CRC

Practical Handbook

of

Materials

Selection

James F. Shackelford
William Alexander
Jun S. Park

CRC Press
Boca Raton   Ann Arbor   London   Tokyo

Library of Congress Cataloging-in-Publication Data

The catalog information is available from the Library of Congress.

0-8493-3709-7

No claim to original U.S. Government works
International Standard Book Number 0-8493-3709-7
Printed in the United States of America 1 2 3 4 5 6 7 8 9 0
Printed on acid-free paper

TABLE OF CONTENTS

Table of Contents (Continued)

Table of Contents (Continued)

Table of Contents (Continued)

Table of Contents (Continued)

PREFACE

The *Practical Handbook of Materials Selection* serves as a companion volume to the *Second Edition* of the *CRC Materials Science and Engineering Handbook*. This more compact volume has been designed to fulfill the goals of providing: i) an introduction to the key professional societies, educational institutions, and employment opportunities in the field of materials science and engineering, ii) an easy to follow organization based on materials properties, and iii) many data sets available in a convenient Selection-Format, in which materials can by compared by property value. The primary sources for data and for verification of existing data are the major professional societies in the materials field, such as ASM International and the American Ceramic Society. From the outset, data storage on magnetic media has been utilized. Relevant tables from existing CRC Press handbooks have been converted to this format and all new data from other sources have been similarly stored.

A special emphasis is given to the "selection by value" presentation. We hope that many users will find this design-approach helpful in the process of materials selection and evaluation.

Every effort has been made to provide a presentation of data in the *Practical Handbook of Materials Selection* that is both easy-to-follow and useful. We have attempted to present data as accurately as possible. This is, however, an ongoing effort and we would greatly appreciate your input. Suggestions and corrections can be sent to:

Dr. James F. Shackelford
College of Engineering
University of California
Davis, CA 95616-5294.

Preface (Continued)

If you are interested in contributing to the upcoming series of CRC materials handbooks related to this volume, you may contact the above address or:

Mr. Joel Claypool
Publisher
CRC Press, Inc.
2000 Corporate Blvd., N.W.
Boca Raton, FL 33431

Finally, we would like to emphasize that, although this book is part of a developing series of reference volumes, it is our intent that it also serve as a stand alone source of scientific and engineering information for many years to come. "Materials science and engineering" continues to be one of the most exciting and dynamic fields in the practice of engineering. We hope that this volume will be a useful reference for those fortunate enough to be involved in that field.

James F. Shackelford

William Alexander

Jun S. Park

ACKNOWLEDGMENTS

We gratefully acknowledge an effective working relationship with CRC Editor Joel Claypool. We appreciate his ongoing enthusiasm and creativity in this effort. Finally, we especially thank our families whose love and support ultimately made this project a reality.

DEDICATION

To Penelope and Scott

and

Li-Li , Cassie Wendy, Cassie, and Bruce

and

Kie Hwa Park and Han Jin Song

Professional Societies in Materials Science and Engineering

AMERICAN CERAMIC SOCIETY (ACerS)

735 Ceramic Place
Westerville, OH 43081-8720
USA

Telephone (614) 890-4700
Fax (614) 899-6109

About the Society:

The American Ceramic Society, founded in 1899, is an international group of scientists, engineers, and industrialists that are active in the creation of now products, applications and research that will advance the usage of ceramic materials. The Society serves more than 14,000 members and subscribers in 80 countries and publishes several periodicals, a bibliographic database *(Ceramic Abstracts)*, and many books and proceedings. An Annual Meeting & Exposition is hold in the spring, other topical meetings are also held.

The Society Headquarters is based in Westerville, Ohio and operates the Ceramic Information Center (CIC), a custom information service, the James I. Mueller Memorial Library, and the Ross C. Purdy Museum of Ceramics, a collection of 2,000 pieces sponsored by Saint-Gobain.

General Publication:

The American Ceramic Society Bulletin, published monthly.

Meetings:

Pacific Coast Regional Meeting
October 19 – 22, 1994
Los Angeles, California.

97th Annual Meeting and Exposition
Meeting: April 30 – May 4, 1995
Exposition: April 30 – May 3, 1995
Cincinnati, Ohio

98th Annual Meeting and Exposition
April 14-18, 1996
Indianapolis, Indiana

ASM INTERNATIONAL (ASM)

Materials Park, Ohio 44073-0002

Telephone (216) 338-5151
Fax (216) 338-4634

About the Society:

ASM International is a society whose mission is to gather, process and disseminate technical information. ASM fosters the understanding and application of engineered materials and their research, design, reliable manufacture, use and economic and social benefits. This is accomplished via a unique global information-sharing network of interaction among members in forums and meetings, education programs, and through publication and electronic media.

General Publication:

Advanced Materials & Processes, published monthly.

Meetings:

ASM/TMS Materials Week '94
3–6 October 1994
O'Hare Exposition Center
Rosemont, Illinois

ASM/TMS Materials Week '95
October 29 – November 2, 1995
O'Hare Exposition Center
Cleveland, Ohio

AMERICAN SOCIETY FOR TESTING AND MATERIALS (ASTM)

1916 Race Street
Philadelphia, Pa. 19103

Telephone (215) 299-5400
Fax (215) 977-9679

About the Society:

The American Society for Testing and Materials (ASTM) is a nonprofit organization devoted to the development of voluntary full consensus standards for materials, products, systems, and services and the promotion of related knowledge.

Each month standards recommendations from the sponsoring ASTM committees are presented on a ballot for Society action. The feature and news sections report on the research, testing, and new activities of the ASTM standards-writing committees. Also included are the legal, governmental, and international events impacting on the standards development process.

News from ASTM committees; discussions of new standards projects; and general features on the development, use, and significance of standards are solicited for publication.

General Publication:

ASTM Standardization News, published monthly.

Meetings:

Monthly meetings.
Dates and locations announced in *ASTM Standardization News*.

THE MATERIALS RESEARCH SOCIETY (MRS)

9800 McKnight Road
Pittsburgh, PA 15237

Telephone (412) 367-3003
Fax (412) 367-4373

About the Society:

The Materials Research Society (MRS) , a non-profit scientific association founded in 1973, promotes interdisciplinary goal-oriented basic research on materials of technological importance. Membership in the Society includes more than 11,600 scientists, engineers, and research managers from industrial, government, and university research laboratories in the United States and nearly 50 countries.

The Society's interdisciplinary approach differs from that of single-discipline professional societies because it promotes information exchange across the many technical fields touching materials development. MRS sponsors two major international annual meetings encompassing approximately 50 topical symposia and also sponsors numerous single-topic scientific meetings. The Society recognizes professional and technical excellence, conducts short courses, and fosters technical interaction in local geographic regions through Sections and University Chapters.

MRS participates in the international arena of materials research through the International Union of Materials Research Societies (IUMRS). MRS is an affiliate of the American Institute of Physics.

General Publication:

MRS Bulletin, published monthly.

Meetings:

MRS Fall Meeting
November 28 – December 2, 1994
Boston, MA

MRS Spring Meeting
April 17 – 21, 1995
San Francisco, CA

THE MINERALS, METALS & MATERIALS SOCIETY (TMS)

420 Commonwealth Drive
Warrendale, PA 15086
USA

Telephone (412) 776-9000
Fax (412) 776-3770
e_mail: tmsgeneral@tms.org

About the Society:

In 1871, a group of visionary metallurgists and mining engineers, realizing the potential of shared knowledge and experience, gathered in Wilkes–Barre, Pennsylvania, and established the Institute of Mining Engineers. The potential in this action has evolved into The Minerals, Metals & Materials Society (TMS) ... a worldwide organization of engineers and researchers promoting professionalism and cooperation, while acting as a bridge for knowledge across the broad spectrum of metallurgy and materials science.

A professional organization by design, TMS both relies on and contributes to the superior qualifications and technical expertise of its membership. TMS views its commitment to a member's continual professional development as the means toward assuring uninterrupted and widely disseminated technological innovation. With more than 12,000 professional and student members worldwide, TMS is able to provide a range of programs, services and benefits tailored to satisfy the level of expertise that characterizes its members.

TMS is an international, as well as an interdisciplinary, society. Its membership spans both the geographical and technological globes. TMS serves as a professional home for metallurgical and materials engineers, scientists, researchers, educators, and administrators in all phases of the minerals, metals, and materials community. And since TMS is the only vertically integrated major materials society in which its members demonstrate a record of specific academic and professional accomplishment, membership reflects prestige and stature within the field.

General Publication:

JOM, published monthly.

The Minerals, Metals & Materials Society (TMS) (Continued)

Meetings:

TMS/ASM Materials Week '94
October 3–6, 1994
O'Hare Exposition Center
Rosemont, Illinois

1995 TMS Annual Meeting
February 12 – 16, 1995
Las Vegas, Nevada

ASM/TMS Materials Week '95
October 29 – November 2, 1995
O'Hare Exposition Center
Cleveland, Ohio

1996 TMS Annual Meeting
February 4-8, 1996
Anaheim, California

1997 TMS Annual Meeting
February 9-13, 1997
Orlando, Florida

SOCIETY OF PLASTICS ENGINEERS (SPE)

Executive Office
14 Fairfield Drive
Brookfield, CT 06804-0403
U.S.A.

Telephone (203) 775-0471
Fax (203) 775-8490

About the Society:

Over 37,000 plastics professionals around the world are now members of the Society of
Plastics Engineers (SPE). These professionals constitute the one greatest source of information and a knowledge of polymers available to the applications engineer. As an international
scientific society, SPE offers the opportunity for professional growth and recognition with
fellow members around the world to create the future of plastics. As a technological
society, SPE functions to provide a forum for members to increase their knowledge of
plastics materials and expand personal contacts. Membership benefits and services are
designed for those who develop, process and produce, engineer, design, purchase, or market
plastics

General Publication:

Plastics Engineering, published monthly.

Meetings:

SPE's Annual Technical Conference (ANTEC), held each spring, is the largest technical
meeting in the world devoted to plastics materials, processing and applications. Over 5,000
plastics professionals attend each year. The conference program of over 500 timely presen-
tations represents each of SPE's 20 technical Divisions. It covers all aspect of plastics
engineering including design, processing, properties, production and product performance.
In conjunction with the ANTEC conference, there is a three-day exhibition of plastics
machinery, materials and equipment.

Accreditation Board for Engineering and Technology (ABET)

ABET Accredited Engineering Programs in Materials Science and Engineering and Related Areas
(for Accreditation Cycle Ending July 1993)

ACCREDITATION BOARD FOR ENGINEERING AND TECHNOLOGY (ABET)

345 East 47th Street
New York, NY 10017

Telephone (212) 705-7685
Fax (212) 838-8062
e_mail: ABET@MTS.CC.Wayne.edu

About the Organization:

Among the purposes of the Accreditation Board for Engineering and Technology are to organize and carry out a comprehensive program of accreditation of pertinent curricula leading to degrees, assist academic institutions in planning their educational programs, promote the intellectual development of those interested in engineering and engineering-related professions, and provide technical assistance to agencies having engineering-related regulatory authority applicable, to accreditation. ABET accomplishes its purposes through standing committees or commissions, one of which is the Engineering Accreditation Commission. The purpose of accrediting is to identify those institutions which offer professional programs in engineering worthy of recognition as such. ABET is recognized by the U.S. Department of Education and Postsecondary Accreditation (COPA) as the sole agency responsible for accreditation of educational programs leading to degrees in engineering.

General Publication:

ABET Accreditation Yearbook, published yearly.

Meetings:

ABET Annual Meeting
October 27-28, 1994
Albuquerque, New Mexico

ABET DEFINITION OF ENGINEERING

It is worthwile to note the central role of materials in the basic definition of engineering from the Accreditation Board for Engineering and Technology

"Engineering is the profession in which a knowledge of the mathematical and natural sciences gained by study, experience, and practice is applied with judgement to develop ways to utilize, economically, the **materials** and forces of nature for the benefit of mankind."

[Boldface our emphasis]

ABET ACCREDITED ENGINEERING PROGRAMS

Ceramic Group (NICE)

Ceramic Engineering

    Alfred University
    Clemson University
    Georgia Institute of Technology
    Illinois at Urbana-Champaign, University of
    Iowa State University
    Missouri-Rolla, University of
    Ohio State University
    Rutgers, The State University of New Jersey
    Washington, University of

Ceramic Engineering Science

    Alfred University

Ceramic Science & Engineering

    Pennsylvania State University

Glass Engineering Science

    Alfred University

ABET Accredited Engineering Programs (Continued)

Geological and Geophysical Group (SME–AIME)

Geological Engineering

Alaska-Fairbanks, University of
Arizona, University of
Colorado School of Mines
Idaho, University of
Michigan Technological University
Minnesota, University of
Mississippi, University of
Missouri-Rolla, University of
Montana College of Mineral Science and Technology
Nevada-Reno, University of
New Mexico State University
North Dakota, University of
Princeton University
South Dakota School of Mines and Technology
Utah, University of

Geophysical Engineering

Colorado School of Mines
Montana College of Mineral Science and Technology

ABET Accredited Engineering Programs (Continued)

Materials Group (TMS Lead Society, with AIChE, ASME, and NICE)
(Programs in this group are accredited according to the program criteria for Metallurgical, Materials, and similarly named engineering programs.)

Materials Engineering

Alabama at Birmingham, University of
Auburn University
Brown University
California, Los Angeles, University of
California Polytechnic State University, San Luis Obispo
Cincinnati, University of
Drexel University
Georgia Institute of Technology
New Mexico Institute of Mining and Technology
Rensselaer Polytechnic Institute
San Jose State University
Wilkes University
Wisconsin-Milwaukee, University of

Materials Science and Engineering

Arizona, University of
California, Davis, University of
Case Western Reserve University
Cornell University
Florida, University of
Johns Hopkins University, The
Kentucky, University of
Lehigh University
Massachusetts Institute of Technology
Michigan State University
Michigan, University of
Minnesota, University of
North Carolina State University at Raleigh
Northwestern University
Pennsylvania, University of
Pittsburgh, University of
Rice University
Tennessee at Knoxville, University of
Utah, University of
Virginia Polytechnic Institute and State University
Washington State University
Wayne State University
Wright State University

Metallurgical Group (TMS Lead Society, with SME–AIME)
(Programs in this group are accredited according to the program criteria for Metallurgical, Materials, and similarly named engineering programs.)

Materials & Metallurgical Engineering

Stevens Institute of Technology

Materials Science and Engineering Option in Metallurgical Engineering

Michigan Technological University

Metallurgical Engineering

Alabama, Tuscaloosa, The University of
Colorado School of Mines
Columbia University
Idaho, University of
Illinois at Chicago, University of
Illinois at Urbana-Champaign, University of
Illinois Institute of Technology
Iowa State University
Missouri-Rolla, University of
Montana College of Mineral Science and Technology
Nevada-Reno, University of
Notre Dame, University of
Ohio State University
Pittsburgh, University of
Polytechnic University
Purdue University at West Lafayette
South Dakota School of Mines and Technology
Texas at El Paso, University of
Utah, University of
Washington, University of
Wisconsin-Madison, University of

Metals Science and Engineering

Pennsylvania State University

Material Science and Engineering

Carnegie–Mellon University

ABET Accredited Engineering Programs (Continued)

Mineral Group (SME–AIME Lead Society, with TMS)
(Programs in this group are accredited according to the program criteria for Metallurgical or Mining and similarly named engineering programs, and/or ABET general criteria.)

Mineral Engineering

California, Berkeley, University of

Mineral Processing Engineering

Michigan Technological University

Mining Group (SME–AIME)
(Programs in this group are accredited according to the program criteria for Mining and similarly named engineering programs.)

Mining Engineering

Alabama, Tuscaloosa, The University of
Alaska-Fairbanks, University of
Arizona, University of
Colorado School of Mines
Columbia University
Idaho, University of
Kentucky, University of
Michigan Technological University
Missouri-Rolla, University of
Montana College of Mineral Science and Technology
Nevada-Reno, University of
South Dakota School of Mines and Technology
Southern Illinois University–Carbondale
Utah, University of
Virginia Polytechnic Institute and State University
West Virginia University

Mining Engineering (Mineral Processing Option)

Pennsylvania State University

Mining Engineering (Mining Option)

Pennsylvania State University

ABET Accredited Engineering Programs (Continued)

Others (EAC of ABET)
(Programs in this group are accredited according to the program criteria for ABET general criteria and program criteria for nontraditional engineering programs.)

    Polymer Science and Engineering

        Case Western Reserve University

    Textile Engineering

        Georgia Institute of Technology
        North Carolina State University at Raleigh

Plastics Group (EAC of ABET, with AIChE, SAE, and SME)
(Programs in this group are accredited according to the general criteria and program criteria for nontraditional engineering programs.)

    Plastics Engineering

        Massachusetts Lowell, University of

Welding Group (TMS Lead Society, with ASME and SME)
(Programs in this group are accredited according to the general criteria and program criteria for nontraditional engineering programs.)

    Welding Engineering

        Ohio State University

Education and Employment Trends in Materials Science and Engineering and Related Areas

BACHELORS DEGREES AWARDED IN ENGINEERING

Year	Degrees Awarded
1976	37,970
1977	40,095
1978	46,091
1979	52,598
1980	58,117
1981	62,935
1982	66,990
1983	72,471
1984	76,931
1985	77,892
1986	78,178
1987	75,735
1988	71,386
1989	68,824
1990	65,967
1991	63,968
1992	63,653
1993	65,001

Source: *data compiled by* A.G. Bormann, Rockwell International, Seal Beach, California *from various sources, including* the Engineering Workforce Commission *and* the College Placement Council.

UNDERGRADUATE DEGREES IN ENGINEERING BY CURRICULUM

Curriculum	1982	1983	1984	1985
Aerospace	1,731	2207	2364	2663
Agricultural	711	704	655	598
Architectural	469	568	341	472
Biomedical	541	577	611	607
Ceramic	260	294	361	283
Chemical	7,039	7499	7685	7244
Civil	10,330	10484	9877	9468
Computer	2,666	2643	3499	4248
Electrical & Electronic	16,094	18590	20495	22135
Engineering Science	1,641	1298	2349	1253
Environmental	254	292	301	224
General Engineering	2,360	1923	2037	1847
Industrial & Manufacturing*	3,695	3808	3923	4330
Engineering Management,*	–	–	–	–
Manufacturing Engineering.*	–	–	–	–
Industrial: All Others*	–	–	–	–
Marine, Naval Arch. & Ocean	698	699	739	736
Materials & Metallurgical	914	1085	1011	988
Mechanical	14,178	16484	17214	17152
Mining & Mineral	1,078	1019	978	926
Nuclear	426	420	434	429
Petroleum	1,256	1420	1587	1550
Systems	346	210	222	442
Other	303	247	248	297
Total Bachelor's Degrees	66,990	72471	76931	77892

Detailed subcategories for industrial engineering programs adopted by EWC in 1991.

Undergraduate Degrees in Engineering by Curriculum (Continued)

Curriculum	1986	1987	1988	1989
Aerospace	2747	2845	2949	3065
Agricultural	638	456	362	330
Architectural	381	400	356	346
Biomedical	618	649	636	677
Ceramic	247	328	368	289
Chemical	6148	5129	4082	3711
Civil	8798	8388	7714	7688
Computer	4999	5012	4275	4398
Electrical & Electronic	24514	25198	24367	22929
Engineering Science	1194	1155	1378	1339
Environmental	182	124	192	138
General Engineering	1385	1315	1085	1058
Industrial & Manufacturing*	4645	4572	4584	4519
Engineering Management,*	–	–	–	–
Manufacturing Engineering.*	–	–	–	–
Industrial: All Others*	–	–	–	–
Marine, Naval Arch. & Ocean	602	537	549	468
Materials & Metallurgical	1011	885	877	842
Mechanical	16702	16056	15610	15369
Mining & Mineral	769	628	404	274
Nuclear	400	324	306	303
Petroleum	1381	1064	612	436
Systems	594	458	474	457
Other	223	212	206	188
Total Bachelor's Degrees	78178	75735	71386	68824

Detailed subcategories for industrial engineering programs adopted by EWC in 1991.

Undergraduate Degrees in Engineering by Curriculum (Continued)

Curriculum	1990	1991	1992	1993
Aerospace	2971	2898	2915	2707
Agricultural	317	267	325	323
Architectural	375	355	445	486
Biomedical	695	665	686	756
Ceramic	348	310	289	286
Chemical	3622	3612	3849	4674
Civil	7587	7748	8413	9196
Computer	4355	4447	4574	4675
Electrical & Electronic	21385	19858	18337	17588
Engineering Science	1045	1110	998	1089
Environmental	137	168	261	388
General Engineering	1239	1115	1120	1087
Industrial & Manufacturing*	4306	–	–	–
Engineering Management,*	–	373	338	279
Manufacturing Engineering.*	–	232	204	168
Industrial: All Others*	–	3690	3541	3242
Marine, Naval Arch. & Ocean	475	410	442	453
Materials & Metallurgical	857	856	875	944
Mechanical	14969	14626	14737	15109
Mining & Mineral	168	182	207	293
Nuclear	264	261	262	257
Petroleum	286	235	237	316
Systems	362	368	382	396
Other	204	200	216	289
Total Bachelor's Degrees	65967	63986	63653	65001

Detailed subcategories for industrial engineering programs adopted by EWC in 1991.

Source: *data compiled by* A.G. Bormann, Rockwell International, Seal Beach, California *from various sources, including* the Engineering Workforce Commission *and* the College Placement Council.

MASTERS AND PROFESSIONAL ENGINEERING DEGREES
BY CURRICULUM

Curriculum	1982	1983	1984	1985
Aerospace	492	512	533	644
Agricultural	144	146	189	224
Architectural	42	30	35	73
Biomedical	199	178	211	272
Ceramic	55	55	87	75
Chemical	1,285	1509	1570	1618
Civil	3,046	3317	3351	3416
Computer	1,371	1420	1533	2232
Electrical & Electronic	4,281	4730	5393	5592
Engineering Science	634	408	434	459
Environmental	412	456	424	382
General Engineering	649	670	887	748
Industrial & Manufacturing*	1,441	1410	1262	1415
Engineering Management,*	–	–	–	–
Manufacturing Engineering.*	–	–	–	–
Industrial: All Others*	–	–	–	–
Marine, Naval Arch. & Ocean	202	163	186	163
Materials & Metallurgical	507	563	583	622
Mechanical	2,573	3001	3160	3315
Mining & Mineral	207	280	242	256
Nuclear	317	311	282	268
Petroleum	147	227	260	240
Systems	490	430	469	380
Other	49	93	105	108
Total	18,543	19909	21226	22502

*Detailed subcategories for industrial engineering programs adopted by EWC in 1991.

Masters and Professional Engineering Degrees by Curriculum (Continued)

Curriculum	1986	1987	1988	1989
Aerospace	652	733	843	865
Agricultural	171	154	183	160
Architectural	48	33	34	29
Biomedical	233	270	257	303
Ceramic	79	102	96	100
Chemical	1430	1314	1274	1220
Civil	3197	3052	3041	3050
Computer	2243	2670	2881	2930
Electrical & Electronic	5926	6780	7335	7520
Engineering Science	641	683	646	675
Environmental	351	337	329	427
General Engineering	513	499	565	601
Industrial & Manufacturing*	1798	1948	2140	2404
Engineering Management,*	–	–	–	–
Manufacturing Engineering.*	–	–	–	–
Industrial: All Others*	–	–	–	–
Marine, Naval Arch. & Ocean	197	132	120	121
Materials & Metallurgical	658	661	623	672
Mechanical	3462	3511	3767	3855
Mining & Mineral	256	296	282	237
Nuclear	284	272	221	245
Petroleum	249	199	220	197
Systems	499	523	636	662
Other	138	121	123	139
Total	23025	24290	25616	26412

*Detailed subcategories for industrial engineering programs adopted by EWC in 1991.

Masters and Professional Engineering Degrees by Curriculum (Continued)

Curriculum	1990	1991	1992	1993
Aerospace	1016	969	1007	1080
Agricultural	189	172	176	145
Architectural	33	40	31	24
Biomedical	310	337	396	446
Ceramic	80	95	91	78
Chemical	1140	1022	1067	1127
Civil	2920	3123	3236	3725
Computer	3221	3398	3388	3540
Electrical & Electronic	7608	7823	7908	8192
Engineering Science	701	742	737	870
Environmental	471	574	783	1052
General Engineering	496	496	536	525
Industrial & Manufacturing*	2387	–	–	–
Engineering Management,*	–	1048	1082	1235
Manufacturing Engineering.*	–	276	350	368
Industrial: All Others*	–	1416	1424	1681
Marine, Naval Arch. & Ocean	146	144	134	128
Materials & Metallurgical	671	690	737	785
Mechanical	3984	3925	4012	4424
Mining & Mineral	192	223	175	219
Nuclear	236	226	208	266
Petroleum	162	117	125	139
Systems	672	738	762	809
Other	120	160	175	246
Total	26755	27754	28540	31104

*Detailed subcategories for industrial engineering programs adopted by EWC in 1991.

Source: data compiled by A.G. Bormann, Rockwell International, Seal Beach, California *from various sources, including* the Engineering Workforce Commission *and* the College Placement Council.

DOCTORATE DEGREES IN ENGINEERING BY CURRICULUM

Curriculum	1982	1983	1984	1985
Aerospace	103	97	131	112
Agricultural	43	51	69	57
Architectural	0	–	–	3
Biomedical	43	50	49	55
Ceramic	12	18	18	14
Chemical	319	379	380	461
Civil	368	390	402	400
Computer	129	102	131	127
Electrical & Electronic	549	628	693	714
Engineering Science	114	170	151	156
Environmental	256	68	79	52
General Engineering	68	76	55	51
Industrial & Manufacturing*	98	108	124	117
Engineering Management,*	–	–	–	–
Manufacturing Engineering.*	–	–	–	–
Industrial: All Others*	–	–	–	–
Marine, Naval Arch. & Ocean	16	18	21	29
Materials & Metallurgical	196	228	241	272
Mechanical	341	399	476	540
Mining & Mineral	32	54	33	45
Nuclear	127	114	115	98
Petroleum	23	14	11	14
Systems	44	56	49	64
Other	6	3	6	2
Total	2,887	3023	3234	3383

*Detailed subcategories for industrial engineering programs adopted by EWC in 1991.

Doctorate Degrees in Engineering by Curriculum (Continued)

Curriculum	1986	1987	1988	1989
Aerospace	114	127	150	167
Agricultural	60	67	61	97
Architectural	–	–	–	–
Biomedical	46	56	70	86
Ceramic	14	30	24	25
Chemical	534	599	657	680
Civil	439	496	518	554
Computer	176	205	262	277
Electrical & Electronic	779	811	1003	1139
Engineering Science	182	217	191	238
Environmental	42	43	63	49
General Engineering	57	63	66	81
Industrial & Manufacturing*	120	158	149	204
Engineering Management,*	–	–	–	–
Manufacturing Engineering.*	–	–	–	–
Industrial: All Others*	–	–	–	–
Marine, Naval Arch. & Ocean	21	21	30	26
Materials & Metallurgical	272	337	310	326
Mechanical	565	691	738	795
Mining & Mineral	59	55	56	74
Nuclear	112	98	102	90
Petroleum	18	19	38	29
Systems	61	69	63	56
Other	15	13	20	24
Total	3686	4175	4571	5017

*Detailed subcategories for industrial engineering programs adopted by EWC in 1991.

Doctorate Degrees in Engineering by Curriculum (Continued)

Curriculum	1990	1991	1992	1993
Aerospace	189	205	239	207
Agricultural	94	76	85	79
Architectural	–	2	–	–
Biomedical	103	120	124	145
Ceramic	38	56	38	39
Chemical	667	692	674	667
Civil	539	567	576	603
Computer	339	414	423	433
Electrical & Electronic	1262	1343	1423	1515
Engineering Science	239	227	247	235
Environmental	51	56	59	66
General Engineering	58	74	75	85
Industrial & Manufacturing*	200	–	–	–
Engineering Management,*	–	20	29	28
Manufacturing Engineering.*	–		3	3
Industrial: All Others*	–	184	199	243
Marine, Naval Arch. & Ocean	21	36	38	38
Materials & Metallurgical	392	414	403	462
Mechanical	900	892	986	1048
Mining & Mineral	79	68	73	50
Nuclear	115	118	125	116
Petroleum	54	35	56	56
Systems	69	65	62	63
Other	15	16	21	17
Total	5424	5680	5958	6198

*Detailed subcategories for industrial engineering programs adopted by EWC in 1991.

Source: *data compiled by* A.G. Bormann, Rockwell International, Seal Beach, California *from various sources, including* the Engineering Workforce Commission *and* the College Placement Council.

STATES AWARDING MOST ENGINEERING DEGREES

BACHELOR'S DEGREES*

School	Number of Degrees (1993)	1992 Rank
California	7,400	(1)
New York	4,960	(2)
Pennsylvania	3,727	(3)
Michigan	3,718	(4)
Texas	3,588	(5)
Total, Top Ten States	23,393	
Total, All States	65,001	
Percent, Top Ten States	36.0%	

* Includes States awarding 3,500 or more Bachelor's degrees.

MASTER'S DEGREES**

School	Number of Degrees (1993)	1992 Rank
California	4,461	(1)
New York	2,613	(2)
Texas	1,964	(4)
Massachusetts	1,778	(3)
Ohio	1,570	(5)
Michigan	1,560	(7)
Total, Top Ten States	13,946	
Total, All States	31,104	
Percent, Top Ten States	44.8%	

** Includes states awarding 1,500 or more MS/PE degrees.

States Awarding Most Engineering Degrees (Continued)

DOCTORATE DEGREES***

School	Number of Degrees (1993)	1992 Rank
California	903	(1)
New York	473	(2)
Illinois	410	(3)
Texas	399	(4)
Pennsylvania	397	(5)
Massachusetts	353	(6)
Total, Top Ton States	2,935	
Total, All States	6,198	
Percent, Top Ten States	47.4%	

*** Includes states awarding 350 or more Doctorate degrees.

Source: *data compiled by* A.G. Bormann, Rockwell International, Seal Beach, California *from various sources, including* the Engineering Workforce Commission *and* the College Placement Council..

SCHOOLS AWARDING MOST ENGINEERING DEGREES

BACHELOR'S DEGREES*

School	Number of Degrees (1993)	1992 Rank
Georgia Tech	1,235	(3)
Penn State	1,212	(1)
Purdue	1,176	(2)
Texas A&M	1,072	(5)
U Illinois	1,068	(4)
N Carolina State/Raleigh	1,064	(7)
Total, Top Schools	6,827	
Total, All Schools	65,001	
Percent, Top Schools	10.5%	

* Includes schools awarding 1,000 or more Bachelor's degrees

MASTER'S DEGREES**

School	Number of Degrees (1993)	1992 Rank
Stanford	910	(1)
Georgia Tech	723	(5)
USC	703	(4)
John Hopkins	681	(2)
MIT	635	(3)
Total, Top Schools	3,652	
Total, All Schools	31,104	
Percent, Top Schools	11.7%	

** Includes schools awarding 600 or more MS/PE degrees

Schools Awarding Most Engineering Degrees (Continued)

School	DOCTORATE*** Number of Degrees (1993)	1992 Rank
MIT	233	(2)
U of Illinois	229	(1)
Stanford	224	(4)
U California/Berkeley	206	(5)
U Michigan/Ann Arbor	195	(3)
Total, Top Schools	1,087	
Total, All Schools	6,198	
Percent, Top Schools	17.5%	

*** Includes schools awarding 180 or more Doctorate degrees

Source: *data compiled by* A.G. Bormann, Rockwell International, Seal Beach, California *from various sources, including* the Engineering Workforce Commission *and* the College Placement Council.

AVERAGE YEARLY SALARY OFFERS IN ENGINEERING

Engineering Field Bachelor's Degree	# Offers	1994	1993	% Change
Aerospace & Aeronautical	178	30,625	31,583	-3.0
Agricultural	28	33,452	31,796	5.2
Architectural	79	30,108	28,296	6.4
Bioengineering & Biomedical	13	29,239	30,874	*
Chemical	555	39,413	39,482	-0.2
Civil	447	29,683	29,211	1.6
Computer	253	33,573	33,963	-1.1
Electrical	842	34,712	34,313	1.2
Environmental	14	33,056	31,569	*
Industrial	247	33,301	32,940	1.1%
Mechanical	993	34,949	34,460	1.4
Metallurgical	84	33,383	34,178	-2.3
Mining	40	32,657	31,826	2.6
Nuclear	5	32,544	34,755	*
Ocean	+	+	N/A	*
Petroleum	40	38,192	38,387	-0.5
Safety	2	24,600	N/A	*
Systems	1	34,200	N/A	*
Textile	3	28,333	37,075	*
Engineering Technology	423	31,777	29,236	8.7
Industrial Technology	16	31,980	28,701	*

N/A - No historic data available
* Not computed for fewer than 20 offers.
+ No offers reported

Source: *data compiled by* A.G. Bormann, Rockwell International, Seal Beach, California *from various sources, including* the Engineering Workforce Commission *and* the College Placement Council.

AVERAGE YEARLY SALARY VS. YEARS OF EDUCATION

Education	Average Salary ($1000s)
High School	38.5
< 4 years College	43.2
College	48.3
Post Graduate	60.8

Source: *Salary Survey Report*, ASM International, Materials Park, Ohio 44073, 1993.

AVERAGE YEARLY SALARY VS. YEARS IN FIELD

Years in Field	Average Salary ($1000s)
0–3	42.9
4–5	44.5
6–10	48.7
11–15	53.7
16–20	58.9
21+	65

Source: *Salary Survey Report*, ASM International, Materials Park, Ohio 44073, 1993.

AVERAGE YEARLY SALARY VS. AGE

Age	Average Salary ($1000s)
<25	32.3
25–29	39.2
30–34	45.1
35–39	49.3
40–44	56.5
45–49	61.4
50–59	65.7
60+	65.1

Source: *Salary Survey Report*, ASM International, Materials Park, Ohio 44073, 1993.

Selecting Metals

Continued

Selecting Metals: Table of Contents (Continued)

Selecting Metals: Table of Contents (Continued)

Electrical Properties   207

Corrosion Properties   211

PERIODIC TABLE OF ELEMENTS IN METALLIC MATERIALS

The Metallic Elements

1 IA	2 IIA	3 IIIB	4 IVB	5 VB	6 VIB	7 VIIB	8 ———	9 VIII	10 ———	11 IB	12 IIB	13 IIIA	14 IVA	15 VA	16 VIA	17 VIIA	18 VIIA
3 Li	4 Be											5 B					
11 Na	12 Mg											13 Al					
19 K	20 Ca	21 Sc	22 Ti	23 V	24 Cr	25 Mn	26 Fe	27 Co	28 Ni	29 Cu	30 Zn	31 Ga					
37 Rb	38 Sr	39 Y	40 Zr	41 Nb	42 Mo	43 Tc	44 Ru	45 Rh	46 Pd	47 Ag	48 Cd	49 In	50 Sn	51 Sb			
55 Cs	56 Ba		72 Hf	73 Ta	74 W	75 Re	76 Os	77 Ir	78 Pt	79 Au	80 Hg	81 Tl	82 Pb	83 Bi			
87 Fr	88 Ra																

57 La	58 Ce	59 Pr	60 Nd	61 Pm	62 Sm	63 Eu	64 Gd	65 Tb	66 Dy	67 Ho	68 Er	69 Tm	70 Yb	71 Lu
89 Ac	90 Th	91 Pa	92 U	93 Np	94 Pu	95 Am	96 Cm	97 Bk	98 Cf	99 Es	100 Fm	101 Md	102 No	103 Lw

Selecting Metals:
Thermodynamic Properties

SELECTING DIFFUSION ACTIVATION ENERGY
IN METALLIC SYSTEMS

Metal	Tracer	Crystal Form	Purity %	Temperature Range °C	Activation Energy, Q kcal $\cdot$ mol^{-1}	Frequency Factor, D_O cm$^2 \cdot$ s^{-1}
Selenium	Hg^{203}	P	99.996	25–100	1.2	—
Zinc	Cu64	S$\perp$c	99.999	338–415	2.0	2.0
Sodium	Au^{198}	P	99.99	1.0–77	2.21	3.34 x 10^{-4}
α-Thallium	Au^{198}	P$\perp$c	99.999	110–260	2.8	2.0 x 10^{-5}
Potassium	Au^{198}	P	99.95	5.6–52.5	3.23	1.29 x10^{-3}
α-Thallium	Au^{198}	P∥c	99.999	110–260	5.2	5.3 x 10^{-4}
Cobalt	S^{35}	P	99.99	1150–1250	5.4	1.3
β-Thallium	Au^{198}	P	99.999	230–310	6.0	5.2 x 10^{-4}
Indium	Au^{198}	S	99.99	25–140	6.7	9 x 10^{-3}
Potassium	Na^{22}	P	99.7	0–62	7.45	0.058
Sodium	K^{42}	P	99.99	0–91	8.43	0.08
Sodium	Rb^{86}	P	99.99	0–85	8.49	0.15
Potassium	Rb^{86}	P	99.95	0.1–59.9	8.78	0.090
Selenium	Fe^{59}	P		40–100	8.88	—
Lithium	Cu64	P	99.98	51–120	9.22	0.47
Potassium	K^{42}	S	99.7	–52–61	9.36	0.16
Phosphorus	P^{32}	P		0–44	9.4	1.07 x 10^{-3}
Lead	Au^{198}	S	99.999	190–320	10.0	8.7 x 10^{-3}
Sodium	Na^{22}	P	99.99	0–98	10.09	0.145
Lithium	Au^{195}	P	92.5	47–153	10.49	0.21
Tin	Au^{198}	S∥c		135–225	11.0	5.8 x 10^{-3}
α-Thallium	Ag^{110}	P∥c	99.999	80–250	11.2	2.7 x 10^{-2}
Indium	Ag^{110}	S∥c	99.99	25–140	11.5	0.11
Selenium	Se^{75}	P		35–140	11.7	1.4 x 10^{-4}

Selecting Diffusion Activation Energy in Metallic Systems (Continued)

Metal	Tracer	Crystal Form	Purity %	Temperature Range °C	Activation Energy, Q kcal $\cdot$ mol^{-1}	Frequency Factor, D_O cm$^2 \cdot$ s^{-1}
α-Thallium	Ag^{110}	P⊥c	99.999	80–250	11.8	3.8 x 10^{-2}
β-Thallium	Ag^{110}	P	99.999	230–310	11.9	4.2 x 10^{-2}
γ-Uranium	Fe^{55}	P	99.99	787–990	12.0	2.69 x 10^{-4}
Tin	Ag^{110}	S‖c		135–225	12.3	7.1 x 10^{-3}
γ-Uranium	Co^{60}	P	99.99	783–989	12.57	3.51 x 10^{-4}
Lithium	Li^{6}	P	99.98	35–178	12.60	0.14
Lithium	Na^{22}	P	92.5	52–176	12.61	0.41
Indium	Ag^{110}	S⊥c	99.99	25–140	12.8	0.52
Lithium	Ag^{110}	P	92.5	65–161	12.83	0.37
Lithium	Ga^{72}	P	99.98	58–173	12.9	0.21
Lithium	Zn^{65}	P	92.5	60–175	12.98	0.57
Aluminum	Mo^{99}	P	99.995	400–630	13.1	1.04 x 10^{-9}
γ-Uranium	Mn^{54}	P	99.99	787–939	13.88	1.81 x 10^{-4}
Lithium	Hg^{203}	P	99.98	58–173	14.18	1.04
Lead	Ag^{110}	P	99.9	200–310	14.4	0.064
Lead	Cu^{64}	S		150–320	14.44	0.046
Tin	Tl^{204}	P	99.999	137–216	14.7	1.2 x 10^{-3}
Lithium	Sn^{113}	P	99.95	108–174	15.0	0.62
Indium	Tl^{204}	S	99.99	49–157	15.5	0.049
Selenium	S^{35}	S‖c		60–90	15.6	1100
γ-Uranium	Ni^{63}	P	99.99	787–1039	15.66	5.36 x10^{-4}
Aluminum	Ni^{63}	P	99.99	360–630	15.7	2.9 x 10^{-8}
Lithium	In^{114}	P	92.5	80–175	15.87	0.39
Lithium	Cd^{115}	P	92.5	80–174	16.05	2.35

Selecting Diffusion Activation Energy in Metallic Systems (Continued)

Metal	Tracer	Crystal Form	Purity %	Temperature Range °C	Activation Energy, Q kcal $\cdot$ mol^{-1}	Frequency Factor, D_0 cm$^2 \cdot$ s^{-1}
α-Praseodymium	Co^{60}	P	99.93	660–780	16.4	4.7×10^{-2}
γ-Uranium	Zr^{95}	P		800–1000	16.5	3.9×10^{-4}
γ–Plutonium	Pu^{238}	P		190–310	16.7	2.1×10^{-5}
Tin	Au^{198}	S⊥c		135–225	17.7	0.16
α-Zirconium	Cr^{51}	P	99.9	700–850	18.0	1.19×10^{-8}
β–Zirconium	Cr^{51}	P	99.9	700–850	18.0	1.19×10^{-8}
Lanthanum	La^{140}	P	99.97	690–850	18.1	2.2×10^{-2}
Zinc	Ga^{72}	S⊥c		240–403	18.15	0.018
Tin	Ag^{110}	S⊥c		135–225	18.4	0.18
Zinc	Ga^{72}	S‖c		240 403	18.4	0.016
Zinc	Sn^{113}	S⊥c		298–400	18.4	0.13
ε-Plutonium	Pu^{238}	P		500–612	18.5	2.0×10^{-2}
Indium	In^{114}	S⊥c	99.99	44–144	18.7	3.7
Indium	In^{114}	S‖c	99.99	44–144	18.7	2.7
Tellurium	Hg^{203}	P		270–440	18.7	3.14×10^{-5}
Cadmium	Zn^{65}	S	99.99	180–300	19.0	0.0016
Zinc	In^{114}	S‖c		271–413	19.10	0.062
Cadmium	Cd^{115}	S	99.95	110–283	19.3	0.14
Zinc	Sn^{113}	S‖c		298–400	19.4	0.15
Aluminum	V^{48}	P	99.995	400–630	19.6	6.05×10^{-8}
Zinc	In^{114}	S⊥c		271–413	19.60	0.14
Aluminum	Nb^{95}	P	99.95	350–480	19.65	1.66×10^{-7}
α-Praseodymium	Au^{195}	P	99.93	650–780	19.7	4.3×10^{-2}
Zinc	Hg^{203}	S‖c		260–413	19.70	0.056

Selecting Diffusion Activation Energy in Metallic Systems (Continued)

Metal	Tracer	Crystal Form	Purity %	Temperature Range °C	Activation Energy, Q kcal $\cdot$ mol^{-1}	Frequency Factor, D_O cm$^2 \cdot$ s^{-1}
Silicon	Fe^{59}	S		1000–1200	20.0	6.2 x 10^{-3}
β-Titanium	C^{14}	P	99.62	1100–1600	20.0	3.02 x 10^{-3}
β-Praseodymium	Au^{195}	P	99.93	800–910	20.1	3.3 x 10^{-2}
Zinc	Cd^{115}	S⊥c	99.999	225–416	20.12	0.117
Zinc	Hg^{203}	S⊥c		260–413	20.18	0.073
Aluminum	Pd^{103}	P	99.995	400–630	20.2	1.92 x 10^{-7}
Zinc	Cd^{115}	S‖c	99.999	225–416	20.54	0.114
β-Thallium	Tl^{204}	S	99.9	230–280	20.7	0.7
Magnesium	Fe^{59}	P	99.95	400–600	21.2	4 x 10^{-6}
Lead	Cd^{115}	S	99.999	150–320	21.23	0.409
Molybdenum	Na^{24}	S		800–1100	21.25	2.95 x 10^{-9}
β-Praseodymium	Ag^{110}	P	99.93	800–900	21.5	3.2 x 10^{-2}
β–Zirconium	Co^{60}	P	99.99	920–1600	21.82	3.26 x 10^{-3}
Zinc	Zn^{65}	S‖c	99.999	240–418	21.9	0.13
Tin	Co^{60}	S,P		140–217	22.0	5.5
α-Zirconium	Sn^{113}	P		300–700	22.0	1.0 x 10^{-8}
β–Zirconium	Sn^{113}	P		300–700	22.0	1 x 10^{-8}
Niobium	K^{42}	S		900–1100	22.10	2.38 x 10^{-7}
α-Thallium	Tl^{204}	S⊥c	99.9	135–230	22.6	0.4
Aluminum	Sm^{153}	P	99.995	450–630	22.88	3.45 x 10^{-7}
Magnesium	Ni^{63}	P	99.95	400–600	22.9	1.2 x 10^{-5}
α-Thallium	Tl^{204}	S‖c	99.9	135–230	22.9	0.4
α-Zirconium	V^{48}	P	99.99	600–850	22.9	1.12 x 10^{-8}
Silicon	Cu^{64}	P		800–1100	23.0	4 x 10^{-2}

Selecting Diffusion Activation Energy in Metallic Systems (Continued)

Metal	Tracer	Crystal Form	Purity %	Temperature Range °C	Activation Energy, Q kcal $\cdot$ mol^{-1}	Frequency Factor, D_O cm$^2 \cdot$ s^{-1}
Zinc	Zn^{65}	S⊥c	99.999	240–418	23.0	0.18
Calcium	Fe^{59}		99.95	500–800	23.3	2.7 x 10^{-3}
δ–Plutonium	Pu^{238}	P		350–440	23.8	4.5 x 10^{-3}
Aluminum	Pr^{142}	P	99.995	520–630	23.87	3.58 x 10^{-7}
γ-Uranium	Cu^{64}	P	99.99	787–1039	24.06	1.96 x 10^{-3}
β-Titanium	P^{32}	P	99.7	950–1600	24.1	3.62x10^{-3}
Copper	Tm^{170}	P	99.999	705–950	24.15	7.28 x 10^{-9}
Lead	Tl^{205}	P	99.999	207–322	24.33	0.511
γ-Uranium	Cr^{51}	P	99.99	797–1037	24.46	5.37 X 10^{-3}
α-Zirconium	Mo^{99}	P		600–850	24.76	6.22 x 10^{-8}
Germanium	Fe^{59}	S		775–930	24.8	0.13
α-Praseodymium	Zn^{65}	P	99.96	766–603	24.8	0.18
Aluminum	Nd^{147}	P	99.995	450–630	25.0	4.8 x 10^{-7}
Molybdenum	K^{42}	S		800–1100	25.04	5.5 x 10^{-9}
Tin	Sn^{113}	S⊥c	99.999	160–226	25.1	10.7
Lithium	Pb^{204}	P	99.95	129–169	25.2	160
Cadmium	Ag^{110}	S	99.99	180–300	25.4	2.21
α-Praseodymium	Ag^{110}	P	99.93	610–730	25.4	0.14
Lead	Pb^{204}	S	99.999	150–320	25.52	0.887
Tin	In^{114}	S‖c	99.998	181–221	25.6	12.2
Tin	Sn^{113}	S‖c	99.999	160–226	25.6	7.7
β-Praseodymium	La^{140}	P	99.96	800–930	25.7	1.8
Tin	In^{114}	S⊥c	99.998	181–221	25.8	34.1
Zinc	Ag^{110}	S‖c	99.999	271–413	26.0	0.32

Selecting Diffusion Activation Energy in Metallic Systems (Continued)

Metal	Tracer	Crystal Form	Purity %	Temperature Range °C	Activation Energy, Q kcal • mol^{-1}	Frequency Factor, D$_O$ cm^2 • s^{-1}
Copper	Lu^{177}	P	99.999	857–1010	26.15	4.3 x 10^{-9}
β-Praseodymium	Ho^{166}	P	99.96	800–930	26.3	9.5
Chromium	C^{14}	P		120–1500	26.5	9.0 x 10^{-3}
Aluminum	Ce^{141}	P	99.995	450–630	26.60	1.9 x 10^{-6}
Copper	Eu^{152}	P	99.999	750–970	26.85	1.17 x 10^{-7}
Aluminum	Au^{198}	S	99.999	423–609	27.0	0.077
Aluminum	La^{140}	P	99.995	500–630	27.0	1.4 x 10^{-6}
Nickel	Sb^{124}	P	99.97	1020–1220	27.0	1.8 x 10^{-5}
β-Praseodymium	Zn^{65}	P	99.96	822–921	27.0	0.63
β–Zirconium	Ta^{182}	P	99.6	900–1200	27.0	5.5 x 10^{-5}
Vanadium	C^{14}	P	99.7	845–1130	27.3	4.9 x 10^{-3}
Magnesium	U^{235}	P	99.95	500–620	27.4	1.6 x 10^{-5}
Copper	Tb^{160}	P	99.999	770–980	27.45	8.96 x 10^{-9}
β–Uranium	Co^{60}	P	99.999	692–763	27.45	1.5 x 10^{-2}
Copper	Pm^{147}	P	99.999	720–955	27.5	3.62 x 10^{-8}
Aluminum	In^{114}	P	99.99	400–600	27.6	0.123
Copper	Ce^{141}	P	99.999	766–947	27.6	2.17 x 10^{-3}
Zinc	Ag^{110}	S⊥c	99.999	271–413	27.6	0.45
Aluminum	Co^{60}	S	99.999	369–655	27.79	0.131
Aluminum	Ag^{110}	S	99.999	371–655	27.83	0.118
Molybdenum	Cs^{134}	S	99.99	1000–1470	28.0	8.7 x 10^{-11}
Magnesium	In^{114}	P	99.9	472–610	28.4	5.2 x 10^{-2}
Aluminum	Sn^{113}	P		400–600	28.5	0.245
γ-Uranium	U^{233}	P	99.99	800–1070	28.5	2.33 x 10^{-3}

Selecting Diffusion Activation Energy in Metallic Systems (Continued)

Metal	Tracer	Crystal Form	Purity %	Temperature Range °C	Activation Energy, Q kcal $\cdot$ mol^{-1}	Frequency Factor, D_O cm$^2 \cdot$ s^{-1}
Magnesium	Ag^{110}	P	99.9	476–621	28.50	0.34
Magnesium	Zn^{65}	P	99.9	467–620	28.6	0.41
Tellurium	Se^{75}	P		320–440	28.6	2.6×10^{-2}
Aluminum	Mn^{54}	P	99.99	450–650	28.8	0.22
Aluminum	Zn^{65}	S	99.999	357–653	28.86	0.259
Calcium	Ni^{63}		99.95	550–800	28.9	1.0×10^{-6}
β–Praseodymium	In^{114}	P	99.96	800–930	28.9	9.6
Aluminum	Ge^{71}	S	99.999	401–653	28.98	0.481
Aluminum	Sb^{124}	P		448–620	29.1	0.09
Aluminum	Ga^{72}	S	99.999	406–652	29.24	0.49
α-Iron	C^{14}	P	99.98	616–844	29.3	2.2
β-Titanium	U^{235}	P	99.9	900–400	29.3	5.1×10^{-4}
β-Praseodymium	Pr^{14}2	P	99.93	800–900	29.4	8.7
Zinc	Cu64	S‖c	99.999	338–415	29.53	2.22
β-Titanium	Ni^{63}	P	99.7	925–1600	29.6	9.2×10^{-3}
Aluminum	Cd^{115}	S	99.999	441–631	29.7	1.04
Zinc	Au^{198}	S⊥c	99.999	315–415	29.72	0.29
Zinc	Au^{198}	S‖c	99.999	315–415	29.73	0.97
Calcium	C^{14}		99.95	550–800	29.8	3.2×10^{-5}
Selenium	S^{35}	S⊥c		60–90	29.9	1700
β–Zirconium	Zr^{95}	P		1100–1500	30.1	2.4×10^{-4}
γ-Uranium	Au^{195}	P	99.99	785–1007	30.4	4.86×10^{-3}
β–Zirconium	U^{235}	P		900–1065	30.5	5.7×10^{-4}
β-Titanium	Co^{60}	P	99.7	900–1600	30.6	1.2×10^{-2}

Selecting Diffusion Activation Energy in Metallic Systems (Continued)

Metal	Tracer	Crystal Form	Purity %	Temperature Range °C	Activation Energy, Q kcal • mol^{-1}	Frequency Factor, D_0 cm^2 • s^{-1}
β–Zirconium	Be[7]	P	99.7	915–1300	31.1	8.33 x 10^{-2}
β-Titanium	Ti[44]	P	99.95	900–1540	31.2	3.58 x 10^{-4}
α-Zirconium	Nb[95]	P	99.99	740–857	31.5	6.6 x 10^{-6}
β-Titanium	Fe[59]	P	99.7	900–1600	31.6	7.8 x 10^{-3}
β-Titanium	Sn[113]	P	99.7	950–1600	31.6	3.8 x 10^{-4}
Niobium	C[14]	P		800–1250	32.0	1.09 x 10^{-5}
Magnesium	Mg[28]	S∥c		467–635	32.2	1.0
β-Titanium	V[48]	P	99.95	900–1545	32.2	3.1 x 10^{-4}
Aluminum	Cu[64]	S	99.999	433–652	32.27	0.647
β-Titanium	Sc[46]	P	99.95	940–1590	32.4	4.0 x 10^{-3}
Magnesium	Mg[28]	S⊥c		467–635	32.5	1.5
β–Zirconium	P[32]	P	99.94	950–1200	33.3	0.33
β-Titanium	Mn[54]	P	99.7	900–1600	33.7	6.1 x 10^{-3}
Aluminum	Al[27]	S		450–650	34.0	1.71
Cobalt	C[14]	P	99.82	600–1400	34.0	0.21
γ-Iron	C[14]	P	99.34	800–1400	34.0	0.15
Nickel	C[14]	P	99.86	600–1400	34.0	0.012
Vanadium	S[35]	P	99.8	1320–1520	34.0	3.1 x 10^{-2}
β–Zirconium	C[14]	P	96.6	1100–1600	34.2	3.57 x 10^{-2}
Calcium	U[235]		99.95	500–700	34.8	1.1 x 10^{-5}
β-Titanium	Cr[51]	P	99.7	950–1600	35.1	5 x 10^{-3}
β–Zirconium	Mo[99]	P		900–1635	35.2	1.99 x 10^{-6}
β-Titanium	Zr[95]	P	98.94	920–1500	35.4	4.7 x 10^{-3}
Tellurium	Te[127]	S∥c	99.9999	300–400	35.5	130

Selecting Diffusion Activation Energy in Metallic Systems (Continued)

Metal	Tracer	Crystal Form	Purity %	Temperature Range °C	Activation Energy, Q kcal · mol^{-1}	Frequency Factor, D_O cm^2 · s^{-1}
α-Titanium	Ti^{44}	P	99.99	700–850	35.9	8.6 x 10^{-6}
Silver	Ge^{77}	P		640–870	36.5	0.084
β–Zirconium	Nb^{95}	P		1230–1635	36.6	7.8 x 10^{-4}
Gold	Hg^{203}	S	99.994	600–1027	37.38	0.116
Copper	Pt^{195}	P		843–997	37.5	4.8 x 10^{-4}
Beryllium	Be^{7}	S⊥c	99.75	565–1065	37.6	0.52
Silver	Tl^{204}	P		640–870	37.9	0.15
Silver	Hg^{20}3	P	99.99	653–948	38.1	0.079
Silver	Pb^{210}	P		700–865	38.1	0.22
Calcium	Ca^{45}		99.95	500–800	38.5	8.3
β–Hafnium	Hf^{181}	P	97.9	1795–1995	38.7	1.2 x10^{-3}
Silver	Te^{125}	P		770–940	38.90	0.47
Silver	Sb^{124}	P	99.999	780–950	39.07	0.234
Beryllium	Ag^{110}	S‖c	99.75	650–900	39.3	0.43
β-Titanium	Nb^{95}	P	99.7	1000–1600	39.3	5.0 x 10^{-3}
Silver	Sn^{113}	S	99.99	592–937	39.30	0.255
Beryllium	Be^{7}	S‖c	99.75	565–1065	39.4	0.62
γ-Uranium	Nb^{95}	P	99.99	791–1102	39.65	4.87 x 10^{-2}
Germanium	In^{114}	S		600–920	39.9	2.9 x 10^{-4}
Silver	S^{35}	S	99.999	600–900	40.0	1.65
α–Uranium	U^{234}	P		580–650	40.0	2 x 10^{-3}
Gold	Ag^{110}	S	99.99	699–1007	40.2	0.072
β-Titanium	Be^{7}	P	99.96	915–1300	40.2	0.8
Tantalum	C^{14}	P		1450–2200	40.3	1.2 x 10^{-2}

Selecting Diffusion Activation Energy in Metallic Systems (Continued)

Metal	Tracer	Crystal Form	Purity %	Temperature Range °C	Activation Energy, Q kcal $\cdot$ mol^{-1}	Frequency Factor, D_O cm^2 $\cdot$ s^{-1}
Silver	In^{114}	S	99.99	592–937	40.80	0.41
Molybdenum	C^{14}	P	99.98	1200–1600	41.0	2.04 x 10^{-2}
Tellurium	Tl^{204}	P		360–430	41.0	320
β–Zirconium	Ce^{141}	P		880–1600	41.4	3.16
Lithium	Sb^{124}	P	99.95	141–176	41.5	1.6 x 10^{10}
Silicon	P^{32}	S		1100–1250	41.5	–
Gold	Co^{60}	P	99.93	702–948	41.6	0.068
Gold	Fe^{59}	P	99.93	701–948	41.6	0.082
Silver	Cd^{115}	S	99.99	592–937	41.69	0.44
Silver	Zn^{65}	S	99.99	640–925	41.7	0.54
Aluminum	Cr^{51}	S	99.999	422–654	41.74	464
Copper	Sb^{124}	S	99.999	600–1000	42.0	0.34
Copper	As^{76}	P		810–1075	42.13	0.20
Gold	Au^{198}	S	99.97	850–1050	42.26	0.107
α-Iron	K^{42}	P	99.92	500–800	42.3	0.036
Copper	Au^{193}	S, P		400–1050	42.6	0.03
β-Titanium	Mo^{99}	P	99.7	900–1600	43.0	8.0 x 10^{-3}
Beryllium	Ag^{110}	S⊥c	99.75	650–900	43.2	1.76
β-Titanium	Ag^{110}	P	99.95	940–1570	43.2	3 x 10^{-3}
Copper	Tl^{204}	S	99.999	785–996	43.3	0.71
γ-Iron	P^{32}	P	99.99	950–1200	43.7	0.01
β-Titanium	W^{185}	P	99.94	900–1250	43.9	3.6 x 10^{-3}
Copper	Hg^{203}	P		–	44.0	0.35
β–Uranium	U^{235}	P		690–750	44.2	2.8 x10^{-3}

Selecting Diffusion Activation Energy in Metallic Systems (Continued)

Metal	Tracer	Crystal Form	Purity %	Temperature Range °C	Activation Energy, Q kcal • mol^{-1}	Frequency Factor, D$_0$ cm^2 • s^{-1}
Copper	Ge^{68}	S	99.998	653–1015	44.76	0.397
Copper	Sn^{113}	P		680–910	45.0	0.11
Lanthanum	Au^{198}	P	99.97	600–800	45.1	1.5
Silver	Ag^{110}	S	99.999	640–955	45.2	0.67
α-Zirconium	Zr^{95}	P	99.95	750–850	45.5	5.6 x 10^{-4}
Copper	Cd^{115}	S	99.98	725–950	45.7	0.935
β–Zirconium	V^{48}	P	99.99	870–1200	45.8	7.59 x 10^{-3}
Copper	Ga^{72}			–	45.90	0.55
Aluminum	Fe^{59}	S	99.99	550–636	46.0	135
Gold	Ni^{63}	P	99.96	880–940	46.0	0.30
Silver	Cu^{64}	P	99.99	717–945	46.1	1.23
Nickel	Be^{7}	P	99.9	1020–1400	46.2	0.019
Copper	Ag^{110}	S, P		580–980	46.5	0.61
Tellurium	Te^{127}	S⊥c	99.9999	300–400	46.7	3.91 x 10^{4}
Silicon	Au^{198}	S		700–1300	47.0	2.75 x 10^{-3}
Carbon	Ni^{63}	⊥c		540–920	47.2	102
Lithium	Bi	P	99.95	141–177	47.3	5.3 x 10^{13}
Copper	Zn^{65}	P	99.999	890–1000	47.50	0.73
α-Zirconium	Fe^{55}	P		750–840	48.0	2.5 x 10^{-2}
β–Zirconium	Fe^{55}	P		750–840	48.0	2.5 x 10^{-2}
Silver	Au^{198}	P	99.99	718–942	48.28	0.85
Silver	Co^{60}	S	99.999	700–940	48.75	1.9
Silver	Fe^{59}	S	99.99	720–930	49.04	2.42
Copper	S^{35}	S	99.999	800–1000	49.2	23

Selecting Diffusion Activation Energy in Metallic Systems (Continued)

Metal	Tracer	Crystal Form	Purity %	Temperature Range °C	Activation Energy, Q kcal $\cdot$ mol^{-1}	Frequency Factor, D_O cm$^2 \cdot$ s^{-1}
Vanadium	P^{32}	P	99.8	1200–1450	49.8	2.45×10^{-2}
Germanium	Sb^{124}	S		720–900	50.2	0.22
Copper	Cu^{67}	S	99.999	698–1061	50.5	0.78
Nickel	Mo^{99}	P		900–1200	51.0	1.6×10^{-3}
Nickel	Pu^{238}	P		1025–1125	51.0	0.5
Niobium	P^{32}	P	99.0	1300–1800	51.5	5.1×10^{-2}
Beryllium	Fe^{59}	S	99.75	700–1076	51.6	0.67
Copper	Fe^{59}	S. P		460–1070	52.0	1.36
α-Iron	Mn^{54}	P	99.97	800–900	52.5	0.35
γ-Iron	S^{35}	P		900–1250	53.0	1.7
Carbon	Ni^{63}	‖c		750–1060	53.3	2.2
Copper	Cr^{51}	S, P		800–1070	53.5	1.02
Tungsten	C^{14}	P	99.51	1200–1600	53.5	8.91×10^{-3}
Copper	Ni^{63}	P		620–1080	53.8	1.1
Molybdenum	Cr^{51}	P		1000–1500	54.0	2.5×10^{-4}
Copper	Co^{60}	S	99.998	701–1077	54.1	1.93
Copper	Pd^{102}	S	99.999	807–1056	54.37	1.71
Silver	Ni^{63}	S	99.99	749–950	54.8	21.9
α-Iron	P^{32}	P		860–900	55.0	2.9
δ-Iron	P^{32}	P	99.99	1370–1460	55.0	2.9
Nickel	Au^{198}	S,P	99.999	700–1075	55.0	0.02
α-Iron	W^{185}	P		755–875	55.1	0.29
α-Iron	V^{48}	P		755–875	55.4	1.43
β–Zirconium	W^{185}	P	99.7	900–1250	55.8	0.41

Selecting Diffusion Activation Energy in Metallic Systems (Continued)

Metal	Tracer	Crystal Form	Purity %	Temperature Range °C	Activation Energy, Q kcal $\cdot$ mol^{-1}	Frequency Factor, D$_0$ cm$^2 \cdot$ s^{-1}
Germanium	Te^{125}	S		770–900	56.0	2.0
α-Iron	Ni^{63}	P	99.97	680–800	56.0	1.3
Silver	Pd^{102}	S	99.999	736–939	56.75	9.56
α-Iron	Cu^{64}	P	99.9	800–1050	57.0	0.57
α-Iron	Cr^{51}	P	99.95	775–875	57.5	2.53
δ-Iron	Fe^{59}	P	99.95	1428–1492	57.5	2.01
γ-Iron	Be^{7}	P	99.9	1100–1350	57.6	0.1
β–Zirconium	V^{48}	P	99.99	1200–1400	57.7	0.32
Beryllium	Ni^{63}	P		800–1250	58.0	0.2
Chromium	Mo^{99}	P		1100–1420	58.0	2.7 x 10^{-3}
Nickel	Sn^{113}	P	99.8	700–1350	58.0	0.83
Nickel	Fe^{59}	P		1020–1263	58.6	0.074
Platinum	Cu^{64}	P		1098–1375	59.5	0.074
Copper	Nb^{95}	P	99.999	807–906	60.06	2.04
Cobalt	Ni^{63}	P		1192–1297	60.2	0.10
α-Iron	Fe^{55}	P	99.92	809–889	60.3	5.4
Yttrium	Y^{90}	S‖c		900–1300	60.3	0.82
Gold	Pt^{195}	P, S	99.98	800–1060	60.9	7.6
δ-Iron	Co^{60}	P	99.995	1428–1521	61.4	6.38
Nickel	Cu^{64}	P	99.95	1050–1360	61.7	0.57
α-Iron	Co^{60}	P	99.995	638–768	62.2	7.19
α-Iron	Au^{198}	P	99.999	800–900	62.4	31
γ-Iron	Mn^{54}	P	99.97	920–1280	62.5	0.16
Cobalt	Fe^{59}	P	99.9	1104–1303	62.7	0.21

Selecting Diffusion Activation Energy in Metallic Systems (Continued)

Metal	Tracer	Crystal Form	Purity %	Temperature Range °C	Activation Energy, Q kcal $\cdot$ mol^{-1}	Frequency Factor, D_O cm$^2 \cdot$ s^{-1}
Palladium	Pd^{103}	S	99.999	1060–1500	63.6	0.205
Carbon	Ag^{110}	⊥c		750–1050	64.3	9280
Vanadium	Cr^{51}	P	99.8	960–1200	64.6	9.54 x10^{-3}
Nickel	Cr^{51}	P	99.95	1100–1270	65.1	1.1
Silver	Ru^{103}	S	99.99	793–945	65.8	180
Nickel	Co^{60}	P	99.97	1149–1390	65.9	1.39
Tungsten	Fe^{59}	P		940–1240	66.0	1.4 x 10^{-2}
Nickel	V^{48}	P	99.99	800–1300	66.5	0.87
α-Iron	Sb^{124}	P		800–900	66.6	1100
γ-Iron	Ni^{63}	P	99.97	930–2050	67.0	0.77
Yttrium	Y^{90}	S⊥c		900–1300	67.1	5.2
Silicon	C^{14}	P		1070–1400	67.2	0.33
Cobalt	Co^{60}	P	99.9	1100–1405	67.7	0.83
γ-Iron	Fe^{59}	P	99.98	1171–1361	67.86	0.49
Nickel	Ni^{63}	P	99.95	1042–1404	68.0	1.9
Platinum	Pt^{195}	P	99.99	1325–1600	68.2	0.33
Germanium	Ge^{71}	S		766–928	68.5	7.8
α-Iron	Ag^{110}	P		748–888	69.0	1950
γ-Iron	V^{48}	P	99.99	1120–1380	69.3	0.28
γ-Iron	Cr^{51}	P	99.99	950–1400	69.7	10.8
Tantalum	S^{35}	P	99.0	1970–2110	70.0	100
α-Zirconium	Ta^{182}	P	99.6	700–800	70.0	100
Niobium	Co^{60}	P	99.85	1500–2100	70.5	0.74
Vanadium	Fe^{59}	P		960–1350	71.0	0.373

Selecting Diffusion Activation Energy in Metallic Systems (Continued)

Metal	Tracer	Crystal Form	Purity %	Temperature Range °C	Activation Energy, Q kcal $\cdot$ mol^{-1}	Frequency Factor, D_O cm$^2 \cdot$ s^{-1}
Tantalum	Fe^{59}	P		930–1240	71.4	0.505
Nickel	W^{185}	P	99.95	1100–1300	71.5	2.0
γ-Iron	Co^{60}	P	99.98	1138–1340	72.9	1.25
α-Iron	Mo^{99}	P		750–875	73.0	7800
Niobium	S^{35}	S	99.9	1100–1500	73.1	2600
Vanadium	V^{48}	S,P	99.99	880–1360	73.65	0.36
Chromium	Cr^{51}	P	99.98	1030–1545	73.7	0.2
Platinum	Co^{60}	P	99.99	900–1050	74.2	19.6
α-Thorium	Pa^{231}	P	99.85	770–910	74.7	126
Molybdenum	U^{235}	P	99.98	1500–2000	76.4	7.6 x 10^{-3}
Niobium	U^{235}	P	99.55	1500–2000	76.8	8.9 x 10^{-3}
Niobium	Fe^{51}	P	99.85	1400–2100	77.7	1.5
Germanium	Tl^{204}	S		800–930	78.4	1700
Niobium	Sn^{113}	P	99.85	1850–2400	78.9	0.14
Chromium	Fe^{59}	P	99.8	980–1420	79.3	0.47
α-Thorium	U^{233}	P	99.85	700–880	79.3	2210
Molybdenum	P^{32}	P	99.97	2000–2200	80.5	0.19
Tantalum	Mo^{99}	P		1750–2220	81.0	1.8 x 10^{-3}
Molybdenum	Ta^{182}	P		1700–2150	83.0	3.5 x 10^{-4}
Niobium	Cr^{51}	S		943–1435	83.5	0.30
Niobium	V^{48}	S	99.99	1000–1400	85.0	2.21
Niobium	Ti^{44}	S		994–1492	86.9	0.099
γ-Iron	W^{185}	P	99.5	1050–1250	90.0	1000
Copper	Mn^{54}	S	99.99	754–950	91.4	10^{7}

Selecting Diffusion Activation Energy in Metallic Systems (Continued)

Metal	Tracer	Crystal Form	Purity %	Temperature Range °C	Activation Energy, Q kcal • mol^{-1}	Frequency Factor, D_O cm^2 • s^{-1}
Niobium	W^{185}	P	99.8	1800–2200	91.7	5 x 10^{-4}
Silicon	Sb^{124}	S		1190–1398	91.7	12.9
Vanadium	V^{48}	S,P	99.99	1360–1830	94.14	214.0
Molybdenum	Re^{186}	P		1700–2100	94.7	0.097
Niobium	Nb^{95}	P, S	99.99	878–2395	96.0	1.1
Molybdenum	Mo^{99}	P		1850–2350	96.9	0.5
Silicon	Ni^{63}	P		450–800	97.5	1000
γ-Iron	Hf^{181}	P	99.99	1110–1360	97.3	3600
Tantalum	Nb^{95}	P, S	99.996	921–2484	98.7	0.23
Tantalum	Ta^{182}	P, S	99.996	1250–2200	98.7	1.24
Niobium	Ta^{182}	P, S	99.997	878–2395	99.3	1.0
Molybdenum	S^{35}	S	99.97	2220–2470	101.0	320
Tungsten	Mo^{99}	P		1700–2100	101.0	0.3
Germanium	Cd^{115}	S		750–950	102.0	1.75 x 10^9
Molybdenum	Co^{60}	P	99.98	1850–2350	106.7	18
Molybdenum	Nb^{95}	P	99.98	1850–2350	108.1	14
Molybdenum	W^{185}	P	99.98	1700–2260	110	1.7
Silicon	Si^{31}	S	99.99999	1225–1400	110.0	1800
Carbon	Th^{228}	‖c		1800–2200	114.7	2.48
Carbon	U^{232}	⊥c		140–2200	115.0	6760
Carbon	U^{232}	‖c		1400–1820	129.5	385
Tungsten	Nb^{95}	P	99.99	1305–2367	137.6	3.01
Tungsten	Ta^{182}	P	99.99	1305–2375	139.9	3.05
Tungsten	W^{185}	P	99.99	1800–2403	140.3	1.88

Selecting Diffusion Activation Energy in Metallic Systems (Continued)

Metal	Tracer	Crystal Form	Purity %	Temperature Range °C	Activation Energy, Q kcal $\cdot$ mol^{-1}	Frequency Factor, D_o cm$^2 \cdot$ s^{-1}
Tungsten	Re^{186}	S		2100–2400	141.0	19.5
Carbon	Th^{228}	$\perp$c		1400–2200	145.4	1.33 x 10^{-5}
Carbon	C^{14}			2000–2200	163	5
α-Thorium	Th^{228}	P	99.85	720–880	716	395

The diffusion coefficient D_T at a temperature T(K) is given by the following:

$$D_T = D_o\, e^{-Q/RT}$$

Abbreviations:

P= polycrystalline
S = single crystal
$\perp$ c = perpendicular to c direction
‖ c = parallel to c direction

Source: data from Askill, J.,in *Handbook of Chemistry and Physics, 55th ed.,*Weast, R.C., Ed., CRC Press, Cleveland,1974, F61.

Selecting Metals: Thermal Properties

SELECTING THERMAL CONDUCTIVITY OF METALS

Metal	Temperature (K)	Thermal Conductivity (watt $\cdot$ cm^{-1} $\cdot$ K^{-1})
Titanium	1	0.0144
Titanium	2	0.0288
Titanium	3	0.0432
Titanium	4	0.0576
Titanium	5	0.0719
Titanium	6	0.0863
Titanium	7	0.101
Zirconium	1	0.111
Titanium	8	0.115
Tantalum	1	0.115
Titanium	9	0.129
Titanium	10	0.144
Molybdenum	1	0.146
Titanium	11	0.158
Titanium	12	0.172
Titanium	13	0.186
Titanium	600	0.194
Titanium	700	0.194
Titanium	500	0.197
Titanium	800	0.197
Titanium	14	0.2
Titanium	900	0.202
Titanium	400	0.204
Zirconium	600	0.207

Selecting Thermal Conductivity of Metals (Continued)

Metal	Temperature (K)	Thermal Conductivity (watt $\bullet$ cm^{-1} $\bullet$ K^{-1})
Titanium	1000	0.207
Zirconium	700	0.209
Zirconium	500	0.21
Titanium	1100	0.213
Titanium	15	0.214
Zirconium	400	0.216
Zirconium	800	0.216
Titanium	300	0.219
Titanium	1200	0.22
Zirconium	2	0.223
Titanium	273	0.224
Zirconium	900	0.226
Zirconium	300	0.227
Titanium	16	0.227
Tantalum	2	0.23
Zirconium	273	0.232
Titanium	1400	0.236
Zirconium	1000	0.237
Titanium	200	0.245
Zirconium	1100	0.248
Niobium	1	0.251
Zirconium	200	0.252
Titanium	1600	0.253
Titanium	18	0.254

Selecting Thermal Conductivity of Metals (Continued)

Metal	Temperature (K)	Thermal Conductivity (watt $\cdot$ cm^{-1} $\cdot$ K^{-1})
Zirconium	1200	0.257
Titanium	1800	0.271
Zirconium	1400	0.275
Titanium	20	0.279
Iron	1200	0.282
Zirconium	1600	0.29
Molybdenum	2	0.292
Iron	1100	0.297
Zirconium	1800	0.302
Iron	1400	0.309
Titanium	100	0.312
Lead	600	0.312
Zirconium	2000	0.313
Titanium	90	0.324
Lead	500	0.325
Iron	1000	0.326
Iron	1600	0.327
Zirconium	100	0.332
Zirconium	3	0.333
Titanium	25	0.337
Lead	400	0.338
Titanium	80	0.339
Tantalum	3	0.345
Zirconium	90	0.35

Selecting Thermal Conductivity of Metals (Continued)

Metal	Temperature (K)	Thermal Conductivity (watt $\cdot$ cm^{-1} $\cdot$ K^{-1})
Lead	300	0.352
Lead	273	0.355
Titanium	70	0.356
Lead	200	0.366
Zirconium	80	0.373
Titanium	60	0.377
Iron	900	0.38
Titanium	30	0.382
Lead	100	0.396
Titanium	50	0.401
Lead	90	0.401
Chromium	1	0.401
Zirconium	70	0.403
Lead	80	0.407
Titanium	35	0.411
Lead	70	0.415
Titanium	45	0.416
Titanium	40	0.422
Lead	60	0.424
Iron	800	0.433
Lead	50	0.435
Molybdenum	3	0.438
Zirconium	4	0.442
Zirconium	60	0.442

Selecting Thermal Conductivity of Metals (Continued)

Metal	Temperature (K)	Thermal Conductivity (watt $\cdot$ cm^{-1} $\cdot$ K^{-1})
Lead	45	0.442
Lead	40	0.451
Tantalum	4	0.459
Lead	35	0.462
Lead	30	0.477
Iron	700	0.487
Zirconium	50	0.497
Niobium	2	0.501
Lead	25	0.507
Niobium	200	0.526
Niobium	273	0.533
Zirconium	45	0.535
Niobium	300	0.537
Iron	600	0.547
Zirconium	5	0.549
Niobium	100	0.552
Niobium	400	0.552
Niobium	90	0.563
Niobium	500	0.567
Tantalum	5	0.571
Tantalum	273	0.574
Tantalum	200	0.575
Tantalum	300	0.575
Tantalum	400	0.578

Selecting Thermal Conductivity of Metals (Continued)

Metal	Temperature (K)	Thermal Conductivity (watt $\cdot$ cm^{-1} $\cdot$ K^{-1})
Zirconium	40	0.58
Niobium	80	0.58
Tantalum	500	0.582
Niobium	600	0.582
Molybdenum	4	0.584
Tantalum	600	0.586
Tantalum	700	0.59
Lead	20	0.59
Tantalum	100	0.592
Tantalum	800	0.594
Tin	500	0.596
Tantalum	90	0.596
Tantalum	900	0.598
Niobium	700	0.598
Tantalum	1000	0.602
Tantalum	80	0.603
Tantalum	1100	0.606
Tantalum	1200	0.61
Niobium	70	0.61
Chromium	1400	0.611
Niobium	800	0.613
Iron	500	0.613
Tantalum	70	0.616
Tantalum	1400	0.618

Selecting Thermal Conductivity of Metals (Continued)

Metal	Temperature (K)	Thermal Conductivity (watt • cm^{-1} • K^{-1})
Tin	400	0.622
Chromium	1200	0.624
Tantalum	1600	0.626
Niobium	900	0.629
Tantalum	1800	0.634
Chromium	1100	0.636
Tantalum	2000	0.64
Nickel	1	0.64
Niobium	1000	0.644
Tantalum	2200	0.647
Zirconium	35	0.65
Tantalum	60	0.651
Zirconium	6	0.652
Nickel	700	0.653
Chromium	1000	0.653
Nickel	600	0.655
Tantalum	2600	0.658
Niobium	1100	0.659
Niobium	60	0.66
Lead	18	0.66
Tantalum	3000	0.665
Tin	300	0.666
Nickel	800	0.674
Niobium	1200	0.675

Selecting Thermal Conductivity of Metals (Continued)

Metal	Temperature (K)	Thermal Conductivity (watt $\cdot$ cm^{-1} $\cdot$ K^{-1})
Chromium	900	0.678
Tantalum	6	0.681
Tin	273	0.682
Iron	400	0.694
Nickel	900	0.696
Niobium	1400	0.705
Chromium	800	0.713
Nickel	1000	0.718
Platinum	500	0.719
Tantalum	50	0.72
Platinum	600	0.72
Nickel	500	0.721
Platinum	400	0.722
Platinum	700	0.723
Platinum	800	0.729
Platinum	300	0.73
Molybdenum	5	0.73
Tin	200	0.733
Platinum	273	0.734
Niobium	1600	0.735
Platinum	900	0.737
Nickel	1100	0.739
Zirconium	30	0.74
Zirconium	7	0.748

Selecting Thermal Conductivity of Metals (Continued)

Metal	Temperature (K)	Thermal Conductivity (watt $\cdot$ cm^{-1} $\cdot$ K^{-1})
Platinum	200	0.748
Platinum	1000	0.748
Niobium	3	0.749
Iron	1	0.75
Chromium	700	0.757
Platinum	1100	0.76
Niobium	50	0.76
Nickel	1200	0.761
Niobium	1800	0.764
Lead	16	0.77
Platinum	1200	0.775
Tantalum	45	0.78
Tantalum	7	0.788
Platinum	100	0.79
Niobium	2000	0.791
Nickel	400	0.801
Chromium	2	0.802
Iron	300	0.803
Nickel	1400	0.804
Chromium	600	0.805
Platinum	1400	0.807
Platinum	90	0.81
Niobium	2200	0.815
Molybdenum	2600	0.825

Selecting Thermal Conductivity of Metals (Continued)

Metal	Temperature (K)	Thermal Conductivity (watt $\cdot$ cm^{-1} $\cdot$ K^{-1})
Iron	273	0.835
Zirconium	8	0.837
Platinum	80	0.84
Niobium	45	0.84
Lead	15	0.84
Platinum	1600	0.842
Chromium	500	0.848
Zirconium	25	0.85
Tin	100	0.85
Molybdenum	2200	0.858
Tantalum	40	0.87
Chromium	400	0.873
Molybdenum	6	0.876
Platinum	1800	0.877
Tin	90	0.88
Molybdenum	2000	0.88
Tantalum	8	0.891
Platinum	70	0.9
Chromium	300	0.903
Nickel	300	0.905
Molybdenum	1800	0.907
Tin	80	0.91
Platinum	2000	0.913
Tungsten	3000	0.915

Selecting Thermal Conductivity of Metals (Continued)

Metal	Temperature (K)	Thermal Conductivity (watt $\cdot$ cm^{-1} $\cdot$ K^{-1})
Zirconium	9	0.916
Cadmium	500	0.92
Tungsten	2600	0.94
Nickel	273	0.94
Lead	14	0.94
Iron	200	0.94
Molybdenum	1600	0.946
Cadmium	400	0.947
Chromium	273	0.948
Tin	70	0.96
Cadmium	300	0.968
Niobium	40	0.97
Cadmium	273	0.975
Tungsten	2200	0.98
Zirconium	10	0.984
Tantalum	9	0.989
Tantalum	35	0.99
Niobium	4	0.993
Cadmium	200	0.993
Molybdenum	1400	0.996
Tungsten	2000	1
Zirconium	20	1.01
Platinum	60	1.01
Molybdenum	7	1.02

Selecting Thermal Conductivity of Metals (Continued)

Metal	Temperature (K)	Thermal Conductivity (watt $\cdot$ cm^{-1} $\cdot$ K^{-1})
Tungsten	1800	1.03
Cadmium	100	1.03
Zirconium	11	1.04
Tin	60	1.04
Cadmium	90	1.04
Zinc	600	1.05
Molybdenum	1200	1.05
Nickel	200	1.06
Cadmium	80	1.06
Tungsten	1600	1.07
Lead	13	1.07
Zirconium	12	1.08
Zirconium	18	1.08
Tantalum	10	1.08
Molybdenum	1100	1.08
Cadmium	70	1.08
Zirconium	13	1.11
Zinc	500	1.11
Tungsten	1400	1.11
Chromium	200	1.11
Zirconium	16	1.12
Molybdenum	1000	1.12
Zirconium	14	1.13
Zirconium	15	1.13

Selecting Thermal Conductivity of Metals (Continued)

Metal	Temperature (K)	Thermal Conductivity (watt $\cdot$ cm^{-1} $\cdot$ K^{-1})
Cadmium	60	1.13
Tungsten	1200	1.15
Tin	50	1.15
Molybdenum	900	1.15
Zinc	400	1.16
Tantalum	11	1.16
Tantalum	30	1.16
Niobium	35	1.16
Molybdenum	8	1.17
Tungsten	1100	1.18
Platinum	50	1.18
Molybdenum	800	1.18
Chromium	3	1.2
Cadmium	50	1.2
Zinc	300	1.21
Tungsten	1000	1.21
Zinc	273	1.22
Molybdenum	700	1.22
Tin	45	1.23
Niobium	5	1.23
Lead	12	1.23
Tungsten	900	1.24
Tantalum	12	1.24
Cadmium	45	1.25

Selecting Thermal Conductivity of Metals (Continued)

Metal	Temperature (K)	Thermal Conductivity (watt $\cdot$ cm^{-1} $\cdot$ K^{-1})
Zinc	200	1.26
Molybdenum	600	1.26
Nickel	2	1.27
Tungsten	800	1.28
Tantalum	13	1.3
Molybdenum	500	1.3
Magnesium	1	1.3
Molybdenum	9	1.31
Zinc	100	1.32
Platinum	45	1.32
Iron	100	1.32
Cadmium	40	1.32
Tungsten	700	1.33
Zinc	90	1.34
Molybdenum	400	1.34
Tin	40	1.35
Tantalum	14	1.36
Tantalum	25	1.36
Zinc	80	1.38
Molybdenum	300	1.38
Tungsten	600	1.39
Molybdenum	273	1.39
Tantalum	15	1.4
Cadmium	35	1.41

Selecting Thermal Conductivity of Metals (Continued)

Metal	Temperature (K)	Thermal Conductivity (watt $\cdot$ cm^{-1} $\cdot$ K^{-1})
Molybdenum	200	1.43
Tantalum	16	1.44
Niobium	30	1.45
Molybdenum	10	1.45
Magnesium	900	1.45
Niobium	6	1.46
Magnesium	800	1.46
Lead	11	1.46
Iron	90	1.46
Tantalum	18	1.47
Tantalum	20	1.47
Magnesium	700	1.47
Zinc	70	1.48
Tungsten	500	1.49
Magnesium	600	1.49
Iron	2	1.49
Tin	35	1.5
Platinum	40	1.51
Magnesium	500	1.51
Magnesium	400	1.53
Magnesium	300	1.56
Cadmium	30	1.56
Magnesium	273	1.57
Nickel	100	1.58

Selecting Thermal Conductivity of Metals (Continued)

Metal	Temperature (K)	Thermal Conductivity (watt $\cdot$ cm^{-1} $\cdot$ K^{-1})
Chromium	100	1.58
Magnesium	200	1.59
Molybdenum	11	1.6
Chromium	4	1.6
Tungsten	400	1.62
Niobium	7	1.67
Iron	80	1.68
Chromium	90	1.68
Magnesium	100	1.69
Zinc	60	1.71
Nickel	90	1.72
Molybdenum	12	1.74
Tin	30	1.76
Tungsten	300	1.78
Magnesium	90	1.78
Lead	10	1.78
Molybdenum	100	1.79
Cadmium	25	1.79
Platinum	35	1.8
Tungsten	273	1.82
Chromium	80	1.82
Niobium	8	1.86
Niobium	25	1.87
Molybdenum	13	1.88

Selecting Thermal Conductivity of Metals (Continued)

Metal	Temperature (K)	Thermal Conductivity (watt $\cdot$ cm^{-1} $\cdot$ K^{-1})
Nickel	3	1.91
Molybdenum	90	1.92
Nickel	80	1.93
Magnesium	80	1.95
Tungsten	200	1.97
Chromium	5	1.99
Molybdenum	14	2.01
Niobium	9	2.04
Iron	70	2.04
Chromium	70	2.08
Molybdenum	80	2.09
Zinc	50	2.13
Aluminum	900	2.13
Molybdenum	15	2.15
Niobium	10	2.18
Aluminum	800	2.2
Nickel	70	2.21
Tin	25	2.22
Magnesium	70	2.23
Iron	3	2.24
Cadmium	20	2.26
Aluminum	700	2.26
Platinum	30	2.28
Molybdenum	16	2.28

Selecting Thermal Conductivity of Metals (Continued)

Metal	Temperature (K)	Thermal Conductivity (watt $\cdot$ cm^{-1} $\cdot$ K^{-1})
Niobium	20	2.29
Niobium	11	2.3
Molybdenum	70	2.3
Lead	9	2.3
Platinum	1	2.31
Aluminum	600	2.32
Tungsten	100	2.35
Aluminum	273	2.36
Aluminum	200	2.37
Aluminum	300	2.37
Aluminum	500	2.37
Chromium	6	2.38
Niobium	12	2.39
Aluminum	400	2.4
Niobium	18	2.42
Tungsten	90	2.44
Niobium	13	2.46
Zinc	45	2.48
Chromium	60	2.48
Niobium	14	2.49
Niobium	16	2.49
Niobium	15	2.5
Molybdenum	18	2.53
Nickel	4	2.54

Selecting Thermal Conductivity of Metals (Continued)

Metal	Temperature (K)	Thermal Conductivity (watt $\cdot$ cm^{-1} $\cdot$ K^{-1})
Tungsten	80	2.56
Magnesium	2	2.59
Molybdenum	60	2.6
Gold	1200	2.62
Cadmium	18	2.62
Nickel	60	2.63
Iron	60	2.65
Gold	1100	2.71
Magnesium	60	2.74
Tungsten	70	2.76
Molybdenum	20	2.77
Chromium	7	2.77
Gold	1000	2.78
Gold	900	2.85
Gold	800	2.92
Zinc	40	2.97
Iron	4	2.97
Gold	700	2.98
Molybdenum	50	3
Aluminum	100	3.0
Gold	600	3.04
Gold	500	3.09
Gold	400	3.12
Chromium	8	3.14

Selecting Thermal Conductivity of Metals (Continued)

Metal	Temperature (K)	Thermal Conductivity (watt • cm^{-1} • K^{-1})
Platinum	25	3.15
Gold	300	3.15
Nickel	5	3.16
Cadmium	16	3.16
Chromium	50	3.17
Tungsten	60	3.18
Gold	273	3.18
Tin	20	3.2
Lead	8	3.2
Molybdenum	25	3.25
Molybdenum	45	3.26
Gold	200	3.27
Nickel	50	3.36
Aluminum	90	3.4
Copper	1200	3.42
Gold	100	3.45
Gold	90	3.48
Copper	1100	3.5
Chromium	9	3.5
Molybdenum	40	3.51
Gold	80	3.52
Molybdenum	30	3.55
Cadmium	15	3.55
Copper	1000	3.57

Selecting Thermal Conductivity of Metals (Continued)

Metal	Temperature (K)	Thermal Conductivity (watt $\cdot$ cm^{-1} $\cdot$ K^{-1})
Silver	1200	3.58
Gold	70	3.58
Molybdenum	35	3.62
Copper	900	3.64
Silver	1100	3.66
Chromium	45	3.67
Iron	5	3.71
Copper	800	3.71
Zinc	35	3.72
Iron	50	3.72
Silver	1000	3.74
Magnesium	50	3.75
Nickel	6	3.77
Copper	700	3.77
Gold	60	3.8
Silver	900	3.82
Copper	600	3.83
Chromium	10	3.85
Magnesium	3	3.88
Copper	500	3.88
Silver	800	3.89
Nickel	45	3.91
Copper	400	3.92
Silver	700	3.97

Selecting Thermal Conductivity of Metals (Continued)

Metal	Temperature (K)	Thermal Conductivity (watt $\cdot$ cm^{-1} $\cdot$ K^{-1})
Copper	300	3.98
Tin	18	4
Aluminum	80	4.0
Copper	273	4.01
Cadmium	14	4.01
Silver	600	4.05
Silver	500	4.13
Copper	200	4.13
Tungsten	50	4.17
Chromium	11	4.18
Silver	400	4.2
Gold	50	4.2
Silver	300	4.27
Silver	273	4.28
Silver	200	4.3
Chromium	40	4.3
Nickel	7	4.36
Gold	1	4.4
Iron	6	4.42
Chromium	12	4.49
Silver	100	4.5
Iron	45	4.5
Magnesium	45	4.57
Silver	90	4.6

Selecting Thermal Conductivity of Metals (Continued)

Metal	Temperature (K)	Thermal Conductivity (watt $\cdot$ cm^{-1} $\cdot$ K^{-1})
Platinum	2	4.6
Gold	45	4.6
Nickel	40	4.63
Cadmium	13	4.67
Silver	80	4.71
Chromium	13	4.78
Copper	100	4.83
Zinc	30	4.9
Platinum	20	4.9
Lead	7	4.9
Nickel	8	4.94
Silver	70	4.97
Aluminum	70	5.0
Chromium	35	5.03
Chromium	14	5.04
Tungsten	45	5.07
Iron	7	5.13
Copper	90	5.14
Magnesium	4	5.15
Gold	40	5.2
Chromium	15	5.27
Tin	16	5.3
Chromium	16	5.48
Nickel	9	5.49

Selecting Thermal Conductivity of Metals (Continued)

Metal	Temperature (K)	Thermal Conductivity (watt $\cdot$ cm^{-1} $\cdot$ K^{-1})
Silver	60	5.5
Iron	40	5.55
Cadmium	12	5.56
Chromium	30	5.58
Nickel	35	5.62
Magnesium	40	5.7
Copper	80	5.7
Iron	8	5.8
Chromium	18	5.81
Nickel	10	6
Chromium	20	6.01
Chromium	25	6.07
Platinum	18	6.1
Gold	35	6.1
Tin	15	6.3
Magnesium	5	6.39
Iron	9	6.45
Nickel	11	6.48
Tungsten	40	6.5
Copper	70	6.7
Aluminum	60	6.7
Platinum	3	6.79
Iron	35	6.81
Zinc	25	6.9

Selecting Thermal Conductivity of Metals (Continued)

Metal	Temperature (K)	Thermal Conductivity (watt $\cdot$ cm^{-1} $\cdot$ K^{-1})
Nickel	12	6.91
Cadmium	11	6.91
Nickel	30	6.95
Silver	50	7
Iron	10	7.05
Nickel	13	7.3
Magnesium	35	7.4
Tin	14	7.6
Platinum	16	7.6
Magnesium	6	7.6
Gold	30	7.6
Iron	11	7.62
Nickel	14	7.64
Aluminum	1	7.8
Nickel	15	7.92
Iron	12	8.13
Iron	30	8.14
Nickel	16	8.15
Nickel	25	8.15
Lead	6	8.2
Silver	45	8.4
Platinum	15	8.4
Nickel	18	8.45
Copper	60	8.5

Selecting Thermal Conductivity of Metals (Continued)

Metal	Temperature (K)	Thermal Conductivity (watt • cm^{-1} • K^{-1})
Nickel	20	8.56
Iron	13	8.58
Magnesium	7	8.75
Platinum	4	8.8
Cadmium	10	8.87
Tungsten	35	8.9
Gold	2	8.9
Iron	14	8.97
Tin	13	9.3
Platinum	14	9.3
Iron	15	9.3
Iron	25	9.36
Magnesium	30	9.5
Iron	16	9.56
Magnesium	8	9.83
Iron	18	9.88
Iron	20	9.97
Aluminum	50	10.0
Platinum	13	10.1
Gold	25	10.2
Silver	40	10.5
Platinum	5	10.5
Zinc	20	10.7
Magnesium	9	10.8

Selecting Thermal Conductivity of Metals (Continued)

Metal	Temperature (K)	Thermal Conductivity (watt $\cdot$ cm^{-1} $\cdot$ K^{-1})
Platinum	12	10.9
Tin	12	11.6
Platinum	11	11.7
Magnesium	10	11.7
Platinum	6	11.8
Magnesium	25	12
Copper	50	12.2
Cadmium	9	12.2
Platinum	10	12.3
Magnesium	11	12.5
Aluminum	45	12.5
Platinum	7	12.6
Platinum	9	12.8
Platinum	8	12.9
Tungsten	30	13.1
Magnesium	12	13.1
Gold	3	13.1
Zinc	18	13.3
Magnesium	13	13.6
Silver	35	13.7
Lead	5	13.8
Magnesium	20	13.9
Magnesium	14	14
Magnesium	15	14.3

Selecting Thermal Conductivity of Metals (Continued)

Metal	Temperature (K)	Thermal Conductivity (watt • cm^{-1} • K^{-1})
Magnesium	18	14.3
Tungsten	1	14.4
Magnesium	16	14.4
Tin	11	14.8
Gold	20	15
Copper	45	15.3
Aluminum	2	15.5
Aluminum	40	16.0
Zinc	16	16.9
Gold	4	17.1
Gold	18	17.7
Cadmium	8	18
Zinc	1	19
Tin	10	19.3
Silver	30	19.3
Zinc	15	19.4
Tungsten	25	20.4
Copper	40	20.5
Gold	5	20.7
Gold	16	20.9
Aluminum	35	21.0
Zinc	14	22.4
Lead	4	22.4
Gold	15	22.6

Selecting Thermal Conductivity of Metals (Continued)

Metal	Temperature (K)	Thermal Conductivity (watt $\cdot$ cm^{-1} $\cdot$ K^{-1})
Aluminum	3	23.2
Gold	6	23.7
Gold	14	24.1
Gold	13	25.5
Tin	9	26
Gold	7	26
Zinc	13	26.1
Gold	12	26.7
Gold	8	27.5
Lead	1	27.7
Gold	11	27.7
Cadmium	7	28
Gold	9	28.2
Gold	10	28.2
Aluminum	30	28.5
Tungsten	2	28.7
Copper	1	28.7
Copper	35	29
Silver	25	29.5
Zinc	12	30.8
Aluminum	4	30.8
Tungsten	20	32.6
Lead	3	34
Tin	8	36

Selecting Thermal Conductivity of Metals (Continued)

Metal	Temperature (K)	Thermal Conductivity (watt $\cdot$ cm^{-1} $\cdot$ K^{-1})
Zinc	11	36.4
Zinc	2	37.9
Aluminum	5	38.1
Silver	1	39.4
Tungsten	18	40
Aluminum	25	40.0
Lead	2	42.4
Tungsten	3	42.6
Copper	30	43
Zinc	10	43.2
Cadmium	6	44.2
Aluminum	6	45.1
Cadmium	1	48.7
Tungsten	16	49.3
Silver	20	51
Aluminum	7	51.5
Zinc	9	51.9
Tin	7	52
Tungsten	15	54.8
Zinc	3	55.5
Tungsten	4	55.6
Aluminum	20	56.5
Copper	2	57.3
Aluminum	8	57.3

Selecting Thermal Conductivity of Metals (Continued)

Metal	Temperature (K)	Thermal Conductivity (watt $\cdot$ cm^{-1} $\cdot$ K^{-1})
Tungsten	14	60.4
Zinc	8	61.8
Aluminum	9	62.2
Aluminum	18	63.5
Silver	18	66
Aluminum	10	66.1
Tungsten	13	66.4
Tungsten	5	67.1
Copper	25	68
Aluminum	16	68.4
Cadmium	5	69
Aluminum	11	69.0
Zinc	4	69.7
Aluminum	15	70.2
Aluminum	12	70.8
Aluminum	14	71.3
Aluminum	13	71.5
Zinc	7	71.7
Tungsten	12	72.4
Tin	6	76
Tungsten	6	76.2
Zinc	5	77.8
Tungsten	11	77.9
Zinc	6	78

Selecting Thermal Conductivity of Metals (Continued)

Metal	Temperature (K)	Thermal Conductivity (watt $\cdot$ cm^{-1} $\cdot$ K^{-1})
Silver	2	78.3
Tungsten	7	82.4
Tungsten	10	82.4
Silver	16	85
Tungsten	9	85.1
Tungsten	8	85.3
Copper	3	85.5
Cadmium	2	89.3
Cadmium	4	92
Silver	15	96
Cadmium	3	104
Copper	20	105
Silver	14	109
Copper	4	113
Silver	3	115
Tin	5	117
Silver	13	124
Copper	18	124
Copper	5	138
Silver	12	139
Copper	16	145
Silver	4	147
Silver	11	154
Copper	15	156

Selecting Thermal Conductivity of Metals (Continued)

Metal	Temperature (K)	Thermal Conductivity (watt $\cdot$ cm^{-1} $\cdot$ K^{-1})
Copper	6	159
Copper	14	166
Silver	10	168
Silver	5	172
Copper	13	176
Copper	7	177
Tin	4	181
Silver	9	181
Copper	12	185
Silver	6	187
Copper	8	189
Silver	8	190
Silver	7	193
Copper	11	193
Copper	9	195
Copper	10	196
Tin	3	297

These data apply only to metals of purity of at least 99.9%.
The third significant figure may not be accurate.

Source: *data from* Ho, C. Y., Powell, R. W., and Liley, P. *E., Thermal Conductictivity of Selected Materials,* NSRDS–NBS–8 and NSRD-NBS-16, Part 2, National Standard Reference Data System–National Bureau of Standards, Part 1, 1966; Part 2, 1968.

SELECTING THERMAL CONDUCTIVITY OF METALS AT TEMPERATURE

Temperature (K)	Metal	Thermal Conductivity (watt $\cdot$ cm^{-1} $\cdot$ K^{-1})
1	Titanium	0.0144
	Zirconium	0.111
	Tantalum	0.115
	Molybdenum	0.146
	Niobium	0.251
	Chromium	0.401
	Nickel	0.64
	Iron	0.75
	Magnesium	1.3
	Platinum	2.31
	Gold	4.4
	Aluminum	7.8
	Tungsten	14.4
	Zinc	19
	Lead	27.7
	Copper	28.7
	Silver	39.4
	Cadmium	48.7
2	Titanium	0.0288
	Zirconium	0.223
	Tantalum	0.23
	Molybdenum	0.292

Selecting Thermal Conductivity of Metals at Temperature(Continued)

Temperature (K)	Metal	Thermal Conductivity (watt $\cdot$ cm^{-1} $\cdot$ K^{-1})
	Niobium	0.501
	Chromium	0.802
	Nickel	1.27
	Iron	1.49
	Magnesium	2.59
	Platinum	4.6
	Gold	8.9
	Aluminum	15.5
	Tungsten	28.7
	Zinc	37.9
	Lead	42.4
	Copper	57.3
	Silver	78.3
	Cadmium	89.3
3	Titanium	0.0432
	Zirconium	0.333
	Tantalum	0.345
	Molybdenum	0.438
	Niobium	0.749
	Chromium	1.2
	Nickel	1.91
	Iron	2.24

Selecting Thermal Conductivity of Metals at Temperature(Continued)

Temperature (K)	Metal	Thermal Conductivity (watt $\cdot$ cm^{-1} $\cdot$ K^{-1})
	Magnesium	3.88
	Platinum	6.79
	Gold	13.1
	Aluminum	23.2
	Lead	34
	Tungsten	42.6
	Zinc	55.5
	Copper	85.5
	Cadmium	104
	Silver	115
	Tin	297
4	Titanium	0.0576
	Zirconium	0.442
	Tantalum	0.459
	Molybdenum	0.584
	Niobium	0.993
	Chromium	1.6
	Nickel	2.54
	Iron	2.97
	Magnesium	5.15
	Platinum	8.8
	Gold	17.1
	Lead	22.4

Selecting Thermal Conductivity of Metals at Temperature(Continued)

Temperature (K)	Metal	Thermal Conductivity (watt $\cdot$ cm^{-1} $\cdot$ K^{-1})
	Aluminum	30.8
	Tungsten	55.6
	Zinc	69.7
	Cadmium	92
	Copper	113
	Silver	147
	Tin	181
5	Titanium	0.0719
	Zirconium	0.549
	Tantalum	0.571
	Molybdenum	0.73
	Niobium	1.23
	Chromium	1.99
	Nickel	3.16
	Iron	3.71
	Magnesium	6.39
	Platinum	10.5
	Lead	13.8
	Gold	20.7
	Aluminum	38.1
	Tungsten	67.1
	Cadmium	69
	Zinc	77.8

Selecting Thermal Conductivity of Metals at Temperature(Continued)

Temperature (K)	Metal	Thermal Conductivity (watt $\cdot$ cm^{-1} $\cdot$ K^{-1})
	Tin	117
	Copper	138
	Silver	172
6	Titanium	0.0863
	Zirconium	0.652
	Tantalum	0.681
	Molybdenum	0.876
	Niobium	1.46
	Chromium	2.38
	Nickel	3.77
	Iron	4.42
	Magnesium	7.6
	Lead	8.2
	Platinum	11.8
	Gold	23.7
	Cadmium	44.2
	Aluminum	45.1
	Tin	76
	Tungsten	76.2
	Zinc	78
	Copper	159
	Silver	187

Selecting Thermal Conductivity of Metals at Temperature(Continued)

Temperature (K)	Metal	Thermal Conductivity (watt $\cdot$ cm^{-1} $\cdot$ K^{-1})
7	Titanium	0.101
	Zirconium	0.748
	Tantalum	0.788
	Molybdenum	1.02
	Niobium	1.67
	Chromium	2.77
	Nickel	4.36
	Lead	4.9
	Iron	5.13
	Magnesium	8.75
	Platinum	12.6
	Gold	26
	Cadmium	28
	Aluminum	51.5
	Tin	52
	Zinc	71.7
	Tungsten	82.4
	Copper	177
	Silver	193
8	Titanium	0.115
	Zirconium	0.837
	Tantalum	0.891
	Molybdenum	1.17

Selecting Thermal Conductivity of Metals at Temperature(Continued)

Temperature (K)	Metal	Thermal Conductivity (watt $\cdot$ cm^{-1} $\cdot$ K^{-1})
	Niobium	1.86
	Chromium	3.14
	Lead	3.2
	Nickel	4.94
	Iron	5.8
	Magnesium	9.83
	Platinum	12.9
	Cadmium	18
	Gold	27.5
	Tin	36
	Aluminum	57.3
	Zinc	61.8
	Tungsten	85.3
	Copper	189
	Silver	190
9	Titanium	0.129
	Zirconium	0.916
	Tantalum	0.989
	Molybdenum	1.31
	Niobium	2.04
	Lead	2.3
	Chromium	3.5
	Nickel	5.49

Selecting Thermal Conductivity of Metals at Temperature(Continued)

Temperature (K)	Metal	Thermal Conductivity (watt $\cdot$ cm^{-1} $\cdot$ K^{-1})
	Iron	6.45
	Magnesium	10.8
	Cadmium	12.2
	Platinum	12.8
	Tin	26
	Gold	28.2
	Zinc	51.9
	Aluminum	62.2
	Tungsten	85.1
	Silver	181
	Copper	195
10	Titanium	0.144
	Zirconium	0.984
	Tantalum	1.08
	Molybdenum	1.45
	Lead	1.78
	Niobium	2.18
	Chromium	3.85
	Nickel	6
	Iron	7.05
	Cadmium	8.87
	Magnesium	11.7
	Platinum	12.3

Selecting Thermal Conductivity of Metals at Temperature(Continued)

Temperature (K)	Metal	Thermal Conductivity (watt • cm^{-1} • K^{-1})
	Tin	19.3
	Gold	28.2
	Zinc	43.2
	Aluminum	66.1
	Tungsten	82.4
	Silver	168
	Copper	196
11	Titanium	0.158
	Zirconium	1.04
	Tantalum	1.16
	Lead	1.46
	Molybdenum	1.6
	Niobium	2.3
	Chromium	4.18
	Nickel	6.48
	Cadmium	6.91
	Iron	7.62
	Platinum	11.7
	Magnesium	12.5
	Tin	14.8
	Gold	27.7
	Zinc	36.4
	Aluminum	69

Practical Handbook of Materials Selection

Selecting Thermal Conductivity of Metals at Temperature(Continued)

Temperature (K)	Metal	Thermal Conductivity (watt $\cdot$ cm^{-1} $\cdot$ K^{-1})
	Tungsten	77.9
	Silver	154
	Copper	193
12	Titanium	0.172
	Zirconium	1.08
	Lead	1.23
	Tantalum	1.24
	Molybdenum	1.74
	Niobium	2.39
	Chromium	4.49
	Cadmium	5.56
	Nickel	6.91
	Iron	8.13
	Platinum	10.9
	Tin	11.6
	Magnesium	13.1
	Gold	26.7
	Zinc	30.8
	Aluminum	70.8
	Tungsten	72.4
	Silver	139
	Copper	185

Selecting Thermal Conductivity of Metals at Temperature(Continued)

Temperature (K)	Metal	Thermal Conductivity (watt $\cdot$ cm^{-1} $\cdot$ K^{-1})
13	Titanium	0.186
	Lead	1.07
	Zirconium	1.11
	Tantalum	1.3
	Molybdenum	1.88
	Niobium	2.46
	Cadmium	4.67
	Chromium	4.78
	Nickel	7.3
	Iron	8.58
	Tin	9.3
	Platinum	10.1
	Magnesium	13.6
	Gold	25.5
	Zinc	26.1
	Tungsten	66.4
	Aluminum	71.5
	Silver	124
	Copper	176
14	Titanium	0.2
	Lead	0.94
	Zirconium	1.13
	Tantalum	1.36

Selecting Thermal Conductivity of Metals at Temperature(Continued)

Temperature (K)	Metal	Thermal Conductivity (watt $\cdot$ cm^{-1} $\cdot$ K^{-1})
	Molybdenum	2.01
	Niobium	2.49
	Cadmium	4.01
	Chromium	5.04
	Tin	7.6
	Nickel	7.64
	Iron	8.97
	Platinum	9.3
	Magnesium	14
	Zinc	22.4
	Gold	24.1
	Tungsten	60.4
	Aluminum	71.3
	Silver	109
	Copper	166
15	Titanium	0.214
	Lead	0.84
	Zirconium	1.13
	Tantalum	1.4
	Molybdenum	2.15
	Niobium	2.5
	Cadmium	3.55
	Chromium	5.27

Selecting Thermal Conductivity of Metals at Temperature(Continued)

Temperature (K)	Metal	Thermal Conductivity (watt $\cdot$ cm^{-1} $\cdot$ K^{-1})
	Tin	6.3
	Nickel	7.92
	Platinum	8.4
	Iron	9.3
	Magnesium	14.3
	Zinc	19.4
	Gold	22.6
	Tungsten	54.8
	Aluminum	70.2
	Silver	96
	Copper	156
16	Titanium	0.227
	Lead	0.77
	Zirconium	1.12
	Tantalum	1.44
	Molybdenum	2.28
	Niobium	2.49
	Cadmium	3.16
	Tin	5.3
	Chromium	5.48
	Platinum	7.6
	Nickel	8.15
	Iron	9.56

Selecting Thermal Conductivity of Metals at Temperature(Continued)

Temperature (K)	Metal	Thermal Conductivity (watt $\cdot$ cm^{-1} $\cdot$ K^{-1})
	Magnesium	14.4
	Zinc	16.9
	Gold	20.9
	Tungsten	49.3
	Aluminum	68.4
	Silver	85
	Copper	145
18	Titanium	0.254
	Lead	0.66
	Zirconium	1.08
	Tantalum	1.47
	Niobium	2.42
	Molybdenum	2.53
	Cadmium	2.62
	Tin	4
	Chromium	5.81
	Platinum	6.1
	Nickel	8.45
	Iron	9.88
	Zinc	13.3
	Magnesium	14.3
	Gold	17.7
	Tungsten	40

Selecting Thermal Conductivity of Metals at Temperature(Continued)

Temperature (K)	Metal	Thermal Conductivity (watt $\cdot$ cm^{-1} $\cdot$ K^{-1})
	Aluminum	63.5
	Silver	66
	Copper	124
20	Titanium	0.279
	Lead	0.59
	Zirconium	1.01
	Tantalum	1.47
	Cadmium	2.26
	Niobium	2.29
	Molybdenum	2.77
	Tin	3.2
	Platinum	4.9
	Chromium	6.01
	Nickel	8.56
	Iron	9.97
	Zinc	10.7
	Magnesium	13.9
	Gold	15
	Tungsten	32.6
	Silver	51
	Aluminum	56.5
	Copper	105

Selecting Thermal Conductivity of Metals at Temperature(Continued)

Temperature (K)	Metal	Thermal Conductivity (watt $\cdot$ cm^{-1} $\cdot$ K^{-1})
25	Titanium	0.337
	Lead	0.507
	Zirconium	0.85
	Tantalum	1.36
	Cadmium	1.79
	Niobium	1.87
	Tin	2.22
	Platinum	3.15
	Molybdenum	3.25
	Chromium	6.07
	Zinc	6.9
	Nickel	8.15
	Iron	9.36
	Gold	10.2
	Magnesium	12
	Tungsten	20.4
	Silver	29.5
	Aluminum	40
	Copper	68
30	Titanium	0.382
	Lead	0.477
	Zirconium	0.74
	Tantalum	1.16

Selecting Thermal Conductivity of Metals at Temperature(Continued)

Temperature (K)	Metal	Thermal Conductivity (watt $\cdot$ cm^{-1} $\cdot$ K^{-1})
	Niobium	1.45
	Cadmium	1.56
	Tin	1.76
	Platinum	2.28
	Molybdenum	3.55
	Zinc	4.9
	Chromium	5.58
	Nickel	6.95
	Gold	7.6
	Iron	8.14
	Magnesium	9.5
	Tungsten	13.1
	Silver	19.3
	Aluminum	28.5
	Copper	43
35	Titanium	0.411
	Lead	0.462
	Zirconium	0.65
	Tantalum	0.99
	Niobium	1.16
	Cadmium	1.41
	Tin	1.5
	Platinum	1.8

Selecting Thermal Conductivity of Metals at Temperature(Continued)

Temperature (K)	Metal	Thermal Conductivity (watt $\cdot$ cm^{-1} $\cdot$ K^{-1})
	Molybdenum	3.62
	Zinc	3.72
	Chromium	5.03
	Nickel	5.62
	Gold	6.1
	Iron	6.81
	Magnesium	7.4
	Tungsten	8.9
	Silver	13.7
	Aluminum	21
	Copper	29
40	Titanium	0.422
	Lead	0.451
	Zirconium	0.58
	Tantalum	0.87
	Niobium	0.97
	Cadmium	1.32
	Tin	1.35
	Platinum	1.51
	Zinc	2.97
	Molybdenum	3.51
	Chromium	4.3
	Nickel	4.63

Selecting Thermal Conductivity of Metals at Temperature(Continued)

Temperature (K)	Metal	Thermal Conductivity (watt $\cdot$ cm^{-1} $\cdot$ K^{-1})
	Gold	5.2
	Iron	5.55
	Magnesium	5.7
	Tungsten	6.5
	Silver	10.5
	Aluminum	16
	Copper	20.5
45	Titanium	0.416
	Lead	0.442
	Zirconium	0.535
	Tantalum	0.78
	Niobium	0.84
	Tin	1.23
	Cadmium	1.25
	Platinum	1.32
	Zinc	2.48
	Molybdenum	3.26
	Chromium	3.67
	Nickel	3.91
	Iron	4.5
	Magnesium	4.57
	Gold	4.6
	Tungsten	5.07

Selecting Thermal Conductivity of Metals at Temperature(Continued)

Temperature (K)	Metal	Thermal Conductivity (watt $\cdot$ cm^{-1} $\cdot$ K^{-1})
	Silver	8.4
	Aluminum	12.5
	Copper	15.3
50	Titanium	0.401
	Lead	0.435
	Zirconium	0.497
	Tantalum	0.72
	Niobium	0.76
	Tin	1.15
	Platinum	1.18
	Cadmium	1.2
	Zinc	2.13
	Molybdenum	3
	Chromium	3.17
	Nickel	3.36
	Iron	3.72
	Magnesium	3.75
	Tungsten	4.17
	Gold	4.2
	Silver	7
	Aluminum	10
	Copper	12.2

Selecting Thermal Conductivity of Metals at Temperature(Continued)

Temperature (K)	Metal	Thermal Conductivity (watt $\cdot$ cm^{-1} $\cdot$ K^{-1})
60	Titanium	0.377
	Lead	0.424
	Zirconium	0.442
	Tantalum	0.651
	Niobium	0.66
	Platinum	1.01
	Tin	1.04
	Cadmium	1.13
	Zinc	1.71
	Chromium	2.48
	Molybdenum	2.6
	Nickel	2.63
	Iron	2.65
	Magnesium	2.74
	Tungsten	3.18
	Gold	3.8
	Silver	5.5
	Aluminum	6.7
	Copper	8.5
70	Titanium	0.356
	Zirconium	0.403
	Lead	0.415
	Niobium	0.61

Selecting Thermal Conductivity of Metals at Temperature(Continued)

Temperature (K)	Metal	Thermal Conductivity (watt • cm^{-1} • K^{-1})
	Tantalum	0.616
	Platinum	0.9
	Tin	0.96
	Cadmium	1.08
	Zinc	1.48
	Iron	2.04
	Chromium	2.08
	Nickel	2.21
	Magnesium	2.23
	Molybdenum	2.3
	Tungsten	2.76
	Gold	3.58
	Silver	4.97
	Aluminum	5
	Copper	6.7
80	Titanium	0.339
	Zirconium	0.373
	Lead	0.407
	Niobium	0.58
	Tantalum	0.603
	Platinum	0.84
	Tin	0.91
	Cadmium	1.06

Selecting Thermal Conductivity of Metals at Temperature(Continued)

Temperature (K)	Metal	Thermal Conductivity (watt $\cdot$ cm^{-1} $\cdot$ K^{-1})
	Zinc	1.38
	Iron	1.68
	Chromium	1.82
	Nickel	1.93
	Magnesium	1.95
	Molybdenum	2.09
	Tungsten	2.56
	Gold	3.52
	Aluminum	4
	Silver	4.71
	Copper	5.7
90	Titanium	0.324
	Zirconium	0.35
	Lead	0.401
	Niobium	0.563
	Tantalum	0.596
	Platinum	0.81
	Tin	0.88
	Cadmium	1.04
	Zinc	1.34
	Iron	1.46
	Chromium	1.68
	Nickel	1.72

Selecting Thermal Conductivity of Metals at Temperature(Continued)

Temperature (K)	Metal	Thermal Conductivity (watt $\cdot$ cm^{-1} $\cdot$ K^{-1})
	Magnesium	1.78
	Molybdenum	1.92
	Tungsten	2.44
	Aluminum	3.4
	Gold	3.48
	Silver	4.6
	Copper	5.14
100	Titanium	0.312
	Zirconium	0.332
	Lead	0.396
	Niobium	0.552
	Tantalum	0.592
	Platinum	0.79
	Tin	0.85
	Cadmium	1.03
	Zinc	1.32
	Iron	1.32
	Nickel	1.58
	Chromium	1.58
	Magnesium	1.69
	Molybdenum	1.79
	Tungsten	2.35
	Aluminum	3

Selecting Thermal Conductivity of Metals at Temperature(Continued)

Temperature (K)	Metal	Thermal Conductivity (watt $\cdot$ cm^{-1} $\cdot$ K^{-1})
	Gold	3.45
	Silver	4.5
	Copper	4.83
200	Titanium	0.245
	Zirconium	0.252
	Lead	0.366
	Niobium	0.526
	Tantalum	0.575
	Tin	0.733
	Platinum	0.748
	Iron	0.94
	Cadmium	0.993
	Nickel	1.06
	Chromium	1.11
	Zinc	1.26
	Molybdenum	1.43
	Magnesium	1.59
	Tungsten	1.97
	Aluminum	2.37
	Gold	3.27
	Copper	4.13
	Silver	4.3

Selecting Thermal Conductivity of Metals at Temperature(Continued)

Temperature (K)	Metal	Thermal Conductivity (watt $\cdot$ cm^{-1} $\cdot$ K^{-1})
273	Titanium	0.224
	Zirconium	0.232
	Lead	0.355
	Niobium	0.533
	Tantalum	0.574
	Tin	0.682
	Platinum	0.734
	Iron	0.835
	Nickel	0.94
	Chromium	0.948
	Cadmium	0.975
	Zinc	1.22
	Molybdenum	1.39
	Magnesium	1.57
	Tungsten	1.82
	Aluminum	2.36
	Gold	3.18
	Copper	4.01
	Silver	4.28
300	Titanium	0.219
	Zirconium	0.227
	Lead	0.352
	Niobium	0.537

Selecting Thermal Conductivity of Metals at Temperature(Continued)

Temperature (K)	Metal	Thermal Conductivity (watt $\cdot$ cm^{-1} $\cdot$ K^{-1})
	Tantalum	0.575
	Tin	0.666
	Platinum	0.73
	Iron	0.803
	Chromium	0.903
	Nickel	0.905
	Cadmium	0.968
	Zinc	1.21
	Molybdenum	1.38
	Magnesium	1.56
	Tungsten	1.78
	Aluminum	2.37
	Gold	3.15
	Copper	3.98
	Silver	4.27
400	Titanium	0.204
	Zirconium	0.216
	Lead	0.338
	Niobium	0.552
	Tantalum	0.578
	Tin	0.622
	Iron	0.694
	Platinum	0.722

Selecting Thermal Conductivity of Metals at Temperature(Continued)

Temperature (K)	Metal	Thermal Conductivity (watt $\cdot$ cm^{-1} $\cdot$ K^{-1})
	Nickel	0.801
	Chromium	0.873
	Cadmium	0.947
	Zinc	1.16
	Molybdenum	1.34
	Magnesium	1.53
	Tungsten	1.62
	Aluminum	2.4
	Gold	3.12
	Copper	3.92
	Silver	4.2
500	Titanium	0.197
	Zirconium	0.21
	Lead	0.325
	Niobium	0.567
	Tantalum	0.582
	Tin	0.596
	Iron	0.613
	Platinum	0.719
	Nickel	0.721
	Chromium	0.848
	Cadmium	0.92
	Zinc	1.11

Selecting Thermal Conductivity of Metals at Temperature(Continued)

Temperature (K)	Metal	Thermal Conductivity (watt $\cdot$ cm^{-1} $\cdot$ K^{-1})
	Molybdenum	1.3
	Tungsten	1.49
	Magnesium	1.51
	Aluminum	2.37
	Gold	3.09
	Copper	3.88
	Silver	4.13
600	Titanium	0.194
	Zirconium	0.207
	Lead	0.312
	Iron	0.547
	Niobium	0.582
	Tantalum	0.586
	Nickel	0.655
	Platinum	0.72
	Chromium	0.805
	Zinc	1.05
	Molybdenum	1.26
	Tungsten	1.39
	Magnesium	1.49
	Aluminum	2.32
	Gold	3.04

Selecting Thermal Conductivity of Metals at Temperature(Continued)

Temperature (K)	Metal	Thermal Conductivity (watt $\cdot$ cm^{-1} $\cdot$ K^{-1})
	Copper	3.83
	Silver	4.05
700	Titanium	0.194
	Zirconium	0.209
	Iron	0.487
	Tantalum	0.59
	Niobium	0.598
	Nickel	0.653
	Platinum	0.723
	Chromium	0.757
	Molybdenum	1.22
	Tungsten	1.33
	Magnesium	1.47
	Aluminum	2.26
	Gold	2.98
	Copper	3.77
	Silver	3.97
800	Titanium	0.197
	Zirconium	0.216
	Iron	0.433
	Tantalum	0.594

Selecting Thermal Conductivity of Metals at Temperature(Continued)

Temperature (K)	Metal	Thermal Conductivity (watt $\cdot$ cm^{-1} $\cdot$ K^{-1})
	Niobium	0.613
	Nickel	0.674
	Chromium	0.713
	Platinum	0.729
	Molybdenum	1.18
	Tungsten	1.28
	Magnesium	1.46
	Aluminum	2.2
	Gold	2.92
	Copper	3.71
	Silver	3.89
900	Titanium	0.202
	Zirconium	0.226
	Iron	0.38
	Tantalum	0.598
	Niobium	0.629
	Chromium	0.678
	Nickel	0.696
	Platinum	0.737
	Molybdenum	1.15
	Tungsten	1.24
	Magnesium	1.45
	Aluminum	2.13

Selecting Thermal Conductivity of Metals at Temperature(Continued)

Temperature (K)	Metal	Thermal Conductivity (watt $\cdot$ cm^{-1} $\cdot$ K^{-1})
	Gold	2.85
	Copper	3.64
	Silver	3.82
1000	Titanium	0.207
	Zirconium	0.237
	Iron	0.326
	Tantalum	0.602
	Niobium	0.644
	Chromium	0.653
	Nickel	0.718
	Platinum	0.748
	Molybdenum	1.12
	Tungsten	1.21
	Gold	2.78
	Copper	3.57
	Silver	3.74
1100	Titanium	0.213
	Zirconium	0.248
	Iron	0.297
	Tantalum	0.606
	Chromium	0.636
	Niobium	0.659
	Nickel	0.739
	Platinum	0.76

Selecting Thermal Conductivity of Metals at Temperature(Continued)

Temperature (K)	Metal	Thermal Conductivity (watt $\cdot$ cm^{-1} $\cdot$ K^{-1})
	Molybdenum	1.08
	Tungsten	1.18
	Gold	2.71
	Copper	3.5
	Silver	3.66
1200	Titanium	0.22
	Zirconium	0.257
	Iron	0.282
	Tantalum	0.61
	Chromium	0.624
	Niobium	0.675
	Nickel	0.761
	Platinum	0.775
	Molybdenum	1.05
	Tungsten	1.15
	Gold	2.62
	Copper	3.42
	Silver	3.58
1400	Titanium	0.236
	Zirconium	0.275
	Iron	0.309
	Chromium	0.611

Selecting Thermal Conductivity of Metals at Temperature(Continued)

Temperature (K)	Metal	Thermal Conductivity (watt $\cdot$ cm^{-1} $\cdot$ K^{-1})
	Tantalum	0.618
	Niobium	0.705
	Nickel	0.804
	Platinum	0.807
	Molybdenum	0.996
	Tungsten	1.11
1600	Titanium	0.253
	Zirconium	0.29
	Iron	0.327
	Tantalum	0.626
	Niobium	0.735
	Platinum	0.842
	Molybdenum	0.946
	Tungsten	1.07
1800	Titanium	0.271
	Zirconium	0.302
	Tantalum	0.634
	Niobium	0.764
	Platinum	0.877
	Molybdenum	0.907
	Tungsten	1.03
2000	Zirconium	0.313
	Tantalum	0.64
	Niobium	0.791

Selecting Thermal Conductivity of Metals at Temperature(Continued)

Temperature (K)	Metal	Thermal Conductivity (watt $\cdot$ cm^{-1} $\cdot$ K^{-1})
	Molybdenum	0.88
	Platinum	0.913
	Tungsten	1
2200	Tantalum	0.647
	Niobium	0.815
	Molybdenum	0.858
	Tungsten	0.98
2600	Tantalum	0.658
	Molybdenum	0.825
	Tungsten	0.94
3000	Tantalum	0.665
	Tungsten	0.915

These data apply only to metals of purity of at least 99.9%.
The third significant figure may not be accurate.

Source: *data from* Ho, C. Y., Powell, R. W., and Liley, P. *E.*, *Thermal Conductivity of Selected Materials*, NSRDS–NBS–8 and NSRD-NBS-16, Part 2 , National Standard Reference Data System–National Bureau of Standards, Part 1, 1966; Part 2, 1968.

SELECTING THERMAL CONDUCTIVITY OF ALLOY CAST IRONS

Description	Thermal Conductivity W/(m • K)
Heat–Resistant High–Nickel Ductile Iron (20 Ni)	13
Corrosion–Resistant High–Nickel Ductile Iron	13.4
Heat–Resistant Gray High–Chromium Iron	20
Abrasion–Resistant Low–C White Iron	22†
Heat–Resistant Gray Nickel–Chromium–Silicon Iron	30
Abrasion–Resistant Martensitic Nickel–Chromium White Iron	30†
Heat–Resistant Gray Medium–Silicon Iron	37
Heat–Resistant Gray High–Nickel Iron	37 to 40
Corrosion–Resistant High–Nickel Gray Iron	38 to 40

† Estimated.

Source: Data from *ASM Metals Reference Book, Second Edition*, American Society for Metals, Metals Park, Ohio 44073, p172, (1984).

SELECTING THERMAL EXPANSION OF TOOL STEELS

Type	Temperature Change from 20 °C to	Thermal Expansion mm/(m•K)
M2	260°C	9.4
T1	200 °C	9.7
T15	200 °C	9.9
M2	100 °C	10.1
H13	100 °C	10.4
W1	100 °C	10.4
A2	260°C	10.6
A2	100 °C	10.7
W1	200 °C	11
T15	425°C	11
M2	425°C	11.2
T1	425°C	11.2
L6	100 °C	11.3
H13	200 °C	11.5
T15	540°C	11.5
T1	540°C	11.7
H11	100 °C	11.9
M2	540°C	11.9
T1	600°C	11.9
H13	425°C	12.2
M2	600°C	12.2
H21	100 °C	12.4
S1	100 °C	12.4
H11	200 °C	12.4

Selecting Thermal Expansion of Tool Steels (Continued)

Type	Temperature Change from 20 °C to	Thermal Expansion mm/(m•K)
H13	540°C	12.4
H26	540°C	12.4
H21	200 °C	12.6
L6	200 °C	12.6
S1	200 °C	12.6
S7	200 °C	12.6
L6	425°C	12.6
S5	425°C	12.6
H11	425°C	12.8
A2	425°C	12.9
H21	425°C	12.9
H11	540°C	12.9
W1	425°C	13.1
H13	600°C	13.1
S7	425°C	13.3
S5	540°C	13.3
H11	600°C	13.3
S7	600°C	13.3
S1	425°C	13.5
H21	540°C	13.5
L6	540°C	13.5
S7	500°C	13.7
L6	600°C	13.7
S5	600°C	13.7

Selecting Thermal Expansion of Tool Steels (Continued)

Type	Temperature Change from 20 °C to	Thermal Expansion mm/(m•K)
W1	500°C	13.8
S1	540°C	13.9
H21	600°C	13.9
A2	540°C	14
A2	600°C	14.2
S1	600°C	14.2
W1	600°C	14.2
L2	425°C	14.4
L2	540°C	14.6
L2	600°C	14.8

Source: data from *ASM Metals Reference Book, Second Edition*, American Society for Metals, Metals Park, Ohio 44073, p242, (1984).

SELECTING THERMAL EXPANSION OF TOOL STEELS AT TEMPERATURE

Temperature Change from 20 °C to	Type	Thermal Expansion mm/(m•K)
100 °C	M2	10.1
	H13	10.4
	W1	10.4
	A2	10.7
	L6	11.3
	H11	11.9
	H21	12.4
	S1	12.4
200 °C	T1	9.7
	T15	9.9
	W1	11
	H13	11.5
	H11	12.4
	H21	12.6
	L6	12.6
	S1	12.6
	S7	12.6
260°C	M2	9.4
	A2	10.6
425°C	T15	11
	M2	11.2
	T1	11.2
	H13	12.2

Selecting Thermal Expansion of Tool Steels at Temperature (Continued)

Temperature Change from 20 °C to	Type	Thermal Expansion mm/(m•K)
	L6	12.6
	S5	12.6
	H11	12.8
	A2	12.9
	H21	12.9
	W1	13.1
	S7	13.3
	S1	13.5
	L2	14.4
500°C	S7	13.7
	W1	13.8
540°C	T15	11.5
	T1	11.7
	M2	11.9
	H13	12.4
	H26	12.4
	H11	12.9
	S5	13.3
	H21	13.5
	L6	13.5
	S1	13.9
	A2	14
	L2	14.6

Selecting Thermal Expansion of Tool Steels at Temperature (Continued)

Temperature Change from 20 °C to	Type	Thermal Expansion mm/(m•K)
600°C	T1	11.9
	M2	12.2
	H13	13.1
	H11	13.3
	S7	13.3
	L6	13.7
	S5	13.7
	H21	13.9
	A2	14.2
	S1	14.2
	W1	14.2
	L2	14.8

Source: data from *ASM Metals Reference Book, Second Edition*, American Society for Metals, Metals Park, Ohio 44073, p242, (1984).

SELECTING THERMAL EXPANSION OF ALLOY CAST IRONS

Description	Thermal Expansion Coefficient mm/(m • °C)
Abrasion–Resistant White Martensitic Nickel–Chromium Iron	8 to 9
Corrosion–Resistant High–Nickel Gray Iron	8.1 to 19.3
Heat–Resistant Gray High–Nickel Iron	8.1 to 19.3
Heat–Resistant Gray High–Chromium Iron	9.3 to 9.9
Corrosion–Resistant High–Chromium Iron	9.4 to 9.9
Heat–Resistant Gray Medium–Silicon Iron	10.8
Heat–Resistant Medium–Silicon Ductile Iron	10.8 to 13.5
Abrasion–Resistant Low–C White Irons	12
Corrosion–Resistant High– Silicon Iron	12.4 to 13.1
Heat–Resistant Gray Nickel–Chromium–Silicon Iron	12.6 to 16.2
Corrosion–Resistant High–Nickel Ductile Iron	12.6 to 18.7
Heat–Resistant Gray High–Aluminum Iron	15.3
Heat–Resistant High–Nickel Ductile (23 Ni)	18.4
Heat–Resistant High–Nickel Ductile (20 Ni)	18.7

Source: Data from *ASM Metals Reference Book, Second Edition*, American Society for Metals, Metals Park, Ohio 44073, p172, (1984).

Selecting Metals: Mechanical Properties

Continued

Selection of Mechanical Properties: Table of Contents (Continued)

SELECTING TENSILE STRENGTH OF TOOL STEELS

Type	Condition	Tensile Strength (MPa)
S7	Annealed	640
L6	Annealed	655
S1	Annealed	690
L2	Annealed	710
S5	Annealed	725
L2	Oil quenched from 855 °C and single tempered at: 650 °C	930
L6	Oil quenched from 845 °C and single tempered at: 650 °C	965
S5	Oil quenched from 870 °C and single tempered at: 650 °C	1035
S7	Fan cooled from 940 °C and single tempered at: 650 °C	1240
L2	Oil quenched from 855 °C and single tempered at: 540 °C	1275
L6	Oil quenched from 845 °C and single tempered at: 540 °C	1345
S1	Oil quenched from 845 °C and single tempered at: 650 °C	1345
S5	Oil quenched from 870 °C and single tempered at: 540 °C	1520
L2	Oil quenched from 855 °C and single tempered at: 425 °C	1550
L6	Oil quenched from 845 °C and single tempered at: 425 °C	1585
S1	Oil quenched from 845 °C and single tempered at: 540 °C	1680
L2	Oil quenched from 855 °C and single tempered at: 315 °C	1790
S1	Oil quenched from 845 °C and single tempered at: 425 °C	1790
S7	Fan cooled from 940 °C and single tempered at: 540 °C	1820
S5	Oil quenched from 870 °C and single tempered at: 425 °C	1895

Selecting Tensile Strength of Tool Steels (Continued)

Type	Condition	Tensile Strength (MPa)
S7	Fan cooled from 940 °C and single tempered at: 425 °C	1895
S7	Fan cooled from 940 °C and single tempered at: 315 °C	1965
L2	Oil quenched from 855 °C and single tempered at: 205 °C	2000
L6	Oil quenched from 845 °C and single tempered at: 315 °C	2000
S1	Oil quenched from 845 °C and single tempered at: 315 °C	2030
S1	Oil quenched from 845 °C and single tempered at: 205 °C	2070
S7	Fan cooled from 940 °C and single tempered at: 205 °C	2170
S5	Oil quenched from 870 °C and single tempered at: 315 °C	2240
S5	Oil quenched from 870 °C and single tempered at: 205 °C	2345

Source: data from *ASM Metals Reference Book, Second Edition*, American Society for Metals, Metals Park, Ohio 44073, p241, (1984).

SELECTING TENSILE STRENGTH OF GRAY CAST IRONS

SAE grade	Maximum Tensile Strength (MPa)
G1800	118
G2500	173
G2500a	173
G3000	207
C3500	241
G4000	276
G3500b	1241
G3500c	1241
G4000d	1276

Grey Cast Iron Bars

ASTM Class	Tensile Strength (MPa)
20	152
25	179
30	214
35	252
40	293
50	362
60	431

Source: data from *ASM Metals Reference Book, Second Edition*, American Society for Metals, Metals Park, Ohio 44073, p166-167, (1984).

SELECTING TENSILE STRENGTH OF DUCTILE IRONS

Specification Number	Grade or Class	Tensile Strength (MPa)
MlL-I-24137(Ships)	Class C	345
MlL-I-24137(Ships)	Class B	379
ASTM A395-76; ASME SA395	60-40-18	414
ASTM A536-72; MIL-1-11466B(MR)	60-40-18	414
SAE J434c	D4018	414
MlL-I-24137(Ships)	Class A	414
ASTM A536-72; MIL-1-11466B(MR)	65-45-12	448
SAE J434c	D4512	448
ASTM A476-70(d); SAE AMS5316	80-60-03	552
ASTM A536-72; MIL-1-11466B(MR)	80-55-06	552
SAE J434c	D5506	552
ASTM A536-72; MIL-1-11466B(MR)	100-70-03	689
SAE J434c	D7003	689
ASTM A536-72; MIL-1-11466B(MR)	120-90-02	827

Source: data from *ASM Metals Reference Book, Second Edition*, American Society for Metals, Metals Park, Ohio 44073, p169, (1984).

SELECTING TENSILE STRENGTHS OF MALLEABLE IRON CASTINGS

Specification Number	grade or class	Tensile Strength (MPa)
ASTM A197		276
ASTM A47, A338; ANSI G48.1; FED QQ–I–666c	32510	345
ASTM A602; SAE J158	M3210(c)	345
ASTM A47, A338; ANSI G48.1; FED QQ–I–666c	35018	365
ASTM A220; ANSI C48.2; MIL–I–11444B	40010	414
ASTM A220; ANSI C48.2; MIL–I–11444B	45008	448
ASTM A220; ANSI C48.2; MIL–I–11444B	45006	448
ASTM A602; SAE J158	M4504(d)	448
ASTM A220; ANSI C48.2; MIL–I–11444B	50005	483
ASTM A602; SAE J158	M5003(d)	517
ASTM A602; SAE J158	M5503(e)	517
ASTM A220; ANSI C48.2; MIL–I–11444B	60004	552
ASTM A220; ANSI C48.2; MIL–I–11444B	70003	586
ASTM A602; SAE J158	M7002(e)	621
ASTM A220; ANSI C48.2; MIL–I–11444B	80002	655
ASTM A220; ANSI C48.2; MIL–I–11444B	90001	724
ASTM A602; SAE J158	M8501(e)	724

Source: data from *ASM Metals Reference Book, Second Edition*, American Society for Metals, Metals Park, Ohio 44073, p171, (1984).

SELECTING TENSILE STRENGTHS OF ALUMINUM CASTING ALLOYS

Alloy AA No.	Temper	Tensile Strength (MPa)
443.0	F	130
208.0	F	145
B443.0	F	159
850.0	T5	160
514.0	F	170
355.0	T71	175
356.0	T51	175
A390.0	F,T5	180
242.0	T21	185
319.0	F	185
308.0	F	195
355.0	T51	195
356.0	T71	195
A390.0	F,T5	200
242.0	T77	205
355.0	T51	210
713.0	T5	210
242.0	T571	220
295.0	T4	220
356.0	T7	220
713.0	T5	220
C443.0	F	228
356.0	T6	230
319.0	F	235

Selecting Tensile Strengths of Aluminum Casting Alloys (Continued)

Alloy AA No.	Temper	Tensile Strength (MPa)
356.0	T7	235
355.0	T6	240
712.0	F	240
295.0	T6	250
319.0	T6	250
336.0	T551	250
355.0	T71	250
A390.0	T7	250
296.0	T4	255
A390.0	T7	260
355.0	T7	265
356.0	T6	265
296.0	T7	270
355.0	T61	270
242.0	T571	275
296.0	T6	275
535.0	F	275
319.0	T6	280
355.0	T7	280
390.0	F	280
A390.0	T6	280
295.0	T62	285
355.0	T6	290
A413.0	F	290

Selecting Tensile Strengths of Aluminum Casting Alloys (Continued)

Alloy AA No.	Temper	Tensile Strength (MPa)
390.0	T5	300
413.0	F	300
355.0	T62	310
383.0	F	310
A390.0	T6	310
518.0	F	310
A360.0	F	320
242.0	T61	325
336.0	T65	325
360.0	F	325
359.0	T61	330
380.0	F	330
384.0, A384.0	F	330
520.0	T4	330
359.0	T62	345
771.0	T6	345
357.0, A357.0	T62	360
201.0	T4	365
354.0	T61	380
206.0, A206.0	T7	435
201.0	T7	460
201.0	T6	485

Source: data from *ASM Metals Reference Book, Second Edition*, American Society for Metals, Metals Park, Ohio 44073, (1984).

SELECTING TENSILE STRENGTHS OF WROUGHT ALUMINUM ALLOYS

Alloy	Temper	Tensile Strength (MPa)
1060	0	69
1050	0	76
1060	H12	83
1350	0	83
1100	0	90
6063	0	90
1060	H14	97
1350	H12	97
6101	H111	97
1050	H14	110
1060	H16	110
1100	H12	110
1350	H14	110
3003	0	110
3105	0	115
Alclad 6061	0	115
1100	H14	125
1350	H16	125
5005	0	125
6061	0	125
1050	H16	130
1060	H18	130
Alclad	H12	130
5457	0	130

Selecting Tensile Strengths of Wrought Aluminum Alloys (Continued)

Alloy	Temper	Tensile Strength (MPa)
5005	H12	140
5005	H32	140
1100	H16	145
4043	0	145
5050	0	145
6070	0	145
3003	H14	150
3105	H12	150
6063	TI	150
6066	0	150
6463	Tl	150
1050	H18	160
5005	H14	160
5005	H34	160
5657	H25	160
1100	H18	165
Alclad 2014	0	170
2219	0	170
3105	H14	170
5050	H32	170
6005	T1	170
6063	T4	170
Alclad 2024	0	180
3003	H16	180

Selecting Tensile Strengths of Wrought Aluminum Alloys (Continued)

Alloy	Temper	Tensile Strength (MPa)
3004	0	180
3105	H25	180
5005	H16	180
5005	H36	180
5457	H25	180
1350	H19	185
2014	0	185
2024	0	185
6063	T5	185
6463	T5	185
7005	0	193
3105	H16	195
5050	H34	195
5052	0	195
5652	0	195
5657	H28, H38	195
3003	H18	200
5005	H18	200
5005	H38	200
5050	H36	205
5457	H28, H38	205
6063	T831	205
Alclad	H32	215
3105	H18	215

Selecting Tensile Strengths of Wrought Aluminum Alloys (Continued)

Alloy	Temper	Tensile Strength (MPa)
5050	H38	220
6151	T6	220
Alclad 7075	0	220
5052	H32	230
5652	H32	230
Alclad 6061	T4, T451	230
7075	0	230
5252	H25	235
6009	T4	235
3004	H34	240
5154	0	240
5154	H112	240
5254	0	240
5254	H112	240
6061	T4, T451	240
6063	T6	240
6463	T6	240
5454	0	250
5454	H112	250
6351	T4	250
6010	T4	255
6063	T83	255
3004	H36	260
5052	H34	260

Selecting Tensile Strengths of Wrought Aluminum Alloys (Continued)

Alloy	Temper	Tensile Strength (MPa)
5086	0	260
5454	H111	260
5454	H311	260
5652	H34	260
6005	T5	260
6205	Tl	260
5086	H112	270
5154	H32	270
5254	H32	270
5052	H36	275
5182	0	275
5454	H32	275
5652	H36	275
3004	H38	285
4043	H18	285
5252	H28, H38	285
5052	H38	290
5056	0	290
5083	0	290
5086	H32, H116, H117	290
5154	H34	290
5254	H34	290
5652	H38	290
Alclad 6061	T6, T651	290

Selecting Tensile Strengths of Wrought Aluminum Alloys (Continued)

Alloy	Temper	Tensile Strength (MPa)
6063	T832	290
5083	H112	305
5454	H34	305
5154	H36	310
5254	H36	310
5456	0	310
5456	H112	310
6061	T6, T651	310
6205	T5	310
6351	T6	310
5083	H113	315
5083	H321	315
5182	H32	315
6070	T4	315
5083	H323, H32	325
5086	H34	325
5456	H111	325
2218	T72	330
5154	H38	330
5254	H38	330
6201	T6	330
6201	T81	330
2036	T4	340
5182	H34	340

Selecting Tensile Strengths of Wrought Aluminum Alloys (Continued)

Alloy	Temper	Tensile Strength (MPa)
5454	H36	340
2218	T71	345
5083	H343, H34	345
6009	T6	345
5456	H321, H116	350
2219	T42	360
2219	T31, T351	360
6066	T4, T451	360
5454	H38	370
7005	T6,T63,T6351	372
2011	T3	380
4032	T6	380
6070	T6	380
7005	T53	393
2219	T37	395
6066	T6, T651	395
6262	T9	400
2011	T8	405
2218	T61	405
2219	T62	415
5056	H38	415
Alclad 2014	T4	420
5182	H19(n)	420
2014	T4	425

Selecting Tensile Strengths of Wrought Aluminum Alloys (Continued)

Alloy	Temper	Tensile Strength (MPa)
Alclad 2014	T3	435
5056	H18	435
Alclad 2024	T4, T351	440
2618	All	440
Alclad 2024	T	450
Alclad 2024	T81, T851	450
2048		455
2219	T81, T851	455
Alclad 2024	T361	460
Alclad 2014	T6	470
2024	T4, T351	470
2219	T87	475
2014	T6	485
2024	T3	485
Alclad 2024	T861	485
2124	T851	490
2024	T361	495
7075	T73	505
7050	T736	515
Alclad 7075	T6,T651	525
7175	T736	525
7475	T61	525
7075	T6,T651	570
7175	T66	595

Source: data from *ASM Metals Reference Book, Second Edition*, American Society for Metals, Metals Park, Ohio 44073, p.299—302, (1984).

SELECTING COMPRESSIVE STRENGTHS OF GRAY CAST IRON BARS

ASTM Class	Compressive Strength (MPa)
20	572
25	669
30	752
35	855
40	965
50	1130
60	1293

Source: data from *ASM Metals Reference Book, Second Edition*, American Society for Metals, Metals Park, Ohio 44073, p166-167, (1984).

SELECTING YIELD STRENGTHS OF TOOL STEELS

Type	Condition	0.2% Yield Strength (MPa)
L6	Annealed	380
S7	Annealed	380
S1	Annealed	415
S5	Annealed	440
L2	Annealed	510
L2	Oil quenched from 855 °C and single tempered at 650 °C	760
L6	Oil quenched from 845 °C and single tempered at 650 °C	830
S7	Fan cooled from 940 °C and single tempered a 650 °C	1035
L6	Oil quenched from 845 °C and single tempered at 540 °C	1100
L2	Oil quenched from 855 °C and single tempered at 540 °C	1170
S5	Oil quenched from 870 °C and single tempered a 650 °C	1170
S1	Oil quenched from 930 °C and single tempered at 650 °C	1240
L2	Oil quenched from 855 °C and single tempered at 425 °C	1380
L6	Oil quenched from 845 °C and single tempered at 425 °C	1380
S5	Oil quenched from 870 °C and single tempered a 540 °C	1380
S7	Fan cooled from 940 °C and single tempered a 540 °C	1380
S7	Fan cooled from 940 °C and single tempered a 425 °C	1410
S7	Fan cooled from 940 °C and single tempered a 205 °C	1450
S1	Oil quenched from 930 °C and single tempered at 540 °C	1525
S7	Fan cooled from 940 °C and single tempered a 315 °C	1585

Selecting Yield Strengths of Tool Steels (Continued)

Type	Condition	0.2% Yield Strength (MPa)
L2	Oil quenched from 855 °C and single tempered at 315 °C	1655
S1	Oil quenched from 930 °C and single tempered at 425 °C	1690
S5	Oil quenched from 870 °C and single tempered a 425 °C	1690
L2	Oil quenched from 855 °C and single tempered at 205 °C	1790
L6	Oil quenched from 845 °C and single tempered at 315 °C	1790
S1	Oil quenched from 930 °C and single tempered at 315 °C	1860
S5	Oil quenched from 870 °C and single tempered a 315 °C	1860
S1	Oil quenched from 930 °C and single tempered at 205 °C	1895
S5	Oil quenched from 870 °C and single tempered a 205 °C	1930

Source: Data from *ASM Metals Reference Book, Second Edition*, American Society for Metals, Metals Park, Ohio 44073, p241, (1984).

SELECTING YIELD STRENGTHS OF DUCTILE IRONS

Specification Number	Grade or Class	Yield Strength (MPa)
MIL-I-24137(Ships)	Class C	172
MIL-I-24137(Ships)	Class B	207
ASTM A395-76; ASME SA395	60-40-18	276
ASTM A536-72, MIL-1-11466B(MR)	60-40-18	276
SAE J434c	D4018	276
ASTM A536-72, MIL-1-11466B(MR)	65-45-12	310
SAE J434c	D4512	310
MIL-I-24137(Ships)	Class A	310
ASTM A536-72, MIL-1-11466B(MR)	80-55-06	379
SAE J434c	D5506	379
ASTM A476-70(d); SAE AMS5316	80-60-03	414
ASTM A536-72, MIL-1-11466B(MR)	100-70-03	483
SAE J434c	D7003	483
ASTM A536-72, MIL-1-11466B(MR)	120-90-02	621

Source: data from *ASM Metals Reference Book, Second Edition*, American Society for Metals, Metals Park, Ohio 44073, p169, (1984).

SELECTING YIELD STRENGTHS OF MALLEABLE IRON CASTINGS

Specification Number	Grade or Class	Yield Strength (MPa)
ASTM A197		207
ASTM A47, A338; ANSI G48.1; FED QQ–I–666c	32510	224
ASTM A602; SAE J158	M3210(c)	224
ASTM A47, A338; ANSI G48.1; FED QQ–I–666c	35018	241
ASTM A220; ANSI C48.2; MIL–I–11444B	40010	276
ASTM A220; ANSI C48.2; MIL–I–11444B	45008	310
ASTM A220; ANSI C48.2; MIL–I–11444B	45006	310
ASTM A602; SAE J158	M4504(d)	310
ASTM A220; ANSI C48.2; MIL–I–11444B	50005	345
ASTM A602; SAE J158	M5003(d)	345
ASTM A602; SAE J158	M5503(e)	379
ASTM A220; ANSI C48.2; MIL–I–11444B	60004	414
ASTM A220; ANSI C48.2; MIL–I–11444B	70003	483
ASTM A602; SAE J158	M7002(e)	483
ASTM A220; ANSI C48.2; MIL–I–11444B	80002	552
ASTM A602; SAE J158	M8501(e)	586
ASTM A220; ANSI C48.2; MIL–I–11444B	90001	621

d Air quenched and tempered
e Liquid quenched and tempered

Source: data from *ASM Metals Reference Book, Second Edition*, American Society for Metals, Metals Park, Ohio 44073, p171, (1984).

SELECTING YIELD STRENGTHS OF CAST ALUMINUM ALLOYS

Alloy AA No.	Temper	Yield Strength (MPa)
443.0	F	55
B443.0	F	62
850.0	T5	75
514.0	F	85
208.0	F	97
295.0	T4	110
308.0	F	110
C443.0	F	110
242.0	T21	125
319.0	F	125
296.0	T4	130
319.0	F	130
A413.0	F	130
296.0	T7	140
356.0	T51	140
413.0	F	140
535.0	F	140
356.0	T71	145
383.0	F	150
713.0	T5	150
713.0	T5	150
242.0	T77	160
355.0	T51	160
295.0	T6	165

Selecting Yield Strengths of Cast Aluminum Alloys (Continued)

Alloy AA No.	Temper	Yield Strength (MPa)
319.0	T6	165
355.0	T51	165
356.0	T6	165
356.0	T7	165
A360.0	F	165
380.0	F	165
384.0, A384.0	F	165
360.0	F	170
712.0	F	170
355.0	T6	175
296.0	T6	180
A390.0	F,T5	180
520.0	T4	180
319.0	T6	185
356.0	T6	185
355.0	T6	190
518.0	F	190
336.0	T551	195
355.0	T71	200
A390.0	F,T5	200
242.0	T571	205
355.0	T7	210
356.0	T7	210
201.0	T4	215

Selecting Yield Strengths of Cast Aluminum Alloys (Continued)

Alloy AA No.	Temper	Yield Strength (MPa)
355.0	T71	215
295.0	T62	220
242.0	T571	235
355.0	T61	240
390.0	F	240
355.0	T7	250
A390.0	T7	250
359.0	T61	255
390.0	T5	260
A390.0	T7	260
771.0	T6	275
355.0	T62	280
A390.0	T6	280
354.0	T61	285
242.0	T61	290
357.0, A357.0	T62	290
359.0	T62	290
336.0	T65	295
A390.0	T6	310
206.0, A206.0	T7	345
201.0	T7	415
201.0	T6	435

Source: data from *ASM Metals Reference Book, Second Edition*, American Society for Metals, Metals Park, Ohio 44073, (1984).

SELECTING YIELD STRENGTHS OF WROUGHT ALUMINUM ALLOYS

Alloy	Yield Strength Temper	(MPa)
1050	O	28
1060	O	28
1350	O	28
1100	O	34
5005	O	41
3003	O	42
5457	O	48
Alclad 6061	O	48
6063	O	48
3105	O	55
5050	O	55
6061	O	55
Alclad 2014	O	69
3004	O	69
4043	O	69
6070	O	69
1060	H12	76
2024	O	76
Alclad 2024	O	76
2219	O	76
6101	Hlll	76
1350	H12	83
6066	O	83
7005	O	83

Selecting Yield Strengths of Wrought Aluminum Alloys (Continued)

Alloy	Yield Strength Temper	(MPa)
1060	H14	90
5052	0	90
5652	0	90
6063	T1	90
6063	T4	90
6463	Tl	90
Alclad 7075	0	95
1350	H14	97
2014	0	97
1050	H14	105
1060	H16	105
1100	H12	105
6005	T1	105
7075	0	105
1350	H16	110
1100	H14	115
5005	H32	115
5086	0	115
5154	0	115
5154	H112	115
5254	0	115
5254	H112	115
5454	0	115
1050	H16	125

Selecting Yield Strengths of Wrought Aluminum Alloys (Continued)

Alloy	Temper	Yield Strength (MPa)
1060	H18	125
Alclad	H12	125
5454	H112	125
3105	H12	130
5005	H12	130
5086	H112	130
6009	T4	130
Alclad 6061	T4, T451	130
1100	H16	140
5005	H34	140
5182	0	140
5657	H25	140
6205	Tl	140
1050	H18	145
3003	H14	145
5050	H32	145
5083	0	145
6061	T4, T451	145
6063	T5	145
6463	T5	145
1100	H18	150
3105	H14	150
5005	H14	150
5056	0	150

Selecting Yield Strengths of Wrought Aluminum Alloys (Continued)

Alloy	Yield Strength Temper	(MPa)
6351	T4	150
3105	H25	160
5456	0	160
5457	H25	160
1350	H19	165
5005	H36	165
5050	H34	165
5456	H112	165
5657	H28, H38	165
3003	H16	170
Alclad	H32	170
3105	H16	170
5005	H16	170
5252	H25	170
6010	T4	170
6070	T4	170
5050	H36	180
5454	H111	180
5454	H311	180
2219	T42	185
3003	H18	185
5005	H38	185
5457	H28, H38	185
6063	T831	185

Selecting Yield Strengths of Wrought Aluminum Alloys (Continued)

Alloy	Temper	Yield Strength (MPa)
2036	T4	195
3105	H18	195
5005	H18	195
5052	H32	195
5083	H112	195
5652	H32	195
6151	T6	195
3004	H34	200
5050	H38	200
5086	H32, H116, H117	205
5154	H32	205
5254	H32	205
5454	H32	205
6066	T4, T451	205
5052	H34	215
5652	H34	215
6063	T6	215
6463	T6	215
3004	H36	230
5083	H113	230
5083	H321	230
5154	H34	230
5254	H34	230
5456	H111	230

Selecting Yield Strengths of Wrought Aluminum Alloys (Continued)

Alloy	Yield Strength Temper	(MPa)
5182	H32	235
5052	H36	240
5252	H28, H38	240
5454	H34	240
5652	H36	240
6005	T5	240
6063	T83	240
2219	T31, T351	250
3004	H38	250
5083	H323, H32	250
5154	H36	250
5254	H36	250
Alclad 2014	T4	255
2218	T72	255
5052	H38	255
5086	H34	255
5456	H321, H116	255
5652	H38	255
Alclad 6061	T6, T651	255
4043	H18	270
5154	H38	270
5254	H38	270
6063	T832	270
Alclad 2014	T3	275

Selecting Yield Strengths of Wrought Aluminum Alloys (Continued)

Alloy	Temper	Yield Strength (MPa)
2218	T71	275
5454	H36	275
6061	T6, T651	275
5083	H343, H34	285
5182	H34	285
6351	T6	285
2014	T4	290
Alclad 2024	T4, T351	290
2219	T62	290
6205	T5	290
2011	T3	295
6201	T6	300
2218	T61	305
2011	T8	310
Alclad 2024	T	310
5454	H38	310
6201	T81	310
2219	T37	315
4032	T6	315
7005	T6,T63,T6351	315
2024	T4, T351	325
6009	T6	325
2024	T3	345
5056	H38	345

Selecting Yield Strengths of Wrought Aluminum Alloys (Continued)

Alloy	Yield Strength Temper	(MPa)
7005	T53	345
2219	T81, T851	350
6070	T6	350
6066	T6, T651	360
Alclad 2024	T361	365
2618	All	370
6262	T9	380
2024	T361	395
2219	T87	395
5182	H19(n)	395
5056	H18	405
2014	T6	415
Alclad 2014	T6	415
Alclad 2024	T81, T851	415
2048		415
7075	T73	435
2124	T851	440
Alclad 2024	T861	455
7050	T736	455
7175	T736	455
Alclad 7075	T6,T651	460
7475	T61	460
7075	T6,T651	505
7175	T66	525

Source: data from *ASM Metals Reference Book, Second Edition*, American Society for Metals, Metals Park, Ohio 44073, p.299–302, (1984).

SELECTING SHEAR STRENGTHS OF WROUGHT ALUMINUM ALLOYS

Alloy AA No.	Temper	Shear Strength (MPa)
1060	0	48
1060	H12	55
1350	0	55
7072	0	55
1050	0	62
1060	H14	62
1100	0	62
1350	H12	62
7072	H12	62
1050	H14	69
1060	H16	69
1100	H12	69
1350	H14	69
6063	0	69
7072	H14	69
1050	H16	76
1060	H18	76
1100	H14	76
1350	H16	76
3003	0	76
5005	0	76
Alclad 6061	0	76
1050	H18	83
1100	H16	83

Selecting Shear Strengths of Wrought Aluminum Alloys (Continued)

Alloy AA No.	Temper	Shear Strength (MPa)
Alclad	H12	83
3105	0	83
5457	0	83
6061	0	83
1100	H18	90
3003	H14	97
3105	H12	97
5005	H12	97
5005	H14	97
5005	H32	97
5005	H34	97
5657	H25	97
6063	T1	97
6066	0	97
6070	0	97
6463	T1	97
1350	H19	105
3003	H16	105
3105	H14	105
3105	H25	105
5005	H16	105
5005	H36	105
5050	0	105
5657	H28, H38	105

Selecting Shear Strengths of Wrought Aluminum Alloys (Continued)

Alloy AA No.	Temper	Shear Strength (MPa)
3003	H18	110
3004	0	110
3105	H16	110
5005	H18	110
5005	H38	110
5457	H25	110
Alclad	H32	115
3105	H18	115
5050	H32	115
6063	T5	115
6463	T5	115
7005	0	117
2014	0	125
Alclad 2014	0	125
2024	0	125
Alclad 2024	0	125
3004	H34	125
5050	H34	125
5052	0	125
5457	H28, H38	125
5652	0	125
6063	T831	125
5050	H36	130
3004	H36	140

Selecting Shear Strengths of Wrought Aluminum Alloys (Continued)

Alloy AA No.	Temper	Shear Strength (MPa)
5050	H38	140
5052	H32	140
5652	H32	140
6151	T6	140
3004	H38	145
5052	H34	145
5252	H25	145
5652	H34	145
5154	0	150
5154	H32	150
5182	0	150
5254	0	150
5254	H32	150
6009	T4	150
Alclad 6061	T4, T451	150
6063	T6	150
6063	T83	150
6463	T6	150
7075	0	150
Alclad 7075	0	150
5052	H36	160
5086	0	160
5252	H28, H38	160
5454	0	160

Selecting Shear Strengths of Wrought Aluminum Alloys (Continued)

Alloy AA No.	Temper	Shear Strength (MPa)
5454	H111	160
5454	H112	160
5454	H311	160
5652	H36	160
5052	H38	165
5154	H34	165
5254	H34	165
5454	H32	165
5652	H38	165
6061	T4, T451	165
5083	0	170
5056	0	180
5154	H36	180
5254	H36	180
5454	H34	180
5086	H34	185
Alclad 6061	T6, T651	185
6063	T832	185
5154	H38	195
5254	H38	195
6066	T4, T451	200
6351	T6	200
2218	T72	205
5456	H321, H116	205

Selecting Shear Strengths of Wrought Aluminum Alloys (Continued)

Alloy AA No.	Temper	Shear Strength (MPa)
6005	T5	205
6061	T6, T651	205
6070	T4	205
6205	T5	205
7005	T6,T63,T6351	214
2011	T3	220
5056	H38	220
7005	T53	221
5056	H18	235
6066	T6, T651	235
6070	T6	235
2011	T8	240
6262	T9	240
Alclad 2014	T3	255
Alclad 2014	T4	255
2014	T4	260
2618	All	260
4032	T6	260
7475	T7351	270
7475	T7651	270
Alclad 2024	T	275
Alclad 2024	T4, T351	275
Alclad 2024	T81, T851	275
Alclad 2014	T6	285

Selecting Shear Strengths of Wrought Aluminum Alloys (Continued)

Alloy AA No.	Temper	Shear Strength (MPa)
2024	T3	285
2024	T4, T351	285
Alclad 2024	T361	285
2014	T6	290
2024	T361	290
Alclad 2024	T861	290
7175	T736	290
7475	T651	295
Alclad 7075	T6,T651	315
7175	T66	325
7075	T6,T651	330

Source: Data from *ASM Metals Reference Book, Second Edition*, American Society for Metals, Metals Park, Ohio 44073, (1984).

SELECTING TORSIONAL SHEAR STRENGTHS OF GRAY CAST IRON BARS

ASTM Class	Torsional Shear Strength (MPa)
20	179
25	220
30	276
35	334
40	393
50	503
60	610

Source: data from *ASM Metals Reference Book, Second Edition*, American Society for Metals, Metals Park, Ohio 44073, p166-167, (1984).

SELECTING HARDNESS OF TOOL STEELS

Type	Condition	Hardness (HRC)
S7	Annealed	95 HRB
L2	Annealed	96 HRB
S1	Annealed	96 HRB
S5	Annealed	96 HRB
L2	Oil quenched from 855 °C and single tempered at 650 °C	30
L6	Oil quenched from 845 °C and single tempered at 315 °C 650 °C	32
S5	Oil quenched from 870 °C and single tempered at 650 °C	37
S7	Fan cooled from 940 °C and single tempered at 650 °C	39
L2	Oil quenched from 855 °C and single tempered at 540 °C	41
L6	Oil quenched from 845 °C and single tempered at 315 °C 540 °C	42
S1	Oil quenched from 930 °C and single tempered at 650 °C	42
L6	Oil quenched from 845 °C and single tempered at 315 °C 425 °C	46
L2	Oil quenched from 855 °C and single tempered at 425 °C	47
S1	Oil quenched from 930 °C and single tempered at 540 °C	47.5
S5	Oil quenched from 870 °C and single tempered at 540 °C	48
S1	Oil quenched from 930 °C and single tempered at 425 °C	50.5
S7	Fan cooled from 940 °C and single tempered at 540 °C	51
L2	Oil quenched from 855 °C and single tempered at 315 °C	52
S5	Oil quenched from 870 °C and single tempered at 425 °C	52
S7	Fan cooled from 940 °C and single tempered at 425 °C	53

Selecting Hardness of Tool Steels (Continued)

Type	Condition	Hardness (HRC)
L2	Oil quenched from 855 °C and single tempered at 205 °C	54
L6	Oil quenched from 845 °C and single tempered at 315 °C	54
S1	Oil quenched from 930 °C and single tempered at 315 °C	54
S7	Fan cooled from 940 °C and single tempered at 315 °C	55
S1	Oil quenched from 930 °C and single tempered at 205 °C	57.5
S5	Oil quenched from 870 °C and single tempered at 315 °C	58
S7	Fan cooled from 940 °C and single tempered at 205 °C	58
S5	Oil quenched from 870 °C and single tempered at 205 °C	59

Source: Data from *ASM Metals Reference Book, Second Edition*, American Society for Metals, Metals Park, Ohio 44073, p241, (1984).

SELECTING HARDNESS OF GRAY CAST IRONS

SAE grade	Hardness (HB)
G2500	170 to 229
G2500a	170 to 229
G1800	187 max
G3000	187 to 241
C3500	207 to 255
G3500b	207 to 255
G3500c	207 to 255
G4000	217 to 269
G4000d	241 to 321

Gray Cast Iron Bars ASTM Class	Hardness (HB)
20	156
25	174
30	210
35	212
40	235
50	262
60	302

Source: data from *ASM Metals Reference Book, Second Edition*, American Society for Metals, Metals Park, Ohio 44073, p166-167, (1984).

SELECTING HARDNESS OF DUCTILE IRONS

Specification Number	Grade or Class	Hardness (HB)
ASTM A395-76; ASME SA395	60-40-18	143-187
SAE J434c	D4512	156-217
SAE J434c	D4018	170 max
MlL-I-24137(Ships)	Class C	175 max
SAE J434c	D5506	187-255
MlL-I-24137(Ships)	Class A	190 max
MlL-I-24137(Ships)	Class B	190 max
ASTM A476-70(d); SAE AMS5316	80-60-03	201 min
SAE J434c	D7003	241-302

Source: data from *ASM Metals Reference Book, Second Edition*, American Society for Metals, Metals Park, Ohio 44073, p169, (1984).

SELECTING HARDNESS OF MALLEABLE IRON CASTINGS

Specification Number	Grade or Class	Hardness (HB)
ASTM A220; ANSI C48.2; MIL–I–11444B	40010	149–197
ASTM A47, A338; ANSI G48.1; FED QQ–I–666c	32510	156 max
ASTM A47, A338; ANSI G48.1; FED QQ–I–666c	35018	156 max
ASTM A197		156 max
ASTM A602; SAE J158	M3210	156 max
ASTM A220; ANSI C48.2; MIL–I–11444B	45008	156–197
ASTM A220; ANSI C48.2; MIL–I–11444B	45006	156–207
ASTM A602; SAE J158	M4504(a)	163–217
ASTM A220; ANSI C48.2; MIL–I–11444B	50005	179–229
ASTM A602; SAE J158	M5003(a)	187–241
ASTM A602; SAE J158	M5503(b)	187–241
ASTM A220; ANSI C48.2; MIL–I–11444B	60004	197–241
ASTM A220; ANSI C48.2; MIL–I–11444B	70003	217–269
ASTM A602; SAE J158	M7002(b)	229–269
ASTM A220; ANSI C48.2; MIL–I–11444B	80002	241–285
ASTM A602; SAE J158	M8501(b)	269–302
ASTM A220; ANSI C48.2; MIL–I–11444B	90001	269–321

[a] Air quenched and tempered
[b] Liquid quenched and tempered

Source: data from *ASM Metals Reference Book, Second Edition*, American Society for Metals, Metals Park, Ohio 44073, p171, (1984).

SELECTING HARDNESS OF WROUGHT ALUMINUM ALLOYS

Alloy AA No.	Temper	Hardness (BHN)
1060	0	19
7072	0	20
1060	H12	23
1100	0	23
6063	0	25
1060	H14	26
1100	H12	28
3003	0	28
5005	0	28
7072	H12	28
1060	H16	30
6061	0	30
1100	H14	32
5457	0	32
7072	H14	32
1060	H18	35
Alclad	H12	35
6070	0	35
5005	H32	36
5050	0	36
1100	H16	38
3003	H14	40
5657	H25	40
5005	H34	41

Selecting Hardness of Wrought Aluminum Alloys (Continued)

Alloy AA No.	Temper	Hardness (BHN)
6063	T1	42
6463	T1	42
6066	0	43
1100	H18	44
2014	0	45
3004	0	45
5005	H36	46
5050	H32	46
2024	0	47
3003	H16	47
5052	0	47
5652	0	47
5457	H25	48
5657	H28, H38	50
5005	H38	51
Alclad	H32	52
5050	H34	53
3003	H18	55
5457	H28, H38	55
5050	H36	58
5154	0	58
5182	0	58
5254	0	58
5052	H32	60

Selecting Hardness of Wrought Aluminum Alloys (Continued)

Alloy AA No.	Temper	Hardness (BHN)
5652	H32	60
6063	T5	60
6463	T5	60
7075	0	60
5454	0	62
5454	H112	62
3004	H34	63
5050	H38	63
5154	H112	63
5254	H112	63
5056	0	65
6061	T4, T451	65
6205	T1	65
5154	H32	67
5254	H32	67
5052	H34	68
5252	H25	68
5652	H34	68
3004	H36	70
5454	H111	70
5454	H311	70
6009	T4	70
6063	T831	70
6151	T6	71

Selecting Hardness of Wrought Aluminum Alloys (Continued)

Alloy AA No.	Temper	Hardness (BHN)
5052	H36	73
5154	H34	73
5254	H34	73
5454	H32	73
5652	H36	73
6063	T6	73
6463	T6	74
5252	H28, H38	75
6010	T4	76
3004	H38	77
5052	H38	77
5652	H38	77
5154	H36	78
5254	H36	78
5154	H38	80
5254	H38	80
5454	H34	81
6063	T83	82
5456	H321, H116	90
6066	T4, T451	90
6070	T4	90
6201	T6	90
2011	T3	95
2218	T72	95

Selecting Hardness of Wrought Aluminum Alloys (Continued)

Alloy AA No.	Temper	Hardness (BHN)
6005	T5	95
6061	T6, T651	95
6063	T832	95
6205	T5	95
6351	T6	95
2011	T8	100
5056	H38	100
2014	T4	105
2218	T71	105
5056	H18	105
2218	T61	115
2024	T3	120
2024	T4, T351	120
4032	T6	120
6066	T6, T651	120
6070	T6	120
6262	T9	120
2024	T361	130
2014	T6	135
7049	T73	135
7175	T736	145
7075	T6,T651	150
7175	T66	150

Source: data from *ASM Metals Reference Book, Second Edition*, American Society for Metals, Metals Park, Ohio 44073, (1984).

SELECTING FATIGUE STRENGTHS OF WROUGHT ALUMINUM ALLOYS

Alloy AA No.	Temper	Fatigue Strength (MPa)
1060	0	21
1060	H12	28
1060	H14	34
1100	0	34
1100	H12	41
1060	H16	45
1060	H18	45
1100	H14	48
1350	H19	48
3003	0	48
Alclad	H12	55
6063	0	55
1100	H16	62
1100	H18	62
3003	H14	62
6061	0	62
6063	T1	62
6070	0	62
3003	H16	69
3003	H18	69
6063	T5	69
6063	T6	69
6463	T1	69
6463	T5	69

Selecting Fatigue Strengths of Wrought Aluminum Alloys (Continued)

Alloy AA No.	Temper	Fatigue Strength (MPa)
6463	T6	69
5050	0	83
2014	0	90
2024	0	90
5050	H32	90
5050	H34	90
6070	T4	90
6262	T9	90
6351	T6	90
3004	0	97
5050	H36	97
5050	H38	97
6005	T1	97
6005	T5	97
6061	T4, T451	97
6061	T6, T651	97
6070	T6	97
2219	T62	105
2219	T81, T851	105
2219	T87	105
Alclad 3004	H32	105
	H34	105

Selecting Fatigue Strengths of Wrought Aluminum Alloys (Continued)

Alloy AA No.	Temper	Fatigue Strength (MPa)
6205	T5	105
3004	H36	110
3004	H38	110
4032	T6	110
5052	0	110
5652	0	110
6066	T6, T651	110
5052	H32	115
5154	0	115
5154	H112	115
5254	0	115
5254	H112	115
5652	H32	115
6009	T4	115
6010	T4	115
2011	T3	125
2011	T8	125
2014	T6	125
2024	T361	125
2036	T4	125
2618	All	125
5052	H34	125
5154	H32	125
5254	H32	125

Selecting Fatigue Strengths of Wrought Aluminum Alloys (Continued)

Alloy AA No.	Temper	Fatigue Strength (MPa)
5652	H34	125
7005	T6,T63,T6351	125
5052	H36	130
5154	H34	130
5254	H34	130
5652	H36	130
2014	T4	140
2024	T3	140
2024	T4, T351	140
5052	H38	140
5056	0	140
5154	H36	140
5182	0	140
5254	H36	140
5652	H38	140
7005	T53	140
5154	H38	145
5254	H38	145
5056	H18	150
5056	H38	150
5083	H321	160
7075	T6,T651	160
7175	T66	160
7175	T736	160

Selecting Fatigue Strengths of Wrought Aluminum Alloys (Continued)

Alloy AA No.	Temper	Fatigue Strength (MPa)
2048		220
7475	T7351	220
7050	T736	240
7049	T73	295

Source: data from *ASM Metals Reference Book, Second Edition*, American Society for Metals, Metals Park, Ohio 44073, (1984).

SELECTING REVERSED BENDING FATIGUE LIMITS OF GRAY CAST IRON BARS

ASTM Class	Reversed Bending Fatigue Limit (MPa)
20	69
25	79
30	97
35	110
40	128
50	148
60	169

Source: data from *ASM Metals Reference Book, Second Edition*, American Society for Metals, Metals Park, Ohio 44073, p166-167, (1984).

SELECTING IMPACT ENERGY OF TOOL STEELS

Type	Condition	Impact Energy (J)
L6	Oil quenched from 845 °C and single tempered at 315 °C	12(b)
L6	Oil quenched from 845 °C and single tempered at 425 °C	18(b)
L2	Oil quenched from 855 °C and single tempered at 315 °C	19(b)
L6	Oil quenched from 845 °C and single tempered at 540 °C	23(b)
L2	Oil quenched from 855 °C and single tempered at 425 °C	26(b)
L2	Oil quenched from 855 °C and single tempered at 205 °C	28(b)
L2	Oil quenched from 855 °C and single tempered at 540 °C	39(b)
L6	Oil quenched from 845 °C and single tempered at 650 °C	81(b)
L2	Oil quenched from 855 °C and single tempered at 650 °C	125(b)
S5	Oil quenched from 870 °C and single tempered at 540 °C	188(c)
S1	Oil quenched from 930 °C and single tempered at 425 °C	203(c)
S5	Oil quenched from 870 °C and single tempered at 205 °C	206(c)
S1	Oil quenched from 930 °C and single tempered at 540 °C	230(c)
S5	Oil quenched from 870 °C and single tempered at 315 °C	232(c)
S1	Oil quenched from 930 °C and single tempered at 315 °C	233(c)
S5	Oil quenched from 870 °C and single tempered at 425 °C	243(c)
S7	Fan cooled from 940 °C and single tempered at 425 °C	243(c)
S7	Fan cooled from 940 °C and single tempered at 205 °C	244(c)
S1	Oil quenched from 930 °C and single tempered at 205 °C	249(c)
S7	Fan cooled from 940 °C and single tempered at 315 °C	309(c)
S7	Fan cooled from 940 °C and single tempered at 540 °C	324(c)
S7	Fan cooled from 940 °C and single tempered at 650 °C	358(c)

(b) Charpy V-notch.
(c) Charpy unnotched.

Source: Data from *ASM Metals Reference Book, Second Edition*, American Society for Metals, Metals Park, Ohio 44073, p241, (1984).

SELECTING TENSILE MODULI OF GRAY CAST IRONS

ASTM Class	Tensile Modulus (GPa)
20	66 to 97
25	79 to 102
30	90 to 113
35	100 to 119
40	110 to 138
50	130 to 157
60	141 to 162

Source: data from *ASM Metals Reference Book, Second Edition*, American Society for Metals, Metals Park, Ohio 44073, p166-167, (1984).

SELECTING TENSILE MODULI OF TREATED DUCTILE IRONS

Treatment	Tension Modulus (GPa)
120 90-02	164
65-45-12	168
80-55-06	168
60-40-18	169

Source: data from *ASM Metals Reference Book, Second Edition*, American Society for Metals, Metals Park, Ohio 44073, p169-170, (1984).

SELECTING COMPRESSION MODULI OF TREATED DUCTILE IRONS

Treatment	Compression Modulus (GPa)
65-45-12	163
60-40-18	164
120 90-02	164
80-55-06	165

Source: data from *ASM Metals Reference Book, Second Edition*, American Society for Metals, Metals Park, Ohio 44073, p169-170, (1984).

SELECTING TORSIONAL MODULI OF GRAY CAST IRONS

ASTM Class	Torsional Modulus (GPa)
20	27 to 39
25	32 to 41
30	36 to 45
35	40 to 48
40	44 to 54
50	50 to 55
60	54 to 59

Source: data from *ASM Metals Reference Book, Second Edition*, American Society for Metals, Metals Park, Ohio 44073, p166-167, (1984).

SELECTING TORSIONAL MODULI OF TREATED DUCTILE IRONS

Treatment	Torsion Modulus (GPa)
60-40-18	63
65-45-12	64
80-55-06	62
120 90-02	63.4

Source: data from *ASM Metals Reference Book, Second Edition*, American Society for Metals, Metals Park, Ohio 44073, p169-170, (1984).

SELECTING COMPRESSION POISSON'S RATIOS OF TREATED DUCTILE IRONS

Treatment	Compression Poisson's Ratio
60-40-18	0.26
120 90-02	0.27
65-45-12	0.31
80-55-06	0.31

Source: data from *ASM Metals Reference Book, Second Edition*, American Society for Metals, Metals Park, Ohio 44073, p169-170, (1984).

SELECTING TORSION POISSON'S RATIOS OF TREATED DUCTILE IRONS

Treatment	Torsion Poisson's Ratio
120 90-02	0.28
60-40-18	0.29
65-45-12	0.29
80-55-06	0.31

Source: data from *ASM Metals Reference Book, Second Edition*, American Society for Metals, Metals Park, Ohio 44073, p169-170, (1984).

SELECTING ELONGATION OF TOOL STEELS

Type	Condition	Elongation (%)
L6	Oil quenched from 845 °C and single tempered at 315 °C	4
S1	Oil quenched from 930 °C and single tempered at 315 °C	4
L2	Oil quenched from 855 °C and single tempered at 205 °C	5
S1	Oil quenched from 930 °C and single tempered at 425 °C	5
S5	Oil quenched from 870 °C and single tempered at 205 °C	5
S5	Oil quenched from 870 °C and single tempered at 315 °C	7
S7	Fan cooled from 940 °C and single tempered at 205 °C	7
L6	Oil quenched from 845 °C and single tempered at 425 °C	8
S1	Oil quenched from 930 °C and single tempered at 540 °C	9
S5	Oil quenched from 870 °C and single tempered at 425 °C	9
S7	Fan cooled from 940 °C and single tempered at 315 °C	9
L2	Oil quenched from 855 °C and single tempered at 315 °C	10
S5	Oil quenched from 870 °C and single tempered at 540 °C	10
S7	Fan cooled from 940 °C and single tempered at 425 °C	10
S7	Fan cooled from 940 °C and single tempered at 540 °C	10
L2	Oil quenched from 855 °C and single tempered at 425 °C	12
L6	Oil quenched from 845 °C and single tempered at 540 °C	12
S1	Oil quenched from 930 °C and single tempered at 650 °C	12
S7	Fan cooled from 940 °C and single tempered at 650 °C	14
L2	Oil quenched from 855 °C and single tempered at 540 °C	15

Selecting Elongation of Tool Steels (Continued)

Type	Condition	Elongation (%)
S5	Oil quenched from 870 °C and single tempered at 650 °C	15
L6	Oil quenched from 845 °C and single tempered at 650 °C	20
S1	Annealed	24
L2	Annealed	25
L2	Oil quenched from 855 °C and single tempered at 650 °C	25
L6	Annealed	25
S5	Annealed	25
S7	Annealed	25

Source: Data from *ASM Metals Reference Book, Second Edition*, American Society for Metals, Metals Park, Ohio 44073, p241, (1984).

SELECTING ELONGATION OF DUCTILE IRONS

Specification Number	Grade or Class	Elongation (%)
ASTM A536-72; MIL-1-11466B(MR)	120-90-02	2
ASTM A476-70(d); SAE AMS5316	80-60-03	3
ASTM A536-72; MIL-1-11466B(MR)	100-70-03	3
SAE J434c	D7003	3
ASTM A536-72; MIL-1-11466B(MR)	80-55-06	6
SAE J434c	D5506	6
MlL-I-24137(Ships)	Class B	7
ASTM A536-72; MIL-1-11466B(MR)	65-45-12	12
SAE J434c	D4512	12
MlL-I-24137(Ships)	Class A	15
ASTM A395-76; ASME SA395	60-40-18	18
ASTM A536-72; MIL-1-11466B(MR)	60-40-18	18
SAE J434c	D4018	18
MlL-I-24137(Ships)	Class C	20

Source: data from *ASM Metals Reference Book, Second Edition*, American Society for Metals, Metals Park, Ohio 44073, p169, (1984).

SELECTING ELONGATION OF MALLEABLE IRON CASTINGS

Specification Number	Grade or Class	Elongation (%)
ASTM A220; ANSI C48.2; MIL-I-11444B	90001	1
ASTM A602; SAE J158	M8501(b)	1
ASTM A220; ANSI C48.2; MIL-I-11444B	80002	2
ASTM A602; SAE J158	M7002(b)	2
ASTM A220; ANSI C48.2; MIL-I-11444B	70003	3
ASTM A602; SAE J158	M5003(a)	3
ASTM A602; SAE J158	M5503(b)	3
ASTM A220; ANSI C48.2; MIL-I-11444B	60004	4
ASTM A602; SAE J158	M4504(a)	4
ASTM A197		5
ASTM A220; ANSI C48.2; MIL-I-11444B	50005	5
ASTM A220; ANSI C48.2; MIL-I-11444B	45006	6
ASTM A220; ANSI C48.2; MIL-I-11444B	45008	8
ASTM A47, A338; ANSI G48.1; FED QQ-I-666c	32510	10
ASTM A220; ANSI C48.2; MIL-I-11444B	40010	10
ASTM A602; SAE J158	M3210	10
ASTM A47, A338; ANSI G48.1; FED QQ-I-666c	35018	18

Source: data from *ASM Metals Reference Book, Second Edition*, American Society for Metals, Metals Park, Ohio 44073, p171, (1984).

SELECTING TOTAL ELONGATION OF CAST ALUMINUM ALLOYS

Alloy AA No.	Temper	Elongation (in 2 in.) (%)
242.0	T571	0.5
242.0	T61	0.5
336.0	T551	0.5
336.0	T65	0.5
355.0	T7	0.5
A390.0	F,T5	<1.0
A390.0	T6	<1.0
A390.0	T7	<1.0
A390.0	T6	<1.0
A390.0	T7	<1.0
242.0	T21	1.0
242.0	T571	1.0
355.0	T61	1.0
390.0	F	1.0
390.0	T5	1.0
A390.0	F,T5	1.0
355.0	T51	1.5
355.0	T71	1.5
355.0	T62	1.5
242.0	T77	2.0
295.0	T62	2.0
308.0	F	2.0
319.0	F	2.0
319.0	T6	2.0

Selecting Total Elongation of Cast Aluminum Alloys (Continued)

Alloy AA No.	Temper	Elongation (in 2 in.) (%)
355.0	T51	2.0
355.0	T7	2.0
356.0	T51	2.0
356.0	T7	2.0
208.0	F	2.5
319.0	F	2.5
384.0, A384.0	F	2.5
413.0	F	2.5
319.0	T6	3.0
355.0	T6	3.0
355.0	T71	3.0
360.0	F	3.0
380.0	F	3.0
713.0	T5	3.0
356.0	T6	3.5
356.0	T71	3.5
383.0	F	3.5
A413.0	F	3.5
355.0	T6	4.0
713.0	T5	4.0
201.0	T7	4.5
296.0	T7	4.5
295.0	T6	5.0
296.0	T6	5.0

Selecting Total Elongation of Cast Aluminum Alloys (Continued)

Alloy AA No.	Temper	Elongation (in 2 in.) (%)
356.0	T6	5.0
A360.0	F	5.0
712.0	F	5.0
518.0	F	5.0—8.0
359.0	T62	5.5
354.0	T61	6.0
356.0	T7	6.0
359.0	T61	6.0
201.0	T6	7
357.0, A357.0	T62	8.0
443.0	F	8.0
295.0	T4	8.5
296.0	T4	9.0
C443.0	F	9.0
514.0	F	9.0
771.0	T6	9.0
B443.0	F	10.0
850.0	T5	10.0
206.0, A206.0	T7	11.7
535.0	F	13
520.0	T4	16
201.0	T4	20

Source: data from *ASM Metals Reference Book, Second Edition*, American Society for Metals, Metals Park, Ohio 44073, (1984).

SELECTING AREA REDUCTION OF TOOL STEELS

Type	Condition	Area Reduction (%)
L6	Oil quenched from 845 °C and single tempered at 315 °C	9
S1	Oil quenched from 930 °C and single tempered at 315 °C	12
L2	Oil quenched from 855 °C and single tempered at 205 °C	15
S1	Oil quenched from 930 °C and single tempered at 425 °C	17
L6	Oil quenched from 845 °C and single tempered at 425 °C	20
S5	Oil quenched from 870 °C and single tempered at 205 °C	20
S7	Fan cooled from 940 °C and single tempered at 205 °C	20
S1	Oil quenched from 930 °C and single tempered at 540 °C	23
S5	Oil quenched from 870 °C and single tempered at 315 °C	24
S7	Fan cooled from 940 °C and single tempered at 315 °C	25
S5	Oil quenched from 870 °C and single tempered at 425 °C	28
S7	Fan cooled from 940 °C and single tempered at 425 °C	29
L2	Oil quenched from 855 °C and single tempered at 315 °C	30
L6	Oil quenched from 845 °C and single tempered at 540 °C	30
S5	Oil quenched from 870 °C and single tempered at 540 °C	30
S7	Fan cooled from 940 °C and single tempered at 540 °C	33
L2	Oil quenched from 855 °C and single tempered at 425 °C	35
S1	Oil quenched from 930 °C and single tempered at 650 °C	37
S5	Oil quenched from 870 °C and single tempered at 650 °C	40
L2	Oil quenched from 855 °C and single tempered at 540 °C	45

Selecting Area Reduction of Tool Steels (Continued)

Type	Condition	Area Reduction (%)
S7	Fan cooled from 940 °C and single tempered at 650 °C	45
L6	Oil quenched from 845 °C and single tempered at 650 °C	48
L2	Annealed	50
S5	Annealed	50
S1	Annealed	52
L2	Oil quenched from 855 °C and single tempered at 650 °C	55
L6	Annealed	55
S7	Annealed	55

Area Reduction in 50 mm or 2 in.

Source: data from *ASM Metals Reference Book, Second Edition*, American Society for Metals, Metals Park, Ohio 44073, p241, (1984).

Selecting Metals:
Electrical Properties

SELECTING ELECTRICAL RESISTIVITY OF ALLOY CAST IRONS

Description	Electrical Resistivity ($\mu\Omega \cdot m$)
Corrosion–Resistant High– Silicon Iron	0.50
Abrasion–Resistant Low–C White Iron	0.53
Heat–Resistant Medium–silicon Ductile Iron	0.58 to 0.87
Abrasion–Resistant Martensitic nickel–chromium White Iron	0.80
Corrosion–Resistant High–nickel gray iron	1.0[a]
Corrosion–Resistant High–nickel ductile iron	1.0[a]
Heat–Resistant High–nickel Ductile Iron (23 Ni)	1.0[a]
Heat–Resistant High–nickel Ductile Iron (20 Ni)	1.02
Heat–Resistant Gray High–nickel Iron	1.4 to 1.7
Heat–Resistant Nickel–chromium–silicon Gray Iron	1.5 to 1.7
Heat–Resistant High–aluminum Gray Iron	2.4

[a] Estimated.

Source: data from *ASM Metals Reference Book, Second Edition*, American Society for Metals, Metals Park, Ohio 44073, (1984).

SELECTING CRITICAL TEMPERATURE OF SUPERCONDUCTIVE ELEMENTS

Element	$T_c(K)$	Element	$T_c(K)$
W	0.0154	Ti	2.332-2.39
Be	0.026	Sb	2.6-2.7[a]
Ir	0.11-0.14	In	3.405
Ti	0.39	Sn	3.721
Ru	0.493	Hg (β)	3.949
Cd	0.518-0.52	Hg (α)	4.154
Zr	0.53	Ta	4.47
Zr (ω)	0.65	La (α)	4.88
Os	0.655	V	5.43-5.31
Zn	0.875	Ga (β)	5.90-6.2
Mo	0.916	La (β)	6.00
Ga	1.0833	Pb	7.23
Al	1.175	Ga (γ)	7.62
Th	1.39	Tc	7.73-7.78
Pa	1.4	Ga (δ)	7.85
Re	1.697	Nb	9.25

[a] Metastable.

Source: data from Roberts, B. W., *Properties of Selected Superconductive Materials* - 1974 Supplement, NBS Technical Note 825, National Bureau of Standards, U.S. Government Printing Office, Washington,D.C., 1974, 10.

Selecting Metals:
Corrosion Properties

SELECTING IRON ALLOYS IN 10% CORROSIVE MEDIUM

Corrosion Rate at 70°F in a 10% Corrosive Medium

Corrosive Medium	1020 Steel 17% Cr Steel	Grey Cast Iron Stainless Steel 301	Ni–Resist Cast Iron Stainless Steel 316	12% C Steel 14% Si Iron
Acetaldehyde	<0.05 —	<0.05 —	— —	— <0.002
Acetic Acid (Aerated)	>0.05 <0.002	>0.05 <0.002	<0.02 <0.002	<0.02 <0.002
Acetic Acid (Air Free)	>0.05 <0.02	>0.05 <0.02	<0.02 <0.002	<0.02 <0.002
Acetic Anhydride	— —	— —	— —	— <0.002
Acetoacetic Acid	>0.05 <0.02	>0.05 <0.02	— <0.02	— <0.02
Acetone	<0.05 <0.02	— <0.02	— <0.02	<0.02 <0.002
Acrolein	<0.02 <0.02	— <0.02	— <0.02	<0.02 <0.02
Alcohol (Ethyl)	<0.02 <0.02	<0.02 <0.02	<0.02 <0.002	<0.02 <0.002
Alcohol (Methyl)	<0.02 <0.02	<0.02 <0.02	<0.02 <0.002	<0.02 <0.002
Alcohol (Allyl)	— —	— —	— —	— <0.02
Allylamine	<0.02 (30%) —	— <0.002 (30%)	— <0.002 (30%)	— <0.002 (30%)
Aluminum Acetate	>0.05 —	>0.05 <0.02	— <0.02	<0.02 <0.02

Selecting Iron Alloys in 10% Corrosive Medium (Continued)

	Corrosion Rate at 70°F in a 10% Corrosive Medium			
	1020 Steel	Grey Cast Iron	Ni–Resist Cast Iron	12% C Steel
Corrosive Medium	17% Cr Steel	Stainless Steel 301	Stainless Steel 316	14% Si Iron
Aluminum Chlorate	— <0.002	— <0.002	— —	— <0.02
Aluminum Chloride	>0.05 >0.05	>0.05 >0.05	>0.05 <0.05	>0.05 <0.002
Aluminum Fluoride	<0.02 >0.05	<0.02 >0.05	— —	>0.05 >0.05
Aluminum Formate	<0.05 <0.02	— <0.02	— <0.02	<0.02 <0.02
Aluminum Hydroxide	<0.02 <0.02	<0.02 <0.02	<0.02 <0.02	<0.02 <0.02
Aluminum Nitrate	>0.05 <0.02	>0.05 <0.02	— <0.02	<0.02 —
Aluminum Potassium Sulfate	>0.05 <0.05	>0.05 <0.02	>0.05 <0.02	>0.05 —
Aluminum Sulfate	>0.05 —	>0.05 <0.02	<0.02 <0.02	>0.05 <0.002
Ammonia	<0.002 <0.002	<0.002 <0.002	<0.002 <0.002	<0.002 <0.02
Ammonium Acetate	— <0.002	— <0.002	<0.002 <0.002	<0.002 <0.002
Ammonium Bicarbonate	<0.02 <0.02	<0.02 <0.02	<0.02 <0.02	<0.02 <0.002
Ammonium Bromide	>0.05 <0.05	>0.05 <0.05	— <0.02	<0.05 <0.002

Selecting Iron Alloys in 10% Corrosive Medium (Continued)

Corrosive Medium	Corrosion Rate at 70°F in a 10% Corrosive Medium			
	1020 Steel	Grey Cast Iron	Ni–Resist Cast Iron	12% C Steel
	17% Cr Steel	Stainless Steel 301	Stainless Steel 316	14% Si Iron
Ammonium Carbonate	<0.02	<0.02	<0.02	<0.02
	<0.02	<0.02	<0.02	<0.002
Ammonium Chloride	<0.05	>0.05	<0.02	<0.05
	<0.05	<0.02	<0.02	<0.002
Ammonium Citrate	>0.05	>0.05	>0.05	—
	<0.02	<0.02	<0.02	—
Ammonium Formate	—	—	—	—
	—	<0.02	<0.02	<0.02
Ammonium Nitrate	<0.002	<0.02	<0.02	<0.02
	<0.002	<0.002	<0.002	<0.002
Ammonium Sulfate	<0.02	<0.05	>0.05	>0.05
	<0.05	<0.05	<0.02	<0.002
Ammonium Sulfite	>0.05	>0.05	>0.05	>0.05
	>0.05	<0.05	<0.02	<0.02
Ammonium Thiocyanate	<0.02	<0.02	<0.02	—
	<0.02	<0.02	<0.02	<0.02
Amyl Acetate	<0.002	—	—	—
	—	<0.002	<0.002	<0.002
Amyl Chloride	>0.05	—	—	—
	—	>0.05	—	<0.02
Aniline	—	—	<0.02	<0.02
	<0.02	<0.02	<0.02	<0.002
Aniline Hydro-chloride	>0.05	>0.05	>0.05	>0.05
	>0.05	>0.05	>0.05	<0.02

Selecting Iron Alloys in 10% Corrosive Medium (Continued)

	Corrosion Rate at 70°F in a 10% Corrosive Medium			
	1020 Steel	Grey Cast Iron	Ni–Resist Cast Iron	12% C Steel
Corrosive Medium	17% Cr Steel	Stainless Steel 301	Stainless Steel 316	14% Si Iron
Antimony Trichloride	>0.05	>0.05	>0.05	>0.05
	>0.05	>0.05	>0.05	<0.002
Barium Carbonate	<0.02	<0.02	<0.02	<0.02
	<0.02	<0.02	<0.02	<0.02
Barium Chloride	<0.02	>0.05	<0.02	<0.05
	<0.02	<0.02	<0.02	<0.02
Barium Nitrate	<0.02	—	—	—
	<0.02	<0.02	<0.02	<0.02
Barium Peroxide	<0.05	—	—	>0.05
	—	<0.02	<0.02	<0.02
Benzaldehyde	>0.05	>0.05	<0.02	—
	—	<0.02	—	<0.02
Benzene	—	—	—	<0.02
	<0.02	<0.02	<0.02	<0.002
Benzoic Acid	>0.05	>0.05	—	<0.02
	<0.02	<0.02	<0.02	<0.02
Boric Acid	<0.05	>0.05	<0.002	<0.02
	<0.02	<0.002	<0.002	<0.02
Bromic Acid	>0.05	>0.05	—	>0.05
	>0.05	>0.05	>0.05	—
Butyric Acid	<0.05	>0.05	>0.05	<0.05
	<0.05	<0.02	<0.02	<0.002
Cadmium Chloride	>0.05	>0.05	>0.05	>0.05
	>0.05	<0.02	<0.02	<0.02

Selecting Iron Alloys in 10% Corrosive Medium (Continued)

	Corrosion Rate at 70°F in a 10% Corrosive Medium			
	1020 Steel	Grey Cast Iron	Ni–Resist Cast Iron	12% C Steel
Corrosive Medium	17% Cr Steel	Stainless Steel 301	Stainless Steel 316	14% Si Iron
Cadmium Sulfate	<0.02	<0.02	—	—
	<0.002	<0.002	<0.002	<0.002
Calcium Acetate	<0.02	<0.05	—	<0.02
	<0.02	<0.02	<0.02	<0.02
Calcium Bicarbonate	<0.02	—	—	—
	—	—	—	—
Calcium Bromide	—	—	—	<0.02
	<0.02	<0.02	<0.02	—
Calcium Chlorate	<0.002	<0.02	<0.05	<0.02
	<0.02	<0.02	<0.02	<0.02
Calcium Chloride	<0.002	<0.02	<0.02	<0.02
	<0.05	<0.02	<0.02	<0.002
Calcium Hydroxide	<0.02	<0.02	<0.02	<0.02
	<0.02	<0.02	<0.02	<0.02
Calcium Hypochlorite	<0.05	<0.05	<0.02	>0.05
	>0.05	<0.05	<0.05	<0.02
Carbon Tetrachloride	—	—	—	>0.05
	<0.002	>0.05	<0.02	<0.002
Carbon Acid (Air Free)	<0.02	—	—	—
	—	<0.02	<0.02	<0.02
Chloroacetic Acid	>0.05	>0.05	>0.05	>0.05
	>0.05	>0.05	>0.05	>0.05
Chlorine Gas	>0.05	>0.05	>0.05	>0.05
	>0.05	—	—	—

Selecting Iron Alloys in 10% Corrosive Medium (Continued)

	Corrosion Rate at 70°F in a 10% Corrosive Medium			
	1020 Steel	Grey Cast Iron	Ni–Resist Cast Iron	12% C Steel
Corrosive Medium	17% Cr Steel	Stainless Steel 301	Stainless Steel 316	14% Si Iron
Chromic Acid	>0.05	<0.05	<0.05	>0.05
	<0.02	<0.02	<0.02	<0.002
Chromic Sulfates	>0.05	—	—	>0.05
	>0.05	<0.02	<0.02	<0.002
Citric Acid	>0.05	>0.05	>0.05	<0.05
	<0.02	<0.02	<0.02	<0.002
Copper Nitrate	>0.05	>0.05	>0.05	<0.02
	<0.02	<0.02	<0.002	<0.002
Copper Sulfate	>0.05	>0.05	>0.05	<0.02
	<0.02	<0.02	<0.02	<0.002
Diethylene Glycol	<0.002 (60%)	—	—	—
	—	—	—	—
Ethyl Chloride	>0.05 (90%)	—	—	>0.05 (90%)
	>0.05 (90%)	>0.05 (90%)	—	—
Ethylene Glycol	<0.02	—	—	—
	—	—	—	<0.02
Ferric Chloride	>0.05	>0.05	>0.05	>0.05
	>0.05	>0.05	>0.05	>0.05
Ferric Nitrate	>0.05	>0.05	—	<0.02
	<0.02	<0.02	<0.02	<0.02
Ferrous Chloride	>0.05	>0.05	>0.05	>0.05
	>0.05	>0.05	>0.05	>0.05
Ferrous Sulfate	>0.05	>0.05	—	<0.02
	<0.02	<0.02	<0.02	<0.02

Selecting Iron Alloys in 10% Corrosive Medium (Continued)

	Corrosion Rate at 70°F in a 10% Corrosive Medium			
	1020 Steel	Grey Cast Iron	Ni–Resist Cast Iron	12% C Steel
Corrosive Medium	17% Cr Steel	Stainless Steel 301	Stainless Steel 316	14% Si Iron
Formaldehyde	<0.05 (40%) <0.002	<0.05 (40%) <0.002 (20%)	<0.05 (40%) <0.02	<0.02 <0.002
Formic Acid	>0.05 <0.05	>0.05 <0.02	>0.05 <0.002	<0.05 <0.002
Furfural	<0.02 (30%) <0.002 (30%)	— <0.002 (30%)	<0.02 (30%) <0.002	<0.02 (80%) <0.02 (20%)
Hydrazine	>0.05 —	>0.05 <0.002	— <0.002	— —
Hydrobromic Acid	>0.05 >0.05	>0.05 >0.05	— >0.05	>0.05 >0.05
Hydrochloric Acid (Areated)	>0.05 >0.05	>0.05 >0.05	>0.05 >0.05	>0.05 <0.02
Hydrochloric Acid (Air Free)	>0.05 >0.05	>0.05 >0.05	<0.05 >0.05	>0.05 <0.02
Hydrofluoric Acid (Areated)	>0.05 —	>0.05 <0.002	<0.002 <0.002	— >0.05
Hydrofluoric Acid (Air Free)	>0.05 >0.05	>0.05 >0.05	<0.002 >0.05	>0.05 >0.05
Hydrogen Chloride	>0.05 90 >0.05 90	>0.05 90 >0.05 90	— —	>0.05 90 <0.02 90
Hydrogen Iodide	<0.05 (1%) —	>0.05 <0.02 1%	— —	<0.05 >0.05
Hydrogen Peroxide	>0.05 (20%) <0.02 (20%)	>0.05 (20%) <0.02 (20%)	— <0.02 (20%)	<0.02 (20%) <0.02 (20%)

Selecting Iron Alloys in 10% Corrosive Medium (Continued)

	Corrosion Rate at 70°F in a 10% Corrosive Medium			
	1020 Steel	Grey Cast Iron	Ni–Resist Cast Iron	12% C Steel
Corrosive Medium	17% Cr Steel	Stainless Steel 301	Stainless Steel 316	14% Si Iron
Hydrogen Sulfide	<0.02 <0.02	<0.02 >0.05	<0.02 <0.002	<0.02 —
Lactic Acid	>0.05 >0.05	>0.05 <0.02	>0.05 <0.02	>0.05 <0.002
Lead Acetate	>0.05 (20%) <0.02	>0.05 <0.02	— <0.02	<0.02 <0.02
Lead Nitrate	>0.05 <0.02	>0.05 <0.02	— <0.02	<0.02 <0.002
Lithium Chloride	<0.02 (30%) —	<0.02 (30%) <0.002 (30%)	<0.002 (30%) <0.002 (30%)	— <0.02 (30%)
Lithium Hydroxide	<0.02 <0.02	<0.02 <0.02	<0.02 <0.02	<0.02 >0.05
Magnesium Chloride	<0.02 <0.05	<0.02 <0.05	<0.02 <0.02	<0.05 <0.002
Magnesium Hydroxide	<0.02 <0.02	<0.02 <0.02	<0.02 <0.02	<0.02 <0.02
Magnesium Sulfate	<0.02 <0.002	>0.05 <0.002	<0.02 <0.002	>0.05 <0.002
Maleic Acid	>0.05 <0.02	>0.05 <0.02	>0.05 <0.02	— <0.02
Malic Acid	>0.05 <0.02	>0.05 <0.002	— <0.002	<0.02 —
Maganous Chloride	>0.05 (40%) —	>0.05 (40%) <0.02 (40%)	<0.05 (40%) <0.02 (40%)	— —

Selecting Iron Alloys in 10% Corrosive Medium (Continued)

Corrosion Rate at 70°F in a 10% Corrosive Medium

Corrosive Medium	1020 Steel 17% Cr Steel	Grey Cast Iron Stainless Steel 301	Ni–Resist Cast Iron Stainless Steel 316	12% C Steel 14% Si Iron
Mercuric Chloride	>0.05 >0.05	>0.05 >0.05	>0.05 >0.05	>0.05 <0.02
Mercurous Nitrate	— <0.02	— <0.02	— <0.02	<0.02 <0.02
Methallylamine	<0.02 <0.02	— <0.02	<0.02 <0.02	<0.02 <0.02
Methanol	<0.02 <0.02	<0.02 <0.02	<0.02 <0.02	<0.02 <0.002
Methyl Ethyl Ketone	<0.02 <0.02	<0.02 <0.02	<0.02 <0.02	<0.02 <0.02
Methyl Isobutyl Ketone	<0.02 <0.02	<0.02 <0.02	<0.02 <0.02	<0.02 <0.02
Methylamine	<0.02 <0.02	<0.02 <0.02	<0.02 <0.02	<0.02 <0.02
Methylene Chloride	— —	— <0.02	— <0.02	— —
Monochloroacetic Acid	>0.05 >0.05	>0.05 <0.05	— <0.05	>0.05 <0.02
Monorthanolamine	<0.02 <0.002	— <0.002	— <0.02	<0.02 —
Monoethalamine	<0.02 <0.02	— <0.02	<0.02 <0.02	<0.02 <0.02
Monoethylamine	<0.02 <0.02	<0.02 <0.02	<0.02 <0.02	<0.02 <0.02

Selecting Iron Alloys in 10% Corrosive Medium (Continued)

Corrosion Rate at 70°F in a 10% Corrosive Medium

Corrosive Medium	1020 Steel / 17% Cr Steel	Grey Cast Iron / Stainless Steel 301	Ni–Resist Cast Iron / Stainless Steel 316	12% C Steel / 14% Si Iron
Monosodium Phosphate	>0.05	>0.05	>0.05	>0.05
	>0.05	<0.02	<0.02	<0.02
Nickel Chloride	>0.05	>0.05	>0.05	>0.05
	>0.05	>0.05	>0.05	<0.02
Nickel Nitrate	<0.02	<0.02	<0.02	<0.02
	<0.02	<0.02	<0.02	<0.002
Nickel Sulfate	>0.05	>0.05	—	—
	—	<0.002	<0.02	<0.002
Nitric Acid	>0.05	>0.05	>0.05	<0.02
	<0.02	<0.002	<0.002	<0.002
Nitric + Sulfuric Acid	—	—	—	—
	—	—	—	<0.02
Nitrous Acid	—	—	—	<0.05
	<0.02	<0.02	<0.02	<0.002
Oleic Acid	—	—	—	<0.02
	<0.02	<0.02	<0.02	<0.002
Oxalic Acid	>0.05	>0.05	>0.05	>0.05
	>0.05	<0.02	<0.02	<0.02
Phosphoric Acid (Areated)	>0.05	>0.05	>0.05	<0.02
	<0.02	<0.02	<0.002	<0.002
Phosphoric Acid (Air Free)	>0.05	>0.05	>0.05	>0.05
	>0.05	<0.02	<0.02	<0.02
Picric Acid	>0.05	>0.05	—	<0.02
	<0.02	<0.02	<0.02	<0.02

Selecting Iron Alloys in 10% Corrosive Medium (Continued)

Corrosion Rate at 70°F in a 10% Corrosive Medium

Corrosive Medium	1020 Steel / 17% Cr Steel	Grey Cast Iron / Stainless Steel 301	Ni–Resist Cast Iron / Stainless Steel 316	12% C Steel / 14% Si Iron
Potassium Bicarbonate	<0.02 <0.02	<0.02 <0.02	<0.02 <0.02	<0.02 <0.02
Potassium Bromide	<0.05 <0.02	<0.05 <0.02	<0.02 <0.02	<0.02 <0.02
Potassium Carbonate	<0.02 <0.02	<0.02 <0.02	<0.02 <0.02	<0.02 <0.02
Potassium Chlorate	<0.02 <0.02	— <0.02	<0.02 <0.02	<0.02 <0.02
Potassium Chromate	<0.02 <0.02	<0.02 <0.02	<0.02 <0.02	<0.02 <0.02
Potassium Cyanide	<0.02 <0.02	>0.05 <0.02	<0.02 <0.02	<0.02 <0.02
Potassium Dichromate	<0.02 <0.02	<0.02 <0.002	<0.02 <0.002	<0.02 <0.002
Potassium Ferricyanide	<0.02 <0.02	<0.02 <0.02	<0.02 <0.02	<0.02 <0.02
Potassium Ferrocyanide	>0.05 <0.02	>0.05 <0.02	<0.02 <0.02	>0.05 <0.02
Potassium Hydroxide	<0.02 <0.02	<0.02 <0.02	<0.02 <0.02	<0.02 >0.05
Potassium Hypochlorite	>0.05 >0.05	>0.05 >0.05	>0.05 <0.05	>0.05 <0.002
Potassium Iodide	<0.02 >0.05	— <0.02	<0.02 <0.02	>0.05 <0.02

Selecting Iron Alloys in 10% Corrosive Medium (Continued)

Corrosion Rate at 70°F in a 10% Corrosive Medium

Corrosive Medium	1020 Steel / 17% Cr Steel	Grey Cast Iron / Stainless Steel 301	Ni–Resist Cast Iron / Stainless Steel 316	12% C Steel / 14% Si Iron
Potassium Nitrate	<0.02 <0.02	<0.02 <0.02	<0.02 <0.02	<0.02 <0.002
Potassium Nitrite	<0.02 <0.02	<0.02 <0.02	<0.02 <0.02	<0.02 <0.02
Potassium Permanganate	<0.02 <0.02	<0.02 <0.02	<0.02 <0.02	<0.002 <0.02
Potassium Silicate	<0.02 <0.02	<0.02 <0.02	<0.02 <0.02	<0.02 <0.02
Propionic Acid	>0.05 —	>0.05 —	— —	— <0.02
Pyridine	<0.02 <0.02	<0.02 <0.02	<0.02 <0.02	<0.02 <0.02
Quinine Sulfate	>0.05 <0.02	>0.05 <0.02	<0.02 <0.02	— <0.02
Silver Bromide	>0.05 >0.05	>0.05 >0.05	>0.05 >0.05	>0.05 —
Silver Chloride	>0.05 >0.05	>0.05 >0.05	— >0.05	>0.05 —
Silver Nitrate	>0.05 <0.02	>0.05 <0.02	— <0.002	<0.02 <0.002
Sodium Acetate	<0.02 <0.02	— <0.02	<0.02 <0.02	<0.02 <0.002
Sodium Bicarbonate	<0.02 <0.02	<0.02 <0.02	<0.02 <0.02	<0.02 <0.002

Selecting Iron Alloys in 10% Corrosive Medium (Continued)

	Corrosion Rate at 70°F in a 10% Corrosive Medium			
	1020 Steel	Grey Cast Iron	Ni–Resist Cast Iron	12% C Steel
Corrosive Medium	17% Cr Steel	Stainless Steel 301	Stainless Steel 316	14% Si Iron
Sodium Bisulfate	>0.05	>0.05	<0.002	<0.002
	<0.002	<0.002	<0.002	<0.002
Sodium Bromide	<0.02	—	<0.02	<0.05
	<0.05	<0.05	<0.05	<0.05
Sodium Carbonate	<0.002	<0.002	<0.002	<0.02
	<0.02	<0.02	<0.02	<0.02
Sodium Chloride	<0.02	<0.02	<0.02	<0.02
	<0.02	<0.02	<0.02	<0.02
Sodium Chromate	<0.02	<0.02	<0.02	<0.02
	<0.02	<0.02	<0.02	<0.02
Sodium Hydroxide	<0.002	<0.02	<0.002	<0.002
	<0.002	<0.002	<0.002	>0.05
Sodium Hypochlorite	>0.05	>0.05	>0.05	>0.05
	>0.05	>0.05	>0.05	—
Sodium Metasilicate	<0.02	<0.02	<0.002	<0.002
	<0.002	<0.002	<0.002	<0.02
Sodium Nitrate	<0.02	<0.02	<0.02	<0.02
	<0.02	<0.002	<0.002	<0.002
Sodium Nitrite	<0.02	<0.02	<0.02	<0.02
	<0.02	<0.02	<0.02	<0.02
Sodium Phosphate	<0.02	<0.02	<0.02	<0.02
	<0.02	<0.02	<0.02	<0.02
Sodium Silicate	<0.02	<0.02	<0.02	<0.02
	<0.02	<0.02	<0.02	<0.02

Selecting Iron Alloys in 10% Corrosive Medium (Continued)

	Corrosion Rate at 70°F in a 10% Corrosive Medium			
	1020 Steel	Grey Cast Iron	Ni–Resist Cast Iron	12% C Steel
Corrosive Medium	17% Cr Steel	Stainless Steel 301	Stainless Steel 316	14% Si Iron
Sodium Sulfate	<0.02 <0.05	<0.02 <0.02	<0.02 <0.002	<0.05 <0.002
Sodium Sulfide	<0.05 >0.05	<0.05 <0.02	— >0.05	>0.05 <0.02
Sodium Sulfite	<0.02 <0.02	>0.05 <0.002	<0.02 <0.002	<0.02 <0.002
Stannic Chloride	>0.05 >0.05	>0.05 >0.05	>0.05 >0.05	>0.05 >0.05
Stannous Chloride	>0.05 >0.05	>0.05 >0.05	>0.05 <0.02	>0.05 <0.002
Strontium Nitrate	>0.05 <0.02	>0.05 <0.02	<0.02 <0.02	<0.02 <0.02
Succinic Acid	<0.02 <0.02	<0.02 <0.02	<0.02 <0.02	<0.02 <0.02
Sulfur Dioxide	>0.05 >0.05	— >0.05	— <0.002	>0.05 —
Sulfuric Acid (Areated)	>0.05 <0.05	>0.05 >0.05	<0.02 <0.002	<0.05 <0.002
Sulfuric Acid (Air Free)	>0.05 >0.05	>0.05 >0.05	<0.02 <0.05	>0.05
Sulfurous Acid	<0.05 >0.05	— <0.02	<0.05 <0.02	>0.05 <0.02
Tannic Acid	>0.05 <0.02	— <0.02	— <0.02	<0.02 <0.002

Selecting Iron Alloys in 10% Corrosive Medium (Continued)

Corrosion Rate at 70°F in a 10% Corrosive Medium

Corrosive Medium	1020 Steel / 17% Cr Steel	Grey Cast Iron / Stainless Steel 301	Ni–Resist Cast Iron / Stainless Steel 316	12% C Steel / 14% Si Iron
Tartaric Acid	>0.05 <0.02	>0.05 <0.002	<0.02 <0.02	<0.02 <0.02
Tetraphosphoric Acid	>0.05 >0.05	>0.05 —	>0.05 —	>0.05 —
Trichloroacetic Acid	>0.05 >0.05	>0.05 >0.05	>0.05 >0.05	>0.05 <0.002
Urea	<0.05 <0.02	— <0.02	— <0.02	<0.02 <0.02
Zinc Chloride	>0.05 —	>0.05 —	<0.02 —	— —
Zinc Sulfate	>0.05 <0.05	>0.05 <0.002	<0.02 <0.02	<0.05 <0.002

10% corrosive medium in 90% water

<0.002 means that corrosion rate is likely to be less than 0.002 inch per year (Excellent).
<0.02 means that corrosion rate is likely to be less than about 0.02 inch per year (Good).
<0.05 means that corrosion rate is likely to be less than about 0.05 inch per year (Fair).
>0.05 means that corrosion rate is likely to be more than 0.05 inch per year (Poor).

Source: data compiled by J.S. Park *from* Earl R. Parker, *Materials Data Book for Engineers and Scientists*, McGraw-Hill Book Company, New York, 1967.

SELECTING IRON ALLOYS IN 100% CORROSIVE MEDIUM

Corrosion Rate at 70°F in a 100% Corrosive Medium

Corrosive Medium	1020 Steel 17% Cr Steel	Grey Cast Iron Stainless Steel 301	Ni–Resist Cast Iron Stainless Steel 316	12% C Steel 14% Si Iron
Acetaldehyde	<0.002 <0.002	<0.002 <0.002	<0.002 <0.002	<0.002 <0.002
Acetic Acid (Aerated)	>0.05 <0.002	>0.05 <0.002	>0.05 <0.002	>0.05 <0.002
Acetic Acid (Air Free)	>0.05 <0.05	>0.05 <0.002	>0.05 <0.02	>0.05 <0.002
Acetic Anhydride	>0.05 <0.05	>0.05 <0.02	<0.02 <0.02	<0.05 <0.002
Acetoacetic Acid	>0.05 <0.02	>0.05 <0.02	— <0.02	— <0.02
Acetone	<0.002 <0.002	<0.002 <0.002	<0.002 <0.002	<0.002 <0.002
Acetylene	<0.002 <0.002	<0.002 <0.002	<0.002 <0.002	<0.002 <0.002
Acrolein	<0.02 <0.02	<0.02 <0.002	<0.02 <0.02	<0.02 <0.02
Acrylonitril	<0.002 <0.002	<0.002 <0.002	<0.002 <0.002	<0.002 <0.002
Alcohol (Ethyl)	<0.002 <0.02	<0.02 <0.02	<0.02 <0.002	<0.02 <0.002
Alcohol (Methyl)	<0.002 <0.02	<0.002 <0.02	<0.002 <0.002	<0.02 <0.002
Alcohol (Allyl)	<0.002 <0.02	<0.02 <0.02	<0.02 <0.02	<0.02 <0.02

Selecting Iron Alloys in 100% Corrosive Medium (Continued)

Corrosion Rate at 70°F in a 100% Corrosive Medium

Corrosive Medium	1020 Steel 17% Cr Steel	Grey Cast Iron Stainless Steel 301	Ni–Resist Cast Iron Stainless Steel 316	12% C Steel 14% Si Iron
Alcohol (Amyl)	<0.02 <0.02	<0.02 <0.02	<0.02 <0.02	<0.02 <0.02
Alcohol (Benzyl)	<0.002 <0.02	— <0.02	— <0.02	<0.02 <0.02
Alcohol (Butyl)	<0.002 <0.002	<0.002 <0.002	— <0.002	<0.002 <0.002
Alcohol (Cetyl)	<0.02 <0.02	— <0.02	— <0.02	<0.02 <0.02
Alcohol (Isopropyl)	<0.002 <0.02	<0.02 <0.02	<0.02 <0.02	<0.02 <0.02
Allylamine	<0.02 <0.02	<0.02 <0.02	<0.02 <0.02	<0.02 <0.02
Allyl Chloride	<0.002 <0.02	<0.02 <0.02	— <0.002	<0.02 <0.002
Allyl Sulfide	<0.02 <0.02	<0.02 <0.02	<0.02 <0.02	<0.02 <0.02
Aluminum Acetate	— —	— <0.02	<0.02 <0.02	<0.02 <0.002
Aluminum Chlorate	— —	— —	— —	— <0.002
Aluminum Chloride	<0.002 <0.002	>0.05 <0.002	>0.05 —	<0.002 <0.02
Aluminum Fluoride	— >0.05	— >0.05	— <0.05	>0.05 >0.05

Selecting Iron Alloys in 100% Corrosive Medium (Continued)

	Corrosion Rate at 70°F in a 100% Corrosive Medium			
	1020 Steel	Grey Cast Iron	Ni–Resist Cast Iron	12% C Steel
Corrosive Medium	17% Cr Steel	Stainless Steel 301	Stainless Steel 316	14% Si Iron
Aluminum Fluosilicate	>0.05	>0.05	—	<0.02
	<0.02	<0.02	<0.02	<0.02
Aluminum Formate	>0.05	—	—	<0.02
	<0.02	<0.02	<0.02	<0.02
Aluminum Hydroxide	—	—	—	—
	<0.02	<0.02	<0.02	—
Aluminum Nitrate	—	—	—	<0.02
	<0.02	<0.02	<0.02	—
Aluminum Potassium Sulfate	—	—	—	<0.05
	>0.05	<0.02	—	<0.002
Aluminum Sulfate	—	—	—	>0.05
	>0.05	<0.02	<0.02	<0.02
Ammonia	<0.002	<0.002	<0.002	<0.002
	<0.002	<0.002	<0.002	<0.02
Ammonium Acetate	<0.002	<0.02	<0.002	<0.002
	<0.002	<0.002	<0.002	<0.02
Ammonium Bicarbonate	<0.002	<0.02	<0.02	—
	—	<0.05	<0.02	<0.002
Ammonium Bromide	>0.05	>0.05	—	>0.05
	—	<0.05	—	—
Ammonium Carbonate	<0.002	<0.02	<0.02	<0.02
	<0.02	<0.02	<0.02	<0.02
Ammonium Chloride	<0.02	—	—	>0.05
	>0.05	>0.05	—	<0.02

Selecting Iron Alloys in 100% Corrosive Medium (Continued)

Corrosion Rate at 70°F in a 100% Corrosive Medium

Corrosive Medium	1020 Steel — 17% Cr Steel	Grey Cast Iron — Stainless Steel 301	Ni–Resist Cast Iron — Stainless Steel 316	12% C Steel — 14% Si Iron
Ammonium Citrate	<0.002 —	— —	— —	— —
Ammonium Formate	— —	— <0.02	— <0.02	— <0.02
Ammonium Nitrate	<0.02 <0.02	<0.05 <0.002	— <0.002	<0.02 —
Ammonium Sulfate	— —	<0.02 —	<0.02 —	— <0.002
Ammonium Sulfite	— —	— <0.05	— <0.02	— —
Amyl Acetate	<0.02 <0.02	<0.02 <0.002	<0.002 <0.002	<0.002 <0.002
Amyl Chloride	<0.02 <0.05	<0.02 <0.002	— <0.002	<0.05 <0.02
Aniline	<0.002 <0.02	<0.002 <0.02	<0.02 <0.02	<0.02 <0.002
Aniline Hydrochloride	>0.05 >0.05	>0.05 >0.05	>0.05 >0.05	>0.05 <0.02
Anthracine	<0.02 <0.02	<0.02 <0.02	<0.02 <0.02	<0.02 <0.02
Antimony Trichloride	<0.05 >0.05	— >0.05	— —	>0.05 —
Barium Carbonate	<0.02 <0.02	<0.02 <0.02	<0.02 <0.02	<0.02 <0.02

Selecting Iron Alloys in 100% Corrosive Medium (Continued)

	Corrosion Rate at 70°F in a 100% Corrosive Medium			
	1020 Steel	Grey Cast Iron	Ni–Resist Cast Iron	12% C Steel
Corrosive Medium	17% Cr Steel	Stainless Steel 301	Stainless Steel 316	14% Si Iron
Barium Chloride	<0.002	<0.02	—	—
	<0.02	<0.05	<0.02	—
Barium Hydroxide	<0.02	<0.02	—	<0.02
	<0.02	<0.02	<0.02	<0.02
Barium Nitrate	<0.02	—	—	—
	—	<0.02	<0.02	<0.02
Barium Oxide	<0.002	—	—	<0.02
	<0.02	<0.02	<0.02	<0.02
Barium Peroxide	<0.002	—	—	—
	—	—	—	—
Benzaldehyde	<0.002	>0.05	<0.002	<0.02
	<0.02	<0.02	<0.02	<0.02
Benzene	<0.02	<0.02	<0.02	<0.02
	<0.02	<0.02	<0.02	<0.002
Benzoic Acid	>0.05	>0.05	<0.02	<0.02
	<0.02	<0.02	<0.02	<0.02
Boric Acid	—	—	<0.02	<0.02
	<0.02	<0.02	<0.02	<0.02
Bromic Acid	>0.05	>0.05	—	>0.05
	>0.05	—	—	—
Bromine (Dry)	<0.05	>0.05	<0.02	>0.05
	>0.05	>0.05	>0.05	>0.05
Bromine (Wet)	>0.05	>0.05	>0.05	>0.05
	>0.05	>0.05	>0.05	>0.05

Selecting Iron Alloys in 100% Corrosive Medium (Continued)

Corrosion Rate at 70°F in a 100% Corrosive Medium

Corrosive Medium	1020 Steel	Grey Cast Iron	Ni–Resist Cast Iron	12% C Steel
	17% Cr Steel	Stainless Steel 301	Stainless Steel 316	14% Si Iron
Butyric Acid	>0.05	—	>0.05	—
	<0.05	<0.02	<0.02	<0.002
Cadmium Chloride	<0.002	—	—	—
	—	—	—	—
Cadmium Sulfate	<0.02	<0.02	—	—
	—	—	—	—
Calcium Acetate	<0.05	<0.05	—	<0.02
	<0.02	<0.02	<0.02	<0.02
Calcium Bicarbonate	<0.02	<0.02	—	<0.02
	<0.02	<0.02	<0.02	<0.02
Calcium Bromide	<0.05	<0.05	—	<0.02
	<0.02	<0.02	<0.02	<0.02
Calcium Chlorate	<0.02	<0.02	<0.02	—
	—	—	—	—
Calcium Chloride	<0.002	<0.002	—	—
	<0.02	<0.02	<0.002	<0.02
Calcium Hydroxide	<0.02	<0.02	—	<0.02
	<0.02	—	—	—
Calcium Hypochlorite	<0.02	<0.02	—	>0.05
	>0.05	—	—	<0.05
Carbon Dioxide	<0.002	<0.002	<0.002	<0.002
	<0.002	<0.002	<0.002	<0.002
Carbon Monoxide	<0.002	<0.002	<0.002	<0.002
	<0.002	<0.002	<0.002	<0.002

Selecting Iron Alloys in 100% Corrosive Medium (Continued)

	Corrosion Rate at 70°F in a 100% Corrosive Medium			
	1020 Steel	Grey Cast Iron	Ni–Resist Cast Iron	12% C Steel
Corrosive Medium	17% Cr Steel	Stainless Steel 301	Stainless Steel 316	14% Si Iron
Carbon Tetrachloride	<0.002 <0.002	<0.05 <0.02	<0.02 <0.02	<0.02 <0.002
Carbon Acid (Air Free)	<0.02 <0.002	<0.05 <0.02	<0.002 <0.02	<0.002 <0.002
Chloroacetic Acid	>0.05 >0.05	>0.05 —	>0.05 —	>0.05 >0.05
Chlorine Gas	<0.02 <0.05	<0.02 <0.002	<0.02 <0.02	<0.05 <0.02
Chlorine Liquid	<0.02 —	— —	— —	— —
Chloroform (Dry)	<0.002 <0.02	<0.002 <0.002	— <0.002	<0.002 —
Chromic Acid	<0.002 —	<0.02 —	<0.02 —	<0.02 <0.02
Chromic Hydroxide	<0.02 <0.02	— <0.02	<0.02 <0.02	<0.02 <0.02
Chromic Sulfates	>0.05 >0.05	— <0.05	— —	>0.05 <0.02
Citric Acid	<0.002 —	— <0.02	>0.05 <0.02	— <0.002
Diethylene Glycol	<0.002 <0.002	— <0.002	— <0.002	— <0.002
Ethyl Chloride	<0.002 <0.002	— <0.002	— <0.002	<0.002 <0.002

Selecting Iron Alloys in 100% Corrosive Medium (Continued)

Corrosion Rate at 70°F in a 100% Corrosive Medium

Corrosive Medium	1020 Steel	Grey Cast Iron	Ni–Resist Cast Iron	12% C Steel
	17% Cr Steel	Stainless Steel 301	Stainless Steel 316	14% Si Iron
Ethylene Glycol	<0.002 <0.02	<0.02 <0.02	<0.02 <0.02	<0.02 <0.02
Ethylene Oxide	<0.002 <0.02	<0.02 <0.02	— <0.02	<0.02 <0.02
Fatty Acids	>0.05 <0.02	>0.05 <0.02	<0.02 <0.002	<0.02 <0.002
Ferric Chloride	<0.02 —	— —	— —	— —
Fluorine	<0.002 <0.002	>0.05 <0.002	— <0.002	>0.05 >0.05
Formaldehyde	<0.002 <0.002	<0.02 <0.002	— <0.002	<0.02 <0.002
Formic Acid	>0.05 <0.05	>0.05 <0.02	>0.05 <0.002	<0.02 <0.002
Furfural	<0.02 —	<0.02 <0.02	<0.02 <0.02	— <0.02
Hydrazine	>0.05 —	— —	— —	— —
Hydrobromic Acid	<0.02 —	<0.02 >0.05	>0.05 —	— >0.05
Hydrocyanic Acid	<0.002 <0.05	<0.02 <0.02	<0.02 <0.02	>0.05 <0.02
Hydrofluoric Acid (Areated)	<0.02 —	>0.05 <0.02	<0.02 <0.02	— >0.05

Selecting Iron Alloys in 100% Corrosive Medium (Continued)

	Corrosion Rate at 70°F in a 100% Corrosive Medium			
	1020 Steel	Grey Cast Iron	Ni–Resist Cast Iron	12% C Steel
Corrosive Medium	17% Cr Steel	Stainless Steel 301	Stainless Steel 316	14% Si Iron
Hydrofluoric Acid (Air Free)	<0.05	>0.05	<0.02	>0.05
	>0.05	>0.05	<0.02	>0.05
Hydrogen Chloride	<0.002	<0.02	<0.002	>0.05
	>0.05	<0.002	<0.002	<0.02
Hydrogen Fluoride	<0.002	—	<0.02	<0.02
	<0.02	<0.002	<0.002	—
Hydrogen Iodide	<0.02	<0.02	<0.02	>0.05
	>0.05	<0.02	<0.02	<0.02
Hydrogen Peroxide	—	—	—	<0.02
	<0.02	<0.02	<0.02	<0.02
Hydrogen Sulfide	<0.02	<0.02	<0.02	<0.02
	<0.05	<0.05	<0.02	<0.02
Lactic Acid	>0.05	>0.05	>0.05	—
	—	<0.02	<0.02	<0.02
Lead Acetate	<0.002	—	—	<0.02
	<0.02	<0.02	<0.02	<0.05
Lead Chromate	<0.02	<0.02	<0.02	<0.02
	<0.02	<0.02	<0.02	<0.02
Lead Nitrate	<0.02	<0.02	—	—
	—	<0.02	<0.02	<0.002
Lead Sulfate	<0.02	<0.02	<0.02	<0.02
	<0.02	<0.02	<0.02	<0.02
Lithium Chloride	<0.002	<0.002	—	—
	—	<0.002	<0.002	<0.02

Selecting Iron Alloys in 100% Corrosive Medium (Continued)

Corrosive Medium	Corrosion Rate at 70°F in a 100% Corrosive Medium			
	1020 Steel	Grey Cast Iron	Ni–Resist Cast Iron	12% C Steel
	17% Cr Steel	Stainless Steel 301	Stainless Steel 316	14% Si Iron
Lithium Hydroxide	<0.002	—	—	—
	—	—	—	—
Magnesium Chloride	<0.002	<0.02	<0.02	—
	—	—	—	>0.05
Magnesium Hydroxide	<0.002	—	<0.02	<0.02
	<0.02	<0.02	<0.02	—
Magnesium Sulfate	<0.02	<0.02	<0.02	<0.05
	<0.02	<0.02	<0.02	<0.002
Maleic Acid	<0.002	—	—	<0.05
	<0.02	<0.02	<0.02	<0.02
Malic Acid	—	—	—	—
	—	<0.002	<0.002	—
Mercuric Chloride	—	—	—	>0.05
	>0.05	>0.05	—	<0.02
Mercurous Nitrate	<0.02	—	—	<0.02
	—	<0.02	<0.02	<0.002
Methallylamine	<0.02	<0.02	<0.02	<0.02
	<0.02	<0.02	<0.02	<0.002
Methanol	<0.002	<0.002	<0.002	<0.002
	<0.002	<0.002	<0.002	<0.002
Methyl Ethyl Ketone	<0.002	<0.002	<0.002	<0.002
	<0.002	<0.002	<0.002	<0.002
Methyl Isobutyl Ketone	<0.02	<0.02	<0.02	<0.02
	<0.02	<0.02	<0.02	<0.02

Selecting Iron Alloys in 100% Corrosive Medium (Continued)

	Corrosion Rate at 70°F in a 100% Corrosive Medium			
	1020 Steel	Grey Cast Iron	Ni–Resist Cast Iron	12% C Steel
Corrosive Medium	17% Cr Steel	Stainless Steel 301	Stainless Steel 316	14% Si Iron
Methylamine	<0.02	<0.02	<0.02	<0.02
	<0.02	<0.02	<0.02	<0.02
Methylene Chloride	<0.02	<0.02	<0.02	<0.02
	<0.02	<0.02	<0.02	<0.02
Monochloroacetic Acid	<0.002	>0.05	<0.05	>0.05
	>0.05	<0.02	<0.02	<0.02
Monorthanolamine	<0.02	<0.02	<0.02	—
	—	<0.02	<0.02	—
Monoethalamine	<0.02	—	<0.02	<0.02
	<0.02	<0.02	<0.02	<0.02
Monoethylamine	<0.02	<0.02	<0.02	<0.02
	<0.02	<0.02	<0.02	<0.02
Nitric Acid	>0.05	>0.05	>0.05	>0.05
	<0.05	<0.002	<0.002	<0.002
Nitric Acid (Red Fuming)	<0.05	>0.05	>0.05	<0.002
	<0.002	<0.002	<0.002	<0.002
Nitric + Hydrochloric Acid	>0.05	>0.05	>0.05	>0.05
	>0.05	>0.05	>0.05	<0.05
Nitric + Hydrofluoric Acid	>0.05	>0.05	>0.05	>0.05
	>0.05	>0.05	>0.05	>0.05
Nitric + Sulfuric Acid	>0.05	>0.05	>0.05	>0.05
	>0.05	>0.05	>0.05	<0.02
Nitrobenzene	<0.002	<0.02	<0.02	<0.02
	<0.02	<0.02	<0.02	<0.002

Selecting Iron Alloys in 100% Corrosive Medium (Continued)

Corrosion Rate at 70°F in a 100% Corrosive Medium

Corrosive Medium	1020 Steel 17% Cr Steel	Grey Cast Iron Stainless Steel 301	Ni–Resist Cast Iron Stainless Steel 316	12% C Steel 14% Si Iron
Nitrocelluolose	<0.02 <0.02	<0.02 <0.02	<0.02 <0.02	<0.02 <0.02
Nitroglycerine	<0.05 <0.02	<0.05 <0.02	<0.02 <0.02	<0.02 <0.05
Nitrotolune	<0.02 <0.02	<0.02 <0.02	<0.02 <0.02	<0.02 <0.02
Nitrous Acid	>0.05 —	— <0.02	— <0.02	— <0.002
Oleic Acid	<0.02 <0.02	<0.02 <0.02	<0.002 <0.02	<0.02 <0.002
Oxalic Acid	>0.05 >0.05	>0.05 >0.05	<0.02 >0.05	>0.05 <0.02
Phenol	<0.002 <0.02	<0.02 <0.02	<0.02 <0.02	<0.02 <0.002
Phosphoric Acid (Areated)	>0.05 —	>0.05 >0.05	>0.05 <0.02	— <0.002
Phosphoric Acid (Air Free)	>0.05 >0.05	>0.05 —	>0.05 —	>0.05 <0.02
Picric Acid	>0.05 <0.02	>0.05 <0.02	>0.05 <0.02	<0.02 <0.02
Potassium Bicarbonate	<0.002 <0.02	— <0.02	— <0.02	— —
Potassium Bromide	>0.05 <0.02	>0.05 <0.05	<0.02 —	<0.002 <0.02

Selecting Iron Alloys in 100% Corrosive Medium (Continued)

	Corrosion Rate at 70°F in a 100% Corrosive Medium			
	1020 Steel	Grey Cast Iron	Ni–Resist Cast Iron	12% C Steel
Corrosive Medium	17% Cr Steel	Stainless Steel 301	Stainless Steel 316	14% Si Iron
Potassium Carbonate	<0.02	<0.02	<0.02	<0.02
	<0.02	<0.02	<0.02	<0.02
Potassium Chlorate	<0.002	—	—	<0.02
	<0.02	<0.02	<0.02	<0.02
Potassium Chromate	—	—	—	<0.02
	<0.02	<0.02	<0.02	—
Potassium Cyanide	<0.002	<0.02	—	<0.02
	<0.02	<0.02	<0.02	<0.02
Potassium Dichromate	—	—	<0.02	<0.02
	<0.02	<0.02	<0.02	—
Potassium Ferricyanide	<0.02	<0.02	<0.02	—
	<0.02	<0.02	<0.02	—
Potassium Hydroxide	<0.002	<0.02	—	<0.002
	<0.002	<0.002	—	>0.05
Potassium Hypochlorite	<0.002	—	—	—
	—	—	<0.02	<0.002
Potassium Iodide	<0.02	—	—	—
	—	<0.02	<0.02	<0.02
Potassium Nitrate	<0.002	<0.02	—	<0.02
	<0.02	<0.02	—	<0.002
Potassium Nitrite	<0.02	<0.02	<0.02	<0.02
	<0.02	<0.02	<0.02	<0.02
Potassium Permanganate	<0.002	<0.02	—	<0.02
	<0.02	<0.02	—	—

Selecting Iron Alloys in 100% Corrosive Medium (Continued)

	Corrosion Rate at 70°F in a 100% Corrosive Medium			
	1020 Steel	Grey Cast Iron	Ni–Resist Cast Iron	12% C Steel
Corrosive Medium	17% Cr Steel	Stainless Steel 301	Stainless Steel 316	14% Si Iron
Potassium Silicate	<0.02 <0.02	<0.02 <0.02	<0.02 <0.02	<0.02 <0.02
Propionic Acid	<0.02 —	— —	— <0.02	— <0.02
Pyridine	<0.02 <0.02	<0.02 <0.02	<0.02 <0.02	<0.02 <0.02
Quinine Sulfate	>0.05 <0.02	>0.05 <0.02	<0.02 <0.02	— <0.02
Salicylic Acid	>0.05 <0.02	>0.05 <0.02	<0.02 <0.02	<0.02 <0.02
Silicon Tetrachloride (Dry)	<0.002 <0.002	<0.002 <0.002	<0.002 <0.002	<0.002 <0.002
Silicon Tetrachloride (Wet)	>0.05 >0.05	>0.05 >0.05	>0.05 —	>0.05 <0.002
Silver Bromide	>0.05 >0.05	>0.05 <0.05	>0.05 —	>0.05 <0.02
Silver Chloride	>0.05 >0.05	>0.05 >0.05	— —	>0.05 <0.02
Silver Nitrate	— —	— —	— <0.02	— —
Sodium Acetate	<0.002 <0.02	<0.002 <0.02	— <0.02	<0.02 <0.02
Sodium Bicarbonate	<0.05 <0.02	<0.05 —	<0.02 —	— —

Selecting Iron Alloys in 100% Corrosive Medium (Continued)

	Corrosion Rate at 70°F in a 100% Corrosive Medium			
	1020 Steel	Grey Cast Iron	Ni–Resist Cast Iron	12% C Steel
Corrosive Medium	17% Cr Steel	Stainless Steel 301	Stainless Steel 316	14% Si Iron
Sodium Bisulfate	<0.002 —	— >0.05	<0.002 —	>0.05 <0.002
Sodium Bromide	<0.02 —	<0.05 —	<0.02 —	— —
Sodium Carbonate	<0.02 <0.02	<0.02 <0.02	<0.02 <0.02	<0.02 <0.02
Sodium Chloride	<0.002 —	<0.02 —	<0.02 —	— —
Sodium Chromate	<0.02 <0.02	<0.02 <0.02	<0.02 <0.02	<0.02 <0.02
Sodium Hydroxide	<0.02 —	— —	<0.02 —	— —
Sodium Hypochlorite	>0.05 >0.05	— >0.05	— >0.05	>0.05 —
Sodium Metasilicate	<0.002 <0.002	<0.02 <0.002	<0.02 <0.002	<0.002 <0.02
Sodium Nitrate	<0.02 <0.002	<0.02 <0.02	<0.02 <0.02	<0.02 <0.002
Sodium Nitrite	<0.002 —	— <0.02	— —	<0.002 —
Sodium Phosphate	<0.02 <0.02	<0.02 <0.02	<0.02 <0.02	<0.02 <0.02
Sodium Silicate	<0.02 <0.02	<0.02 <0.02	<0.02 <0.02	<0.02 <0.02

Selecting Iron Alloys in 100% Corrosive Medium (Continued)

Corrosion Rate at 70°F in a 100% Corrosive Medium

Corrosive Medium	1020 Steel	Grey Cast Iron	Ni–Resist Cast Iron	12% C Steel
	17% Cr Steel	Stainless Steel 301	Stainless Steel 316	14% Si Iron
Sodium Sulfate	<0.02 >0.05	<0.02 <0.002	<0.02 <0.002	>0.05 <0.002
Sodium Sulfide	<0.02 >0.05	<0.02 >0.05	— —	<0.02 <0.02
Stannic Chloride	<0.002 —	— —	— —	— —
Stannous Chloride	<0.02 <0.05	<0.02 <0.05	<0.02 —	— —
Strontium Nitrate	>0.05 <0.02	>0.05 <0.02	— <0.02	— <0.02
Succinic Acid	<0.02 <0.02	<0.02 <0.02	<0.02 <0.02	<0.02 —
Sulfur Dioxide	<0.002 <0.02	<0.02 <0.02	<0.02 <0.02	<0.02 >0.05
Sulfur Trioxide	<0.02 <0.02	<0.02 <0.02	<0.02 <0.02	<0.02 >0.05
Sulfuric Acid (Areated)	<0.02 >0.05	<0.02 <0.02	<0.02 <0.02	>0.05 <0.02
Sulfuric Acid (Air Free)	<0.02 <0.05	<0.02 <0.05	<0.02 <0.02	<0.05
Sulfuric Acid (Fuming)	<0.02 <0.002	<0.02 <0.02	<0.05 <0.02	<0.002 <0.02
Sulfurous Acid	>0.05 >0.05	>0.05 >0.05	>0.05 <0.002	>0.05 <0.02

Selecting Iron Alloys in 100% Corrosive Medium (Continued)

Corrosion Rate at 70°F in a 100% Corrosive Medium

Corrosive Medium	1020 Steel / 17% Cr Steel	Grey Cast Iron / Stainless Steel 301	Ni–Resist Cast Iron / Stainless Steel 316	12% C Steel / 14% Si Iron
Tannic Acid	<0.002 <0.02	<0.02 <0.02	— <0.02	<0.02 <0.002
Tartaric Acid	<0.05 —	>0.05 —	— —	— <0.02
Tetraphosphoric Acid	>0.05 >0.05	>0.05 <0.02	<0.05 <0.02	>0.05 <0.05
Trichloroacetic Acid	>0.05 >0.05	>0.05 >0.05	>0.05 >0.05	>0.05 <0.002
Trichloro-ethylene	<0.002 <0.02	<0.02 <0.02	<0.02 <0.02	<0.02 <0.002
Zinc Chloride	<0.002 >0.05	<0.02 —	<0.02 —	>0.05 —

Water-free, dry or maximum concentration of corrosive medium. Quantitatively

<0.002 means that corrosion rate is likely to be less than 0.002 inch per year (Excellent).
<0.02 means that corrosion rate is likely to be less than about 0.02 inch per year (Good).
<0.05 means that corrosion rate is likely to be less than about 0.05 inch per year (Fair).
>0.05 means that corrosion rate is likely to be more than 0.05 inch per year (Poor).

Source: data compiled by J.S. Park *from* Earl R. Parker, *Materials Data Book for Engineers and Scientists*, McGraw-Hill Book Company, New York, 1967.

SELECTING NONFERROUS METALS FOR USE IN
A 10% CORROSIVE MEDIUM

Corrosion Rate at 70°F in a 10% Corrosive Medium

Corrosive Medium	Copper, Sn-Braze, Al-Braze / Inconel	70-30 Brass / Hastelloy	Silicon Bronze / Aluminum	Monel / Lead	Nickel / Titanium
Acetaldehyde	<0.002 —	<0.02 —	<0.02 <0.02	<0.002 <0.02	<0.002 —
Acetic Acid (Aerated)	>0.05 <0.02	>0.05 <0.002	>0.05 <0.02	<0.02 >0.05	<0.05 <0.002
Acetic Acid (Air Free)	<0.002 <0.02	>0.05 <0.002	>0.05 <0.002	<0.02 >0.05	<0.02 <0.002
Acetic Anhydride	— —	— <0.002	— —	— —	— —
Acetoacetic Acid	— —	— <0.02	— <0.02	<0.02 —	<0.02 —
Acetone	<0.002 <0.002	<0.002 <0.002	<0.002 <0.02	<0.002 <0.002	<0.002 <0.002
Acrolein	<0.02 —	<0.02 —	<0.02 <0.02	— <0.02	— —
Alcohol (Ethyl)	<0.002 <0.002	<0.002 <0.002	<0.002 <0.02	<0.002 <0.002	<0.002 <0.002
Alcohol (Methyl)	<0.02 <0.002	<0.02 <0.002	<0.02 —	<0.002 <0.02	<0.002 —
Alcohol (Benzyl)	— —	— <0.02	— —	— —	— —
Alcohol (Butyl)	— —	— —	— <0.002	— —	— —
Aluminum Acetate	<0.02 <0.02	— <0.02	<0.02 <0.002	<0.02 <0.002	<0.02 —

Selecting Nonferrous Metals for use in a 10% Corrosive Medium (Continued)

Corrosion Rate at 70°F in a 10% Corrosive Medium

Corrosive Medium	Copper, Sn-Braze, Al-Braze / Inconel	70-30 Brass / Hastelloy	Silicon Bronze / Aluminum	Monel / Lead	Nickel / Titanium
Aluminum Chlorate	— <0.02	— <0.02	— —	<0.02 <0.02	<0.02 <0.002
Aluminum Chloride	<0.02 >0.05	>0.05 <0.002	<0.02 >0.05	<0.02 >0.05	<0.05 >0.05
Aluminum Fluoride	<0.02 —	>0.05 <0.02	<0.02 <0.002	<0.002 <0.02	<0.02 —
Aluminum Formate	— <0.02	— <0.02	<0.02 <0.02	<0.02 —	<0.02 —
Aluminum Hydroxide	<0.02 —	<0.02 <0.02	<0.02 <0.02	<0.02 <0.02	<0.02 <0.002
Aluminum Nitrate	— <0.02	— <0.02	— <0.02	<0.02 <0.02	<0.02 <0.002
Aluminum Potassium Sulfate	<0.02 —	>0.05 <0.02	<0.02 <0.02	<0.02 <0.002	<0.02 —
Aluminum Sulfate	<0.02 <0.02	<0.02 <0.002	<0.02 <0.002	<0.02 <0.02	<0.02 <0.002
Ammonia	>0.05 <0.002	>0.05 <0.002	>0.05 <0.002	>0.05 <0.02	>0.05 <0.002
Ammonium Acetate	— <0.002	— <0.002	— <0.002	<0.002 —	<0.002 —
Ammonium Bicarbonate	>0.05 —	>0.05 —	>0.05 <0.02	— <0.02	— —
Ammonium Bromide	>0.05 —	>0.05 <0.02	>0.05 >0.05	<0.02 >0.05	<0.02 —

Selecting Nonferrous Metals for use in a 10% Corrosive Medium
(Continued)

Corrosion Rate at 70°F in a 10% Corrosive Medium

Corrosive Medium	Copper, Sn-Braze, Al-Braze ——— Inconel	70-30 Brass ——— Hastelloy	Silicon Bronze ——— Aluminum	Monel ——— Lead	Nickel ——— Titanium
Ammonium Carbonate	>0.05 >0.05	>0.05 >0.05	>0.05 <0.02	<0.02 <0.02	>0.05 —
Ammonium Chloride	>0.05 <0.02	>0.05 <0.002	>0.05 >0.05	<0.02 >0.05	<0.02 <0.002
Ammonium Citrate	>0.05 <0.02	>0.05 <0.02	>0.05 <0.02	<0.02 —	<0.02 <0.002
Ammonium Formate	— <0.02	— <0.002	— <0.02	<0.02 —	<0.02 <0.002
Ammonium Nitrate	>0.05 —	>0.05 <0.02	>0.05 <0.02	>0.05 >0.05	<0.02 <0.05
Ammonium Sulfate	<0.05 <0.02	>0.05 <0.02	<0.02 >0.05	<0.02 <0.02	<0.02 <0.002
Ammonium Sulfite	>0.05 >0.05	>0.05 —	>0.05 —	>0.05 —	>0.05 —
Ammonium Thiocyanate	>0.05 —	>0.05 —	>0.05 —	<0.02 —	<0.02 —
Amyl Acetate	<0.02 —	<0.02 <0.002	<0.02 —	<0.02 —	— —
Amyl Chloride	<0.02 —	— —	— —	<0.02 —	<0.02 —
Aniline	— —	— —	— —	<0.02 —	<0.02 —
Aniline Hydrochloride	>0.05 >0.05	>0.05 <0.02	>0.05 >0.05	>0.05 >0.05	<0.05 <0.002

Selecting Nonferrous Metals for use in a 10% Corrosive Medium (Continued)

Corrosion Rate at 70°F in a 10% Corrosive Medium

Corrosive Medium	Copper, Sn-Braze, Al-Braze _____ Inconel	70-30 Brass _____ Hastelloy	Silicon Bronze _____ Aluminum	Monel _____ Lead	Nickel _____ Titanium
Antimony Trichloride	>0.05 —	>0.05 >0.05	>0.05 >0.05	>0.05 <0.02	>0.05 —
Barium Carbonate	<0.02 —	<0.02 —	<0.02 —	<0.02 —	<0.02 —
Barium Chloride	<0.02 <0.02	>0.05 <0.02	<0.02 <0.02	<0.02 <0.02	<0.02 <0.002
Barium Hydroxide	>0.05 <0.02	>0.05 <0.02	>0.05 >0.05	<0.02 >0.05	<0.002 —
Barium Nitrate	>0.05 <0.02	>0.05 <0.02	>0.05 <0.02	— <0.02	<0.02 —
Barium Peroxide	>0.05 —	>0.05 —	>0.05 >0.05	<0.02 >0.05	<0.02 —
Benzaldehyde	>0.05 —	>0.05 —	>0.05 <0.02	— >0.05	— —
Benzene	<0.002 <0.002	<0.02 <0.02	<0.02 <0.02	<0.002 <0.02	<0.002 <0.002
Benzoic Acid	<0.02 <0.02	<0.02 <0.002	<0.02 <0.02	<0.02 >0.05	<0.02 <0.002
Boric Acid	<0.02 <0.02	<0.02 <0.002	<0.02 <0.05	<0.02 <0.02	<0.02 <0.002
Bromic Acid	>0.05 >0.05	>0.05 —	>0.05 >0.05	>0.05 <0.02	>0.05 —
Butyric Acid	<0.05 <0.05	<0.05 <0.002	<0.02 <0.02	<0.05 >0.05	<0.05 <0.002

Selecting Nonferrous Metals for use in a 10% Corrosive Medium
(Continued)

Corrosion Rate at 70°F in a 10% Corrosive Medium

Corrosive Medium	Copper, Sn-Braze, Al-Braze / Inconel	70-30 Brass / Hastelloy	Silicon Bronze / Aluminum	Monel / Lead	Nickel / Titanium
Cadmium Chloride	<0.02 —	>0.05 <0.02	<0.02 >0.05	<0.02 —	<0.02 —
Cadmium Sulfate	<0.02 <0.002	<0.02 <0.002	<0.02 <0.02	<0.002 <0.002	<0.002 —
Calcium Acetate	<0.02 <0.02	<0.02 <0.02	<0.02 —	<0.02 <0.02	<0.02 <0.002
Calcium Bromide	<0.02 <0.02	<0.02 <0.02	<0.02 <0.05	<0.02 <0.02	<0.02 —
Calcium Chlorate	<0.02 <0.02	>0.05 <0.02	<0.02 <0.02	<0.02 <0.02	<0.02 —
Calcium Chloride	<0.002 <0.002	<0.02 <0.002	<0.02 <0.002	<0.002 >0.05	<0.002 <0.002
Calcium Hydroxide	<0.02 <0.02	<0.02 <0.002	<0.02 >0.05	<0.02 >0.05	<0.02 —
Calcium Hypochlorite	<0.02 >0.05	<0.02 <0.02	<0.02 >0.05	>0.05 <0.05	>0.05 <0.002
Carbon Tetrachloride	— <0.002	— <0.002	— —	<0.02 —	<0.02 —
Carbon Acid (Air Free)	<0.02 <0.02	— <0.002	<0.02 <0.02	<0.02 —	<0.02 —
Chloroacetic Acid	>0.05 —	>0.05 <0.02	— >0.05	<0.02 >0.05	— —
Chromic Acid	>0.05 <0.02	>0.05 <0.02	>0.05 >0.05	>0.05 <0.02	>0.05 <0.002

Selecting Nonferrous Metals for use in a 10% Corrosive Medium (Continued)

Corrosion Rate at 70°F in a 10% Corrosive Medium

Corrosive Medium	Copper, Sn-Braze, Al-Braze	70-30 Brass	Silicon Bronze	Monel	Nickel
	Inconel	Hastelloy	Aluminum	Lead	Titanium
Chromic Sulfates	<0.02	<0.02	<0.02	—	—
	—	<0.02	—	<0.02	—
Citric Acid	<0.05	>0.05	<0.05	<0.02	<0.02
	<0.02	<0.002	<0.02	<0.02	<0.002
Copper Nitrate	>0.05	>0.05	>0.05	>0.05	>0.05
	>0.05	<0.02	>0.05	—	—
Copper Sulfate	>0.05	>0.05	<0.02	<0.02	<0.02
	<0.02	<0.002	>0.05	<0.02	—
Ethyl Chloride	<0.02	—	—	<0.02	—
	—	—	—	—	—
Ethylene Glycol	<0.02	—	—	—	—
	—	—	<0.002	—	—
Ferric Chloride	>0.05	>0.05	>0.05	>0.05	>0.05
	<0.05	<0.002	>0.05	>0.05	<0.002
Ferric Nitrate	>0.05	>0.05	>0.05	>0.05	>0.05
	>0.05	<0.002	>0.05	<0.002	<0.002
Ferrous Chloride	<0.02	>0.05	<0.05	>0.05	<0.05
	>0.05	<0.02	>0.05	>0.05	<0.002
Ferrous Sulfate	<0.02	>0.05	<0.02	—	>0.05
	<0.02	<0.02	<0.002	<0.02	<0.002
Formaldehyde	<0.002	<0.002	<0.002	<0.002	<0.002
	<0.002	<0.02	<0.02	<0.02	<0.002
Formic Acid	<0.02	<0.05	<0.02	<0.02	<0.02
	<0.02	<0.002	<0.02	>0.05	<0.02

Selecting Nonferrous Metals for use in a 10% Corrosive Medium
(Continued)

Corrosion Rate at 70°F in a 10% Corrosive Medium

Corrosive Medium	Copper, Sn-Braze, Al-Braze	70-30 Brass	Silicon Bronze	Monel	Nickel
	Inconel	Hastelloy	Aluminum	Lead	Titanium
Furfural	<0.02	<0.02	<0.02	<0.02	<0.02
	<0.02	<0.02	—	—	—
Hydrazine	>0.05	>0.05	>0.05	—	—
	—	—	—	>0.05	—
Hydrobromic Acid	>0.05	>0.05	<0.02	>0.05	>0.05
	—	<0.02	>0.05	>0.05	—
Hydrochloric Acid (Areated)	>0.05	>0.05	>0.05	>0.05	>0.05
	>0.05	<0.02	>0.05	<0.02	<0.02
Hydrochloric Acid (Air Free)	>0.05	>0.05	<0.02	>0.05	>0.05
	>0.05	<0.02	>0.05	<0.02	<0.02
Hydrocyanic Acid	>0.05	>0.05	>0.05	>0.05	—
	—	—	<0.02	>0.05	—
Hydrofluoric Acid (Areated)	<0.02	>0.05	>0.05	<0.02	<0.02
	<0.02	<0.02	>0.05	>0.05	>0.05
Hydrofluoric Acid (Air Free)	<0.02	>0.05	<0.02	<0.02	<0.02
	<0.02	<0.02	>0.05	<0.002	>0.05
Hydrogen Iodide	—	—	—	<0.02	—
	—	—	—	—	—
Hydrogen Peroxide	>0.05	>0.05	>0.05	<0.02	<0.02
	<0.02	<0.002	<0.002	>0.05	<0.002
Hydrogen Sulfide	<0.02	<0.02	<0.02	—	—
	<0.02	—	—	—	—
Lactic Acid	<0.002	<0.05	<0.05	>0.05	<0.02
	<0.02	<0.02	<0.02	>0.05	<0.002

Selecting Nonferrous Metals for use in a 10% Corrosive Medium (Continued)

Corrosion Rate at 70°F in a 10% Corrosive Medium

Corrosive Medium	Copper, Sn-Braze, Al-Braze / Inconel	70-30 Brass / Hastelloy	Silicon Bronze / Aluminum	Monel / Lead	Nickel / Titanium
Lead Acetate	<0.05 <0.02	 <0.02	— —	<0.02 —	<0.02 <0.002
Lead Chromate	— —	— —	— >0.05	— —	— —
Lead Nitrate	— —	— —	— >0.05	— —	<0.02 —
Lead Sulfate	— —	— —	— >0.05	— —	<0.02 —
Lithium Chloride	<0.02 30 <0.002 30	<0.02 30 <0.002 30	<0.02 30 <0.05	<0.002 30 <0.02	<0.002 30 —
Lithium Hydroxide	>0.05 <0.02	>0.05 <0.02	>0.05 >0.05	<0.02 >0.05	<0.02 —
Magnesium Chloride	<0.02 <0.002	<0.02 <0.002	<0.02 >0.05	<0.002 >0.05	<0.002 <0.002
Magnesium Hydroxide	<0.02 —	<0.02 <0.02	<0.02 >0.05	<0.02 >0.05	— —
Magnesium Sulfate	<0.002 <0.02	<0.02 <0.002	<0.002 <0.02	<0.02 <0.02	<0.02 —
Maleic Acid	<0.02 <0.02	<0.02 <0.002	<0.02 <0.02	<0.05 —	<0.02 —
Malic Acid	— <0.002	— —	— <0.02	<0.02 —	<0.02 —
Maganous Chloride	— —	— <0.02	— —	— —	— <0.002

Selecting Nonferrous Metals for use in a 10% Corrosive Medium (Continued)

Corrosion Rate at 70°F in a 10% Corrosive Medium

Corrosive Medium	Copper, Sn-Braze, Al-Braze	70-30 Brass	Silicon Bronze	Monel	Nickel
	Inconel	Hastelloy	Aluminum	Lead	Titanium
Mercuric Chloride	>0.05	>0.05	>0.05	>0.05	<0.05
	>0.05	<0.02	>0.05	<0.05	<0.002
Mercurous Nitrate	>0.05	>0.05	>0.05	<0.02	—
	—	<0.02	>0.05	—	—
Methanol	<0.02	<0.02	<0.02	<0.002	<0.002
	<0.002	<0.002	—	<0.02	—
Methyl Ethyl Ketone	<0.02	<0.02	<0.02	<0.02	<0.02
	<0.02	<0.02	<0.02	<0.02	<0.002
Methyl Isobutyl Ketone	<0.02	<0.02	<0.02	<0.02	<0.02
	<0.02	<0.02	<0.02	<0.02	<0.002
Methylamine	—	—	—	—	—
	—	—	<0.02	—	—
Methylene Chloride	<0.02	—	<0.02	—	—
	—	<0.02	>0.05	—	—
Monochloroacetic Acid	>0.05	>0.05	>0.05	—	<0.02
	<0.02	—	>0.05	>0.05	—
Monoethalamine	—	—	—	—	—
	—	—	<0.02	—	—
Monoethylamine	—	—	—	—	—
	—	—	<0.02	—	—
Monosodium Phosphate	<0.02	<0.02	<0.02	<0.02	<0.02
	<0.02	<0.02	>0.05	<0.02	—
Nickel Chloride	>0.05	>0.05	>0.05	<0.02	—
	—	<0.002	>0.05	—	<0.02

Selecting Nonferrous Metals for use in a 10% Corrosive Medium (Continued)

Corrosion Rate at 70°F in a 10% Corrosive Medium

Corrosive Medium	Copper, Sn-Braze, Al-Braze	70-30 Brass	Silicon Bronze	Monel	Nickel
	Inconel	Hastelloy	Aluminum	Lead	Titanium
Nickel Nitrate	<0.05	<0.05	<0.05	>0.05	>0.05
	>0.05	<0.02	>0.05	—	—
Nickel Sulfate	<0.02	<0.05	<0.02	—	<0.02
	<0.02	<0.02	>0.05	<0.02	—
Nitric Acid	>0.05	>0.05	>0.05	>0.05	>0.05
	<0.02	<0.002	>0.05	>0.05	<0.002
Nitric + Sulfuric Acid	>0.05	>0.05	>0.05	>0.05	>0.05
	>0.05	—	>0.05	>0.05	—
Nitrous Acid	—	—	—	—	>0.05
	—	—	<0.05	—	—
Oleic Acid	—	>0.05	—	—	—
	—	—	—	—	—
Oxalic Acid	<0.02	<0.02	<0.02	<0.02	<0.02
	<0.02	<0.02	<0.02	>0.05	<0.02
Phenol	—	—	—	<0.002	—
	—	—	—	—	—
Phosphoric Acid (Areated)	>0.05	>0.05	>0.05	<0.05	<0.05
	<0.02	<0.002	>0.05	<0.02	<0.02
Phosphoric Acid (Air Free)	<0.02	<0.02	<0.02	<0.02	<0.02
	<0.02	<0.002	>0.05	<0.002	—
Picric Acid	>0.05	>0.05	>0.05	<0.05	>0.05
	—	<0.02	>0.05	>0.05	—
Potassium Bicarbonate	<0.02	<0.02	<0.02	<0.02	<0.02
	<0.02	<0.02	>0.05	>0.05	—

Selecting Nonferrous Metals for use in a 10% Corrosive Medium
(Continued)

Corrosion Rate at 70°F in a 10% Corrosive Medium

Corrosive Medium	Copper, Sn-Braze, Al-Braze / Inconel	70-30 Brass / Hastelloy	Silicon Bronze / Aluminum	Monel / Lead	Nickel / Titanium
Potassium Bromide	<0.02 <0.02	<0.02 <0.002	<0.02 <0.02	<0.02 <0.02	<0.02 <0.002
Potassium Carbonate	<0.02 <0.02	<0.02 <0.02	<0.02 >0.05	<0.02 >0.05	<0.02 <0.002
Potassium Chlorate	<0.02 <0.05	<0.02 <0.02	<0.02 <0.02	<0.05 <0.02	<0.02 <0.002
Potassium Chromate	<0.02 <0.002	<0.02 <0.002	<0.02 <0.02	<0.02 <0.02	<0.002 —
Potassium Cyanide	>0.05 <0.02	>0.05 <0.02	>0.05 >0.05	<0.02 >0.05	<0.02 —
Potassium Dichromate	<0.02 <0.02	<0.02 <0.02	<0.02 <0.002	<0.02 <0.02	<0.02 <0.002
Potassium Ferricyanide	<0.02 —	<0.02 <0.02	<0.02 <0.02	<0.02 <0.02	<0.02 —
Potassium Ferrocyanide	<0.02 <0.02	<0.02 <0.02	<0.02 <0.002	<0.02 <0.02	<0.02 —
Potassium Hydroxide	<0.02 <0.02	<0.02 <0.02	<0.02 >0.05	<0.002 >0.05	<0.002 <0.002
Potassium Hypochlorite	<0.02 <0.05	>0.05 <0.02	>0.05 >0.05	<0.05 <0.02	<0.05 <0.002
Potassium Iodide	<0.02 <0.02	— <0.02	<0.02 <0.02	<0.02 >0.05	<0.02 <0.002
Potassium Nitrate	<0.02 <0.02	<0.02 <0.02	<0.02 <0.002	<0.02 <0.02	<0.02 <0.002

Selecting Nonferrous Metals for use in a 10% Corrosive Medium (Continued)

Corrosion Rate at 70°F in a 10% Corrosive Medium

Corrosive Medium	Copper, Sn-Braze, Al-Braze / Inconel	70-30 Brass / Hastelloy	Silicon Bronze / Aluminum	Monel / Lead	Nickel / Titanium
Potassium Nitrite	<0.02 / <0.02	<0.02 / <0.02	<0.02 / <0.02	<0.02 / <0.02	<0.02 / <0.002
Potassium Permanganate	<0.02 / <0.02	<0.02 / <0.002	<0.02 / <0.02	<0.05 / <0.05	<0.02 / —
Potassium Silicate	<0.02 / <0.02	<0.02 / <0.02	<0.02 / >0.05	<0.02 / —	<0.02 / —
Propionic Acid	<0.02 / —	<0.02 / —	<0.02 / <0.02	<0.02 / >0.05	<0.02 / —
Pyridine	<0.02 / <0.02	<0.02 / <0.02	<0.02 / <0.02	<0.02 / <0.02	<0.02 / —
Quinine Sulfate	<0.02 / <0.02	<0.02 / <0.02	<0.02 / —	<0.02 / —	<0.02 / —
Salicylic Acid	— / —	— / —	— / >0.05	<0.02 / —	<0.02 / —
Silver Bromide	>0.05 / —	>0.05 / <0.002	>0.05 / >0.05	— / —	— / —
Silver Chloride	>0.05 / —	>0.05 / <0.02	>0.05 / >0.05	— / —	— / <0.002
Silver Nitrate	>0.05 / <0.02	>0.05 / <0.002	>0.05 / >0.05	>0.05 / >0.05	>0.05 / —
Sodium Acetate	<0.02 / <0.02	<0.02 / <0.02	<0.02 / <0.02	<0.05 / —	<0.02 / —
Sodium Bicarbonate	<0.02 / <0.02	<0.02 / <0.02	<0.02 / >0.05	<0.02 / <0.02	<0.02 / —

Selecting Nonferrous Metals for use in a 10% Corrosive Medium (Continued)

Corrosion Rate at 70°F in a 10% Corrosive Medium

Corrosive Medium	Copper, Sn-Braze, Al-Braze / Inconel	70-30 Brass / Hastelloy	Silicon Bronze / Aluminum	Monel / Lead	Nickel / Titanium
Sodium Bisulfate	— <0.02	>0.05 <0.02	<0.02 >0.05	<0.02 <0.02	<0.02 —
Sodium Bromide	<0.02 <0.02	<0.05 <0.02	<0.02 <0.05	<0.02 —	<0.02 —
Sodium Carbonate	<0.02 <0.02	>0.05 <0.02	<0.02 >0.05	<0.02 <0.02	<0.02 —
Sodium Chloride	<0.02 <0.002	<0.05 <0.02	<0.02 <0.05	<0.002 <0.02	<0.002 <0.002
Sodium Chromate	<0.02 <0.02	<0.02 <0.02	<0.02 <0.02	<0.02 <0.02	<0.02 —
Sodium Hydroxide	<0.002 <0.002	>0.05 <0.002	<0.02 >0.05	<0.002 <0.02	<0.002 <0.002
Sodium Hypochlorite	>0.05 >0.05	>0.05 <0.002	<0.02 >0.05	>0.05 >0.05	>0.05 <0.002
Sodium Metasilicate	<0.02 <0.002	<0.02 <0.002	<0.02 >0.05	<0.002 —	<0.002 —
Sodium Nitrate	<0.02 <0.002	<0.05 <0.02	<0.02 <0.002	<0.02 >0.05	<0.02 —
Sodium Nitrite	<0.02 <0.02	<0.02 <0.02	<0.02 <0.02	<0.02 <0.02	<0.02 <0.002
Sodium Phosphate	<0.02 <0.02	<0.02 <0.02	<0.02 >0.05	<0.02 <0.02	<0.02 —
Sodium Silicate	<0.02 <0.02	<0.02 <0.02	<0.02 >0.05	<0.02 >0.05	<0.02 —

Selecting Nonferrous Metals for use in a 10% Corrosive Medium (Continued)

Corrosion Rate at 70°F in a 10% Corrosive Medium

Corrosive Medium	Copper, Sn-Braze, Al-Braze / Inconel	70-30 Brass / Hastelloy	Silicon Bronze / Aluminum	Monel / Lead	Nickel / Titanium
Sodium Sulfate	<0.02 <0.02	<0.02 <0.02	<0.02 <0.002	<0.02 <0.02	<0.02 —
Sodium Sulfide	>0.05 <0.02	<0.05 <0.02	>0.05 >0.05	<0.02 <0.002	<0.02 <0.002
Sodium Sulfite	<0.02 <0.02	>0.05 <0.02	<0.02 <0.02	<0.02 <0.02	<0.02 —
Stannic Chloride	>0.05 >0.05	>0.05 <0.02	>0.05 >0.05	>0.05 >0.05	>0.05 <0.002
Stannous Chloride	>0.05 >0.05	>0.05 <0.02	<0.02 >0.05	>0.05 >0.05	<0.05 —
Strontium Nitrate	<0.02 <0.02	<0.02 <0.02	<0.02 <0.02	<0.02 —	<0.02 —
Succinic Acid	<0.02 <0.02	<0.02 <0.02	<0.02 <0.02	<0.02 <0.02	<0.02 <0.002
Sulfur Dioxide	<0.02 <0.02	>0.05 <0.002	— >0.05	>0.05 —	>0.05 —
Sulfuric Acid (Areated)	>0.05 >0.05	>0.05 <0.002	>0.05 >0.05	<0.05 <0.002	<0.05 <0.02
Sulfuric Acid (Air Free)	<0.02 <0.05	<0.05 <0.002	<0.02 >0.05	<0.002 <0.002	<0.02 —
Sulfurous Acid	<0.02 <0.05	<0.02 <0.02	<0.02 <0.02	>0.05 <0.02	<0.05 <0.002
Tannic Acid	<0.02 —	— <0.02	<0.02 <0.02	<0.02 >0.05	— <0.002

Selecting Nonferrous Metals for use in a 10% Corrosive Medium (Continued)

Corrosion Rate at 70°F in a 10% Corrosive Medium

Corrosive Medium	Copper, Sn-Braze, Al-Braze / Inconel	70-30 Brass / Hastelloy	Silicon Bronze / Aluminum	Monel / Lead	Nickel / Titanium
Tartaric Acid	<0.02 <0.02	<0.05 <0.02	<0.05 <0.02	<0.02 <0.02	<0.02 <0.002
Tetraphosphoric Acid	— —	>0.05 —	>0.05 >0.05	— >0.05	— —
Trichloroacetic Acid	>0.05 —	>0.05 <0.02	— >0.05	— >0.05	— <0.002
Urea	<0.02 <0.02	<0.02 <0.02	<0.02 <0.02	<0.02 —	<0.02 —
Zinc Chloride	<0.02 —	>0.05 <0.02	<0.02 >0.05	<0.02 <0.02	<0.02 <0.002
Zinc Sulfate	<0.02 <0.002	<0.05 <0.02	<0.02 <0.05	<0.02 <0.02	<0.02 —

10% corrosive medium in 90% water

<0.002 means that corrosion rate is likely to be less than 0.002 inch per year (Excellent).
<0.02 means that corrosion rate is likely to be less than about 0.02 inch per year (Good).
<0.05 means that corrosion rate is likely to be less than about 0.05 inch per year (Fair).
>0.05 means that corrosion rate is likely to be more than 0.05 inch per year (Poor).

Source: data compiled by J.S. Park *from* Earl R. Parker, *Materials Data Book for Engineers and Scientists*, McGraw-Hill Book Company, New York, 1967.

SELECTING NONFERROUS METALS FOR USE IN A 100% CORROSIVE MEDIUM

Corrosion Rate at 70°F in a 100% Corrosive Medium

Corrosive Medium	Copper, Sn-Braze, Al-Braze / Inconel	70-30 Brass / Hastelloy	Silicon Bronze / Aluminum	Monel / Lead	Nickel / Titanium
Acetaldehyde	<0.002 <0.002	<0.002 <0.002	<0.002 <0.002	<0.002 <0.002	<0.002 <0.002
Acetic Acid (Aerated)	<0.02 <0.02	>0.05 <0.002	>0.05 <0.002	<0.02 <0.05	>0.05 <0.002
Acetic Acid (Air Free)	<0.002 <0.02	>0.05 <0.002	<0.02 <0.002	<0.02 <0.02	<0.02 <0.002
Acetic Anhydride	<0.02 <0.02	>0.05 <0.002	<0.02 <0.002	<0.02 <0.002	<0.02 <0.002
Acetoacetic Acid	— —	— <0.02	— <0.02	<0.02 <0.02	<0.02 —
Acetone	<0.002 <0.002	<0.002 <0.002	<0.002 <0.002	<0.002 <0.02	<0.002 <0.002
Acetylene	<0.002 <0.002	<0.002 <0.002	<0.002 <0.002	<0.002 <0.002	<0.002 <0.002
Acrolein	<0.02 <0.02	<0.02 <0.02	<0.02 <0.02	<0.02 —	<0.02 <0.02
Acrylonitril	<0.002 <0.002	<0.002 <0.002	<0.002 <0.002	<0.002 <0.002	<0.002 <0.002
Alcohol (Ethyl)	<0.002 <0.002	<0.002 <0.002	<0.002 <0.02	<0.002 <0.002	<0.002 <0.002
Alcohol (Methyl)	<0.02 <0.002	<0.02 <0.002	<0.02 <0.02	<0.002 <0.02	<0.002 —
Alcohol (Allyl)	<0.02 <0.02	<0.02 <0.02	<0.02 <0.02	<0.02 <0.02	<0.02 <0.002

Selecting Nonferrous Metals for use in a 100% Corrosive Medium (Continued)

Corrosion Rate at 70°F in a 100% Corrosive Medium

Corrosive Medium	Copper, Sn-Braze, Al-Braze / Inconel	70-30 Brass / Hastelloy	Silicon Bronze / Aluminum	Monel / Lead	Nickel / Titanium
Alcohol (Amyl)	<0.002 —	— —	<0.02 <0.002	— —	— <0.002
Alcohol (Benzyl)	<0.02 <0.02	<0.02 <0.02	<0.02 <0.02	<0.02 <0.02	<0.02 <0.002
Alcohol (Butyl)	<0.002 <0.002	<0.002 —	<0.002 <0.002	<0.002 —	<0.002 <0.002
Alcohol (Cetyl)	<0.02 <0.02	— —	— <0.02	<0.02 <0.02	<0.02 <0.002
Alcohol (Isopropyl)	<0.02 <0.02	<0.02 <0.02	<0.02 <0.02	<0.02 <0.002	<0.02 —
Allylamine	>0.05 —	>0.05 —	>0.05 —	— —	— —
Allyl Chloride	<0.02 <0.02	<0.02 <0.02	<0.02 >0.05	<0.02 <0.05	<0.02 —
Allyl Sulfide	>0.05 —	>0.05 —	>0.05 <0.02	— >0.05	— —
Aluminum Acetate	<0.02 —	<0.02 <0.02	<0.02 <0.002	— <0.002	— <0.002
Aluminum Chlorate	— <0.02	— <0.02	— —	<0.02 <0.02	<0.02 —
Aluminum Chloride	<0.02 —	>0.05 <0.002	<0.02 <0.02	— —	<0.02 —
Aluminum Fluosilicate	<0.02 <0.02	<0.02 <0.02	<0.02 —	<0.02 <0.02	<0.02 —

Selecting Nonferrous Metals for use in a 100% Corrosive Medium (Continued)

Corrosion Rate at 70°F in a 100% Corrosive Medium

Corrosive Medium	Copper, Sn-Braze, Al-Braze / Inconel	70-30 Brass / Hastelloy	Silicon Bronze / Aluminum	Monel / Lead	Nickel / Titanium
Aluminum Formate	<0.02	—	<0.02	<0.02	<0.02
	<0.02	<0.02	<0.02	<0.02	<0.002
Aluminum Hydroxide	—	—	<0.02	—	—
	—	—	—	—	<0.002
Aluminum Nitrate	—	—	—	—	—
	—	—	<0.02	—	<0.002
Aluminum Potassium Sulfate	<0.02	>0.05	<0.02	—	—
	—	—	<0.02	<0.02	<0.002
Aluminum Sulfate	<0.002	<0.05	<0.02	<0.02	<0.02
	—	<0.02	>0.05	—	—
Ammonia	<0.002	<0.002	<0.002	<0.002	<0.002
	<0.002	<0.002	<0.002	<0.02	<0.002
Ammonium Acetate	>0.05	>0.05	>0.05	<0.002	<0.002
	<0.002	<0.002	<0.002	—	—
Ammonium Bicarbonate	—	—	—	—	—
	—	—	<0.02	—	—
Ammonium Carbonate	—	—	<0.02	<0.02	<0.02
	<0.02	—	<0.02	—	—
Ammonium Chloride	>0.05	>0.05	>0.05	<0.02	<0.02
	<0.02	<0.02	<0.02	<0.02	—
Ammonium Citrate	—	—	—	—	—
	<0.02	—	<0.02	—	<0.002
Ammonium Formate	—	—	—	—	—
	<0.02	—	—	—	<0.002

Selecting Nonferrous Metals for use in a 100% Corrosive Medium
(Continued)

Corrosion Rate at 70°F in a 100% Corrosive Medium

Corrosive Medium	Copper, Sn-Braze, Al-Braze / Inconel	70-30 Brass / Hastelloy	Silicon Bronze / Aluminum	Monel / Lead	Nickel / Titanium
Ammonium Nitrate	>0.05 —	>0.05 —	>0.05 <0.02	<0.02 —	<0.02 —
Ammonium Sulfate	<0.02 —	<0.02 <0.02	<0.02 <0.02	<0.02 <0.02	<0.02 —
Ammonium Sulfite	>0.05 —	>0.05 —	>0.05 —	— —	— —
Ammonium Thiocyanate	— —	— —	— —	<0.02 —	<0.02 —
Amyl Acetate	<0.02 <0.02	<0.02 <0.002	<0.02 <0.002	<0.02 <0.02	<0.02 <0.002
Amyl Chloride	<0.002 —	<0.02 <0.02	<0.002 <0.02	<0.02 >0.05	<0.02 —
Aniline	>0.05 —	>0.05 <0.02	— <0.02	<0.02 >0.05	<0.02 —
Aniline Hydrochloride	— —	— <0.05	— >0.05	— —	— —
Anthracine	<0.02 <0.02	<0.02 <0.02	<0.02 <0.02	<0.02 <0.02	<0.02 <0.002
Antimony Trichloride	<0.05 —	— <0.002	— <0.02	— <0.002	<0.02 —
Barium Carbonate	<0.02 —	<0.02 <0.02	<0.02 >0.05	<0.02 >0.05	<0.02 —
Barium Chloride	<0.02 <0.02	<0.02 <0.02	<0.02 >0.05	<0.02 —	<0.02 —

Selecting Nonferrous Metals for use in a 100% Corrosive Medium (Continued)

Corrosion Rate at 70°F in a 100% Corrosive Medium

Corrosive Medium	Copper, Sn-Braze, Al-Braze / Inconel	70-30 Brass / Hastelloy	Silicon Bronze / Aluminum	Monel / Lead	Nickel / Titanium
Barium Hydroxide	—	—	—	<0.02	<0.02
	<0.02	<0.02	>0.05	>0.05	—
Barium Nitrate	—	—	—	—	—
	<0.02	<0.02	—	—	—
Barium Oxide	—	—	—	<0.02	—
	<0.02	<0.02	—	—	—
Benzaldehyde	<0.02	<0.02	<0.02	<0.02	<0.02
	<0.02	<0.02	<0.002	>0.05	—
Benzene	<0.02	<0.02	<0.02	<0.02	<0.02
	<0.02	<0.02	<0.02	<0.02	<0.002
Benzoic Acid	<0.02	<0.02	<0.02	<0.02	<0.02
	—	—	<0.02	>0.05	<0.002
Boric Acid	<0.02	<0.02	<0.02	<0.02	<0.02
	<0.02	<0.002	<0.02	<0.02	—
Bromic Acid	>0.05	>0.05	>0.05	>0.05	>0.05
	>0.05	—	—	<0.02	—
Bromine (Dry)	<0.02	<0.02	<0.02	<0.002	<0.002
	<0.002	<0.002	<0.02	<0.002	>0.05
Bromine (Wet)	>0.05	>0.05	>0.05	>0.05	>0.05
	>0.05	<0.002	>0.05	>0.05	>0.05
Butyric Acid	<0.02	—	<0.02	<0.02	<0.05
	<0.05	<0.002	<0.002	>0.05	<0.002
Calcium Acetate	<0.02	<0.02	<0.02	<0.02	<0.02
	<0.02	<0.02	<0.05	<0.02	<0.002

Selecting Nonferrous Metals for use in a 100% Corrosive Medium
(Continued)

Corrosion Rate at 70°F in a 100% Corrosive Medium

Corrosive Medium	Copper, Sn-Braze, Al-Braze / Inconel	70-30 Brass / Hastelloy	Silicon Bronze / Aluminum	Monel / Lead	Nickel / Titanium
Calcium Bicarbonate	<0.02 <0.02	<0.02 <0.02	<0.02 <0.02	<0.02 <0.05	<0.02 <0.002
Calcium Bromide	<0.02 <0.02	<0.02 <0.02	<0.02 <0.05	<0.02 <0.02	<0.02 <0.05
Calcium Chlorate	— —	<0.02 <0.02	— —	— —	— <0.002
Calcium Chloride	<0.02 <0.02	<0.02 <0.002	<0.02 >0.05	<0.02 —	<0.02 —
Calcium Hydroxide	— <0.02	— —	— >0.05	<0.02 —	<0.02 —
Calcium Hypochlorite	— —	— <0.02	— —	— <0.002	— —
Carbon Dioxide	<0.002 <0.002	<0.002 <0.002	<0.002 <0.002	<0.002 <0.002	<0.002 <0.002
Carbon Monoxide	<0.002 <0.002	<0.002 <0.002	<0.002 <0.002	<0.002 <0.002	<0.002 <0.002
Carbon Tetrachloride	<0.002 <0.002	<0.05 <0.002	<0.002 <0.02	<0.002 <0.002	<0.002 <0.002
Carbon Acid (Air Free)	<0.02 <0.002	>0.05 <0.002	<0.02 <0.002	<0.05 >0.05	<0.02 —
Chloroacetic Acid	>0.05 <0.05	>0.05 <0.002	<0.05 >0.05	<0.05 >0.05	<0.02 <0.002
Chlorine Gas	<0.02 <0.02	>0.05 <0.02	<0.02 <0.02	<0.02 <0.02	<0.002 >0.05

Selecting Nonferrous Metals for use in a 100% Corrosive Medium (Continued)

Corrosive Medium	Corrosion Rate at 70°F in a 100% Corrosive Medium				
	Copper, Sn-Braze, Al-Braze	70-30 Brass	Silicon Bronze	Monel	Nickel
	Inconel	Hastelloy	Aluminum	Lead	Titanium
Chlorine Liquid	—	—	—	<0.02	—
	—	—	—	<0.02	—
Chloroform (Dry)	<0.002	<0.02	<0.02	<0.002	<0.002
	<0.002	<0.02	<0.02	<0.02	—
Chromic Acid	—	>0.05	—	—	—
	—	<0.02	>0.05	—	—
Chromic Hydroxide	<0.02	<0.02	<0.02	<0.02	<0.02
	<0.02	<0.02	<0.02	<0.02	—
Chromic Sulfates	<0.05	—	—	<0.05	—
	—	<0.02	<0.05	<0.02	—
Citric Acid	<0.02	<0.02	<0.02	<0.02	<0.02
	<0.02	<0.002	<0.02	>0.05	—
Copper Nitrate	>0.05	>0.05	<0.05	—	—
	—	<0.02	—	—	—
Copper Sulfate	>0.05	>0.05	>0.05	—	—
	—	<0.002	>0.05	<0.02	—
Diethylene Glycol	<0.002	<0.002	<0.002	<0.02	<0.02
	<0.02	<0.02	<0.02	<0.02	<0.002
Ethyl Chloride	<0.002	<0.002	<0.002	<0.02	<0.002
	<0.002	<0.02	<0.002	<0.02	<0.002
Ethylene Glycol	<0.02	<0.02	<0.02	<0.02	<0.02
	<0.02	—	<0.002	<0.05	—
Ethylene Oxide	>0.05	>0.05	>0.05	<0.02	<0.02
	<0.02	<0.002	<0.002	<0.02	<0.002

Selecting Nonferrous Metals for use in a 100% Corrosive Medium
(Continued)

Corrosion Rate at 70°F in a 100% Corrosive Medium

Corrosive Medium	Copper, Sn-Braze, Al-Braze	70-30 Brass	Silicon Bronze	Monel	Nickel
	Inconel	Hastelloy	Aluminum	Lead	Titanium
Fatty Acids	<0.05	<0.05	<0.05	<0.02	<0.02
	<0.02	<0.002	<0.002	>0.05	<0.002
Ferric Chloride	<0.02	<0.02	<0.02	>0.05	—
	>0.05	<0.02	>0.05	—	—
Ferric Nitrate	—	—	—	—	—
	—	—	—	<0.002	—
Ferrous Chloride	<0.02	—	<0.02	—	—
	—	<0.02	—	—	—
Ferrous Sulfate	<0.02	<0.05	<0.02	<0.02	<0.02
	—	<0.02	—	—	—
Fluorine	<0.002	<0.02	>0.05	<0.002	<0.002
	<0.002	<0.02	>0.05	<0.02	—
Formaldehyde	<0.002	<0.02	<0.02	<0.002	<0.002
	<0.02	<0.02	<0.002	<0.02	<0.002
Formic Acid	<0.02	<0.02	<0.02	—	<0.02
	<0.02	<0.002	<0.02	>0.05	<0.02
Furfural	<0.02	<0.02	<0.02	<0.02	<0.02
	<0.02	<0.02	<0.02	<0.02	<0.002
Hydrazine	—	—	—	>0.05	<0.002
	<0.002	<0.002	<0.002	>0.05	—
Hydrobromic Acid	<0.02	>0.05	<0.02	—	<0.02
	—	—	>0.05	—	—
Hydrocyanic Acid	<0.02	<0.02	<0.02	<0.02	<0.02
	<0.02	<0.02	<0.002	<0.02	—

Selecting Nonferrous Metals for use in a 100% Corrosive Medium (Continued)

Corrosion Rate at 70°F in a 100% Corrosive Medium

Corrosive Medium	Copper, Sn-Braze, Al-Braze	70-30 Brass	Silicon Bronze	Monel	Nickel
	Inconel	Hastelloy	Aluminum	Lead	Titanium
Hydrofluoric Acid (Areated)	<0.02	—	—	<0.02	<0.02
	<0.02	<0.02	—	—	—
Hydrofluoric Acid (Air Free)	<0.02	<0.02	<0.02	<0.02	<0.02
	<0.02	<0.05	—	>0.05	>0.05
Hydrogen Chloride	<0.02	<0.02	<0.02	<0.002	<0.002
	<0.002	<0.002	>0.05	<0.02	—
Hydrogen Fluoride	<0.02	<0.02	<0.02	<0.02	<0.002
	<0.02	<0.02	<0.02	>0.05	<0.002
Hydrogen Iodide	<0.02	>0.05	<0.02	—	<0.02
	—	<0.02	>0.05	—	—
Hydrogen Peroxide	>0.05	>0.05	>0.05	<0.002	<0.02
	<0.02	<0.002	<0.002	<0.002	>0.05
Hydrogen Sulfide	<0.02	<0.02	<0.02	<0.02	<0.02
	<0.02	<0.002	<0.002	<0.02	<0.002
Lactic Acid	<0.02	<0.05	<0.02	—	—
	—	<0.02	<0.02	>0.05	<0.002
Lead Acetate	—	<0.05	<0.02	<0.02	—
	—	>0.05	>0.05	—	
Lead Chromate	<0.02	<0.02	<0.02	<0.02	<0.02
	<0.02	<0.02	—	<0.02	—
Lead Nitrate	—	—	—	<0.02	<0.02
	<0.02	<0.02	—	<0.02	—
Lead Sulfate	<0.02	<0.02	<0.02	<0.02	<0.02
	<0.02	<0.02	—	<0.02	—

Selecting Nonferrous Metals for use in a 100% Corrosive Medium
(Continued)

Corrosion Rate at 70°F in a 100% Corrosive Medium

Corrosive Medium	Copper, Sn-Braze, Al-Braze / Inconel	70-30 Brass / Hastelloy	Silicon Bronze / Aluminum	Monel / Lead	Nickel / Titanium
Lithium Chloride	—	—	—	<0.002	—
	—	—	—	<0.02	—
Lithium Hydroxide	—	—	—	<0.02	<0.02
	<0.02	<0.02	>0.05	—	—
Magnesium Chloride	<0.02	—	<0.02	<0.02	<0.02
	<0.02	<0.002	—	>0.05	<0.002
Magnesium Hydroxide	<0.02	<0.02	<0.02	<0.02	<0.02
	—	—	>0.05	—	—
Magnesium Sulfate	<0.02	<0.02	<0.02	<0.02	<0.02
	<0.02	<0.002	<0.02	—	—
Maleic Acid	<0.02	—	—	—	—
	—	<0.02	—	—	—
Malic Acid	—	—	—	—	<0.02
	<0.02	—	<0.002	—	<0.002
Mercuric Chloride	>0.05	>0.05	>0.05	—	—
	—	—	—	—	—
Mercurous Nitrate	>0.05	>0.05	—	—	—
	—	<0.02	>0.05	>0.05	—
Mercury	>0.05	>0.05	>0.05	<0.02	<0.02
	<0.02	<0.02	>0.05	>0.05	—
Methallylamine	>0.05	>0.05	>0.05	<0.05	<0.02
	<0.02	<0.02	<0.02	—	—
Methanol	<0.02	<0.02	<0.02	<0.002	<0.002
	<0.002	<0.02	<0.02	<0.02	—

Selecting Nonferrous Metals for use in a 100% Corrosive Medium (Continued)

Corrosive Medium	Corrosion Rate at 70°F in a 100% Corrosive Medium				
	Copper, Sn-Braze, Al-Braze	70-30 Brass	Silicon Bronze	Monel	Nickel
	Inconel	Hastelloy	Aluminum	Lead	Titanium
Methyl Ethyl Ketone	<0.002	<0.002	<0.002	<0.002	<0.002
	<0.002	<0.002	<0.002	<0.002	<0.002
Methyl Isobutyl Ketone	<0.02	<0.02	<0.02	<0.02	<0.02
	<0.02	<0.002	<0.002	<0.002	<0.002
Methylamine	>0.05	>0.05	>0.05	—	—
	—	—	<0.02	—	—
Methylene Chloride	<0.002	<0.002	<0.02	<0.002	<0.02
	<0.02	—	<0.002	<0.02	—
Monochloroacetic Acid	>0.05	>0.05	>0.05	<0.05	<0.02
	<0.02	<0.002	>0.05	>0.05	<0.002
Monorthanolamine	>0.05	>0.05	>0.05	<0.02	<0.02
	<0.02	—	<0.02	—	—
Monoethalamine	>0.05	>0.05	>0.05	—	—
	—	—	<0.02	—	—
Monoethylamine	>0.05	>0.05	>0.05	—	—
	—	—	<0.02	—	—
Nickel Chloride	—	—	<0.02	<0.02	—
	<0.02	<0.002	>0.05	<0.02	—
Nickel Nitrate	—	—	—	<0.02	<0.02
	<0.02	<0.02	—	<0.02	—
Nickel Sulfate	<0.02	<0.02	<0.02	<0.02	—
	<0.02	<0.02	>0.05	<0.02	—
Nitric Acid	>0.05	>0.05	>0.05	>0.05	>0.05
	—	—	<0.02	>0.05	—

Selecting Nonferrous Metals for use in a 100% Corrosive Medium
(Continued)

Corrosion Rate at 70°F in a 100% Corrosive Medium

Corrosive Medium	Copper, Sn-Braze, Al-Braze / Inconel	70-30 Brass / Hastelloy	Silicon Bronze / Aluminum	Monel / Lead	Nickel / Titanium
Nitric Acid (Red Fuming)	>0.05 <0.02	>0.05 <0.02	>0.05 <0.002	>0.05 —	>0.05 <0.002
Nitric + Hydrochloric Acid	>0.05 >0.05	>0.05 >0.05	>0.05 >0.05	>0.05 >0.05	>0.05 <0.02
Nitric + Hydrofluoric Acid	>0.05 —	— <0.05	— —	— —	— >0.05
Nitric + Sulfuric Acid	>0.05 >0.05	>0.05 —	>0.05 >0.05	>0.05 >0.05	>0.05 —
Nitrobenzene	<0.02 <0.02	<0.02 <0.02	<0.02 <0.02	<0.02 <0.02	<0.02 —
Nitrocelluolose	— <0.02	<0.02 —	<0.02 <0.002	<0.002 <0.002	<0.02 —
Nitroglycerine	<0.02 <0.02	<0.02 —	<0.02 <0.002	<0.02 <0.05	— —
Nitrotolune	<0.02 <0.02	<0.02 —	<0.02 <0.02	<0.02 <0.02	<0.02 —
Nitrous Acid	>0.05 —	>0.05 —	>0.05 —	>0.05 >0.05	>0.05 —
Oleic Acid	<0.002 <0.002	<0.02 <0.02	<0.02 <0.002	<0.002 >0.05	<0.002 <0.002
Oxalic Acid	<0.05 <0.02	<0.05 <0.02	<0.02 <0.02	<0.02 >0.05	<0.05 —
Phenol	<0.002 <0.002	<0.002 <0.002	<0.002 <0.002	<0.002 <0.02	<0.002 —

Selecting Nonferrous Metals for use in a 100% Corrosive Medium (Continued)

Corrosion Rate at 70°F in a 100% Corrosive Medium

Corrosive Medium	Copper, Sn-Braze, Al-Braze Inconel	70-30 Brass Hastelloy	Silicon Bronze Aluminum	Monel Lead	Nickel Titanium
Phosphoric Acid (Areated)	>0.05 >0.05	>0.05 <0.002	>0.05 <0.02	— <0.02	>0.05 >0.05
Phosphoric Acid (Air Free)	— —	>0.05 <0.002	— >0.05	— <0.02	— >0.05
Picric Acid	>0.05 <0.02	>0.05 <0.02	>0.05 <0.02	>0.05 <0.02	<0.02 —
Potassium Bicarbonate	<0.02 —	<0.02 —	<0.02 <0.02	— —	— —
Potassium Bromide	<0.02 <0.02	<0.02 <0.02	<0.02 —	<0.02 <0.02	<0.02 —
Potassium Carbonate	<0.02 <0.02	<0.02 <0.02	<0.02 >0.05	<0.02 >0.05	<0.02 —
Potassium Chlorate	<0.05 —	<0.05 —	<0.05 <0.02	— —	— —
Potassium Chromate	— —	<0.02 —	<0.02 <0.02	— —	— —
Potassium Cyanide	>0.05 <0.02	>0.05 —	>0.05 —	<0.02 —	<0.02 >0.05
Potassium Dichromate	— —	— —	— <0.02	— —	— —
Potassium Ferricyanide	<0.02 —	— —	— —	— —	— —
Potassium Ferrocyanide	— —	— —	— <0.02	— —	— —

Selecting Nonferrous Metals for use in a 100% Corrosive Medium
(Continued)

Corrosion Rate at 70°F in a 100% Corrosive Medium

Corrosive Medium	Copper, Sn-Braze, Al-Braze / Inconel	70-30 Brass / Hastelloy	Silicon Bronze / Aluminum	Monel / Lead	Nickel / Titanium
Potassium Hydroxide	—	—	>0.05	—	—
	—	—	—	>0.05	—
Potassium Hypochlorite	—	—	—	—	—
	—	<0.02	—	—	—
Potassium Iodide	<0.02	—	<0.02	<0.02	<0.02
	<0.02	<0.02	—	—	<0.002
Potassium Nitrate	<0.002	<0.02	<0.02	<0.02	<0.02
	—	—	<0.02	—	—
Potassium Nitrite	<0.02	<0.02	<0.02	<0.02	<0.02
	<0.02	<0.02	<0.02	<0.02	<0.002
Potassium Permanganate	<0.02	—	<0.02	—	—
	—	<0.002	<0.02	>0.05	—
Potassium Silicate	<0.02	<0.02	<0.02	<0.02	<0.02
	<0.02	<0.02	<0.02	—	—
Propionic Acid	<0.02	—	—	<0.02	—
	—	—	<0.02	—	>0.05
Pyridine	<0.02	<0.02	<0.02	<0.02	<0.02
	<0.02	<0.02	<0.02	<0.02	—
Quinine Sulfate	<0.02	<0.02	<0.02	<0.02	<0.02
	<0.02	<0.02	—	—	<0.002
Salicylic Acid	<0.02	—	<0.02	<0.02	<0.02
	<0.02	<0.02	<0.02	<0.02	—
Silicon Tetrachloride (Dry)	<0.002	<0.002	<0.002	<0.002	<0.002
	<0.002	<0.02	<0.02	<0.02	—

Selecting Nonferrous Metals for use in a 100% Corrosive Medium (Continued)

Corrosion Rate at 70°F in a 100% Corrosive Medium

Corrosive Medium	Copper, Sn-Braze, Al-Braze / Inconel	70-30 Brass / Hastelloy	Silicon Bronze / Aluminum	Monel / Lead	Nickel / Titanium
Silicon Tetrachloride (Wet)	>0.05	>0.05	>0.05	>0.05	>0.05
	—	<0.02	>0.05	—	—
Silver Bromide	—	—	—	<0.02	<0.02
	—	—	—	—	<0.002
Silver Chloride	<0.02	—	—	—	—
	—	—	—	—	—
Sodium Acetate	<0.02	—	—	<0.02	<0.02
	<0.02	—	<0.002	<0.02	—
Sodium Bicarbonate	<0.02	—	—	—	—
	—	—	<0.02	—	—
Sodium Bisulfate	<0.02	<0.05	<0.02	<0.02	<0.02
	<0.02	<0.02	—	—	—
Sodium Bromide	<0.05	—	—	—	—
	—	—	—	—	—
Sodium Carbonate	—	—	<0.02	<0.02	<0.02
	<0.02	<0.02	—	—	—
Sodium Chromate	<0.02	<0.02	<0.02	<0.02	<0.02
	<0.02	<0.02	<0.02	<0.02	—
Sodium Hydroxide	—	—	—	<0.002	<0.002
	<0.002	<0.002	—	—	—
Sodium Hypochlorite	—	>0.05	>0.05	<0.02	—
	—	<0.05	>0.05	>0.05	<0.002
Sodium Metasilicate	<0.02	<0.02	<0.02	<0.002	<0.002
	<0.002	<0.002	<0.02	—	—

Selecting Nonferrous Metals for use in a 100% Corrosive Medium (Continued)

Corrosion Rate at 70°F in a 100% Corrosive Medium

Corrosive Medium	Copper, Sn-Braze, Al-Braze / Inconel	70-30 Brass / Hastelloy	Silicon Bronze / Aluminum	Monel / Lead	Nickel / Titanium
Sodium Nitrate	<0.05 / —	<0.05 / —	<0.02 / <0.02	<0.02 / —	<0.02 / —
Sodium Nitrite	— / <0.02	— / —	— / —	<0.002 / —	<0.02 / —
Sodium Phosphate	<0.02 / <0.02	<0.02 / <0.02	<0.02 / —	<0.02 / <0.02	<0.02 / —
Sodium Silicate	<0.02 / <0.02	<0.02 / <0.02	<0.02 / <0.002	<0.02 / —	<0.02 / —
Sodium Sulfate	<0.02 / <0.02	>0.05 / <0.002	<0.02 / —	<0.02 / <0.02	<0.02 / —
Sodium Sulfide	>0.05 / —	>0.05 / —	>0.05 / >0.05	— / <0.002	— / —
Sodium Sulfite	<0.05 / <0.02	>0.05 / —	<0.02 / —	<0.02 / <0.02	— / —
Stannic Chloride	— / —	— / <0.02	>0.05 / —	— / —	— / —
Stannous Chloride	— / <0.02	— / <0.02	<0.02 / —	<0.02 / —	<0.02 / —
Strontium Nitrate	<0.02 / <0.02	<0.02 / <0.02	<0.02 / <0.02	<0.02 / —	<0.02 / —
Succinic Acid	<0.02 / <0.02	<0.02 / —	<0.02 / <0.02	<0.02 / <0.02	<0.02 / <0.002

Selecting Nonferrous Metals for use in a 100% Corrosive Medium (Continued)

Corrosion Rate at 70°F in a 100% Corrosive Medium

Corrosive Medium	Copper, Sn-Braze, Al-Braze / Inconel	70-30 Brass / Hastelloy	Silicon Bronze / Aluminum	Monel / Lead	Nickel / Titanium
Sulfur Dioxide	<0.02 <0.02	<0.05 <0.02	<0.02 <0.02	<0.02 <0.02	<0.02 —
Sulfur Trioxide	<0.02 <0.02	<0.02 <0.02	<0.02 <0.02	<0.02 <0.02	<0.02 —
Sulfuric Acid (Areated)	>0.05 >0.05	>0.05 <0.02	>0.05 >0.05	>0.05 >0.05	>0.05 >0.05
Sulfuric Acid (Air Free)	— —	— <0.02	— >0.05	>0.05 >0.05	>0.05 >0.05
Sulfuric Acid (Fuming)	>0.05 <0.02	>0.05 <0.002	>0.05 <0.02	>0.05 >0.05	>0.05 —
Sulfurous Acid	<0.05 <0.02	>0.05 <0.02	<0.02 <0.02	>0.05 <0.02	>0.05 <0.002
Tannic Acid	<0.02 <0.02	<0.05 —	<0.02 >0.05	<0.02 >0.05	<0.02 <0.002
Tartaric Acid	<0.02 —	— <0.02	<0.02 —	— >0.05	— <0.002
Tetraphosphoric Acid	<0.05 <0.02	<0.05 <0.02	<0.05 >0.05	<0.05 >0.05	>0.05 —
Trichloroacetic Acid	>0.05 —	>0.05 <0.02	<0.05 >0.05	>0.05 >0.05	<0.02 >0.05
Trichloroethylene	<0.002 <0.02	<0.02 <0.002	<0.02 <0.002	<0.002 >0.05	<0.002 <0.002

Selecting Nonferrous Metals for use in a 100% Corrosive Medium (Continued)

Corrosion Rate at 70°F in a 100% Corrosive Medium

Corrosive Medium	Copper, Sn-Braze, Al-Braze / Inconel	70-30 Brass / Hastelloy	Silicon Bronze / Aluminum	Monel / Lead	Nickel / Titanium
Urea	—	—	—	—	—
	—	—	<0.02	—	—
Zinc Chloride	—	—	>0.05	<0.02	<0.02
	<0.02	<0.02	—	<0.02	—
Zinc Sulfate	<0.02	<0.02	<0.02	—	—
	—	—	—	—	—

Water-free, Dry or Maximum concentration of corrosive medium. Quantitatively

<0.002 means that corrosion rate is likely to be less than 0.002 inch per year (Excellent).
<0.02 means that corrosion rate is likely to be less than about 0.02 inch per year (Good).
<0.05 means that corrosion rate is likely to be less than about 0.05 inch per year (Fair).
>0.05 means that corrosion rate is likely to be more than 0.05 inch per year (Poor).

Source: data compiled by J.S. Park *from* Earl R. Parker, *Materials Data Book for Engineers and Scientists*, McGraw-Hill Book Company, New York, 1967.

SELECTING CORROSION RATES OF METALS

Metal	Corrosive Environment	Corrosion Rate (Mils Penetration per Year)
Silicon iron	Acetic, 5% (Non–oxidizing)	0–0.2
Iron	Sodium Hydroxide, 5%	0–0.2
Nickel alloys	Sodium Hydroxide, 5%	0–0.2
Stainless steel	Sodium Hydroxide, 5%	0–0.2
Nickel alloys	Fresh Water	0–0.2
Silicon iron	Fresh Water	0–0.2
Stainless steel	Fresh Water	0–0.2
Copper alloys	Normal Outdoor Air (Urban Exposure)	0–0.2
Lead	Normal Outdoor Air (Urban Exposure)	0–0.2
Nickel alloys	Normal Outdoor Air (Urban Exposure)	0–0.2
Silicon iron	Normal Outdoor Air (Urban Exposure)	0–0.2
Stainless steel	Normal Outdoor Air (Urban Exposure)	0–0.2
Tin	Normal Outdoor Air (Urban Exposure)	0–0.2
Stainless steel	Acetic, 5% (Non–oxidizing)	0–0.5
Tin	Fresh Water	0–0.5
Aluminum	Normal Outdoor Air (Urban Exposure)	0–0.5
Zinc	Normal Outdoor Air (Urban Exposure)	0–0.5
Copper alloys	Fresh Water	0–1
Nickel alloys	Sea Water	0–1
Silicon iron	Sodium Hydroxide, 5%	0–10
Stainless steel	Sulfuric, 5% (Non–oxidizing)	0–2
Stainless steel	Nitric, 5% (Oxidizing)	0–2
Silicon iron	Sulfuric, 5% (Non–oxidizing)	0–20
Silicon iron	Nitric, 5% (Oxidizing)	0–20

Selecting Corrosion Rates of Metals (Continued)

Metal	Corrosive Environment	Corrosion Rate (Mils Penetration per Year)
Stainless steel	Sea Water	0–200**
Silicon iron	Sea Water	0–3
Gold	Sulfuric, 5% (Non–oxidizing)	<0.1
Platinum	Sulfuric, 5% (Non–oxidizing)	<0.1
Tantalum	Sulfuric, 5% (Non–oxidizing)	<0.1
Zirconium	Sulfuric, 5% (Non–oxidizing)	<0.1
Gold	Acetic, 5% (Non–oxidizing)	<0.1
Molybdenum	Acetic, 5% (Non–oxidizing)	<0.1
Platinum	Acetic, 5% (Non–oxidizing)	<0.1
Silver	Acetic, 5% (Non–oxidizing)	<0.1
Tantalum	Acetic, 5% (Non–oxidizing)	<0.1
Titanium	Acetic, 5% (Non–oxidizing)	<0.1
Zirconium	Acetic, 5% (Non–oxidizing)	<0.1
Gold	Nitric, 5% (Oxidizing)	<0.1
Platinum	Nitric, 5% (Oxidizing)	<0.1
Tantalum	Nitric, 5% (Oxidizing)	<0.1
Zirconium	Nitric, 5% (Oxidizing)	<0.1
Gold	Sodium Hydroxide, 5%	<0.1
Molybdenum	Sodium Hydroxide, 5%	<0.1
Platinum	Sodium Hydroxide, 5%	<0.1
Silver	Sodium Hydroxide, 5%	<0.1
Zirconium	Sodium Hydroxide, 5%	<0.1
Gold	Fresh Water	<0.1
Molybdenum	Fresh Water	<0.1

Selecting Corrosion Rates of Metals (Continued)

Metal	Corrosive Environment	Corrosion Rate (Mils Penetration per Year)
Platinum	Fresh Water	<0.1
Silver	Fresh Water	<0.1
Tantalum	Fresh Water	<0.1
Titanium	Fresh Water	<0.1
Zirconium	Fresh Water	<0.1
Aluminum	Fresh Water	0.1
Gold	Sea Water	<0.1
Molybdenum	Sea Water	<0.1
Platinum	Sea Water	<0.1
Silver	Sea Water	<0.1
Tantalum	Sea Water	<0.1
Titanium	Sea Water	<0.1
Zirconium	Sea Water	<0.1
Tin	Sea Water	0.1
Gold	Normal Outdoor Air (Urban Exposure)	<0.1
Molybdenum	Normal Outdoor Air (Urban Exposure)	<0.1
Platinum	Normal Outdoor Air (Urban Exposure)	<0.1
Silver	Normal Outdoor Air (Urban Exposure)	<0.1
Tantalum	Normal Outdoor Air (Urban Exposure)	<0.1
Titanium	Normal Outdoor Air (Urban Exposure)	<0.1
Zirconium	Normal Outdoor Air (Urban Exposure)	<0.1
Titanium	Sulfuric, 5% (Non–oxidizing)	0.1–1
Titanium	Nitric, 5% (Oxidizing)	0.1–1
Iron	Fresh Water	0.1–10*

Selecting Corrosion Rates of Metals (Continued)

Metal	Corrosive Environment	Corrosion Rate (Mils Penetration per Year)
Iron	Sea Water	0.1–10*
Nickel alloys	Sulfuric, 5% (Non–oxidizing)	0.1–1500
Nickel alloys	Nitric, 5% (Oxidizing)	0.1–1500
Lead	Fresh Water	0.1–2
Titanium	Sodium Hydroxide, 5%	<0.2
Lead	Sea Water	0.2–15
Copper alloys	Sea Water	0.2–15*
Zinc	Fresh Water	0.5–10
Zinc	Sea Water	0.5–10*
Aluminum	Acetic, 5% (Non–oxidizing)	0.5–5
Tantalum	Sodium Hydroxide, 5%	<1
Aluminum	Sea Water	1–50
Iron	Normal Outdoor Air (Urban Exposure)	1–8
Nickel alloys	Acetic, 5% (Non–oxidizing)	2–10*
Copper alloys	Acetic, 5% (Non–oxidizing)	2–15*
Copper alloys	Sodium Hydroxide, 5%	2–5
Tin	Acetic, 5% (Non–oxidizing)	2–500*
Tin	Sodium Hydroxide, 5%	5–20
Lead	Sodium Hydroxide, 5%	5–500*
Lead	Acetic, 5% (Non–oxidizing)	10–150*
Iron	Acetic, 5% (Non–oxidizing)	10–400
Zinc	Sodium Hydroxide, 5%	15–200
Aluminum	Sulfuric, 5% (Non–oxidizing)	15–80
Aluminum	Nitric, 5% (Oxidizing)	15–80

Selecting Corrosion Rates of Metals (Continued)

Metal	Corrosive Environment	Corrosion Rate (Mils Penetration per Year)
Tin	Sulfuric, 5% (Non–oxidizing)	100–400
Tin	Nitric, 5% (Oxidizing)	100–400
Lead	Sulfuric, 5% (Non–oxidizing)	100–6000
Lead	Nitric, 5% (Oxidizing)	100–6000
Copper alloys	Sulfuric, 5% (Non–oxidizing)	150–1500
Copper alloys	Nitric, 5% (Oxidizing)	150–1500
Zinc	Acetic, 5% (Non–oxidizing)	600–800
Iron	Sulfuric, 5% (Non–oxidizing)	1000–10000
Iron	Nitric, 5% (Oxidizing)	1000–10000
Aluminum	Sodium Hydroxide, 5%	13000
Molybdenum	Sulfuric, 5% (Non–oxidizing)	high
Silver	Sulfuric, 5% (Non–oxidizing)	high
Zinc	Sulfuric, 5% (Non–oxidizing)	high
Molybdenum	Nitric, 5% (Oxidizing)	high
Silver	Nitric, 5% (Oxidizing)	high
Zinc	Nitric, 5% (Oxidizing)	high

* Aeration leads to the higher rates in the range.
** Aeration leads to passivity, scarcity of dissolved air to activity.

Corrosion Rate Ranges Expressed in Mils Penetration per Year (1 Mil = 0.001 in)
Note: The corrosion–rate ranges for the solutions are based on temperature up to 212 °F.

Source: data compiled by J.S. Park *from* R. E. Bolz and G. L. Tuve, *CRC Handbook of Tables for Applied Engineering Science, 2nd edition*, CRC Press, Inc., Boca Raton, Florida, (1987).

SELECTING CORROSION RATES OF METALS
IN CORROSIVE ENVIRONMENTS

Corrosive Environment	Metal	Corrosion Rate (Mils Penetration per Year)
Sulfuric, 5% (Non–oxidizing)	Stainless steel	0–2
	Silicon iron	0–20
	Gold	<0.1
	Platinum	<0.1
	Tantalum	<0.1
	Zirconium	<0.1
	Titanium	0.1–1
	Nickel alloys	0.1–1500
	Aluminum	15–80
	Tin	100–400
	Lead	100–6000
	Copper alloys	150–1500
	Iron	1000–10000
	Molybdenum	high
	Silver	high
	Zinc	high
Acetic, 5% (Non–oxidizing)	Gold	<0.1
	Molybdenum	<0.1
	Platinum	<0.1
	Silver	<0.1
	Tantalum	<0.1
	Titanium	<0.1
	Zirconium	<0.1
	Silicon iron	0–0.2

Selecting Corrosion Rates of Metals in Corrosive Environments(Continued)

Corrosive Environment	Metal	Corrosion Rate (Mils Penetration per Year)
	Stainless steel	0–0.5
	Aluminum	0.5–5
	Nickel alloys	2–10*
	Copper alloys	2–15*
	Tin	2–500*
	Lead	10–150*
	Iron	10–400
	Zinc	600–800
Nitric, 5% (Oxidizing)	Stainless steel	0–2
	Silicon iron	0–20
	Gold	<0.1
	Platinum	<0.1
	Tantalum	<0.1
	Zirconium	<0.1
	Titanium	0.1–1
	Nickel alloys	0.1–1500
	Aluminum	15–80
	Tin	100–400
	Lead	100–6000
	Copper alloys	150–1500
	Iron	1000–10000
	Molybdenum	high
	Silver	high
	Zinc	high

Selecting Corrosion Rates of Metals in Corrosive Environments(Continued)

Corrosive Environment	Metal	Corrosion Rate (Mils Penetration per Year)
Sodium Hydroxide, 5%	Iron	0–0.2
	Nickel alloys	0–0.2
	Stainless steel	0–0.2
	Silicon iron	0–10
	Gold	<0.1
	Molybdenum	<0.1
	Platinum	<0.1
	Silver	<0.1
	Zirconium	<0.1
	Titanium	<0.2
	Tantalum	<1
	Copper alloys	2–5
	Tin	5–20
	Lead	5–500[*]
	Zinc	15–200
	Aluminum	13000
Fresh Water	Nickel alloys	0–0.2
	Silicon iron	0–0.2
	Stainless steel	0–0.2
	Tin	0–0.5
	Gold	<0.1
	Molybdenum	<0.1
	Platinum	<0.1
	Silver	<0.1

Selecting Corrosion Rates of Metals in Corrosive Environments(Continued)

Corrosive Environment	Metal	Corrosion Rate (Mils Penetration per Year)
	Tantalum	<0.1
	Titanium	<0.1
	Zirconium	<0.1
	Copper alloys	0–1
	Aluminum	0.1
	Iron	0.1–10[*]
	Lead	0.1–2
	Zinc	0.5–10
Sea Water	Nickel alloys	0–1
	Stainless steel	0–200[**]
	Silicon iron	0–3
	Gold	<0.1
	Molybdenum	<0.1
	Platinum	<0.1
	Silver	<0.1
	Tantalum	<0.1
	Titanium	<0.1
	Zirconium	<0.1
	Tin	0.1
	Iron	0.1–10[*]
	Lead	0.2–15
	Copper alloys	0.2–15[*]
	Zinc	0.5–10[*]
	Aluminum	1–50

Selecting Corrosion Rates of Metals in Corrosive Environments(Continued)

Corrosive Environment	Metal	Corrosion Rate (Mils Penetration per Year)
Normal Outdoor Air		
(Urban Exposure)	Copper alloys	0–0.2
	Lead	0–0.2
	Nickel alloys	0–0.2
	Silicon iron	0–0.2
	Stainless steel	0–0.2
	Tin	0–0.2
	Aluminum	0–0.5
	Zinc	0–0.5
	Gold	<0.1
	Molybdenum	<0.1
	Platinum	<0.1
	Silver	<0.1
	Tantalum	<0.1
	Titanium	<0.1
	Zirconium	<0.1
	Iron	1–8

[*] Aeration leads to the higher rates in the range.
[**] Aeration leads to passivity, scarcity of dissolved air to activity.

Corrosion Rate Ranges Expressed in Mils Penetration per Year (1 Mil = 0.001 in)
Note: The corrosion–rate ranges for the solutions are based on temperature up to 212 °F.

Source: data compiled by J.S. Park from R. E. Bolz and G. L. Tuve, CRC Handbook of Tables for Applied Engineering Science, 2nd edition, CRC Press, Inc., Boca Raton, Florida, (1987).

Selecting Ceramic Materials and Glasses

(Continued)

Selecting Ceramic Materials and Glasses (Continued)

Selecting Ceramic Materials and Glasses (Continued)

PERIODIC TABLE OF ELEMENTS IN CERAMIC MATERIALS

The Elements in Ceramic Materials

1 IA	2 IIA	3 IIIB	4 IVB	5 VB	6 VIB	7 VIIB	8 ------	9 VIII	10 ------	11 IB	12 IIB	13 IIIA	14 IVA	15 VA	16 VIA	17 VIIA	18 VIIA
3 Li	4 Be											5 B	6 C	7 N	8 O		
11 Na	12 Mg											13 Al	14 Si	14 Si	15 P		
19 K	20 Ca	21 Sc	22 Ti	23 V	24 Cr	25 Mn	26 Fe	27 Co	28 Ni	29 Cu	30 Zn	31 Ga	32 Ge				
37 Rb	38 Sr	39 Y	40 Zr	41 Nb	42 Mo	43 Tc	44 Ru	45 Rh	46 Pd	47 Ag	48 Cd	49 In	50 Sn	51 Sb			
55 Cs	56 Ba	57 La	72 Hf	73 Ta	74 W	75 Re	76 Os	77 Ir	78 Pt	79 Au	80 Hg	81 Tl	82 Pb	83 Bi			
87 Fr	88 Ra	89 Ac															

57 La	58 Ce	59 Pr	60 Nd	61 Pm	62 Sm	63 Eu	64 Gd	65 Tb	66 Dy	67 Ho	68 Er	69 Tm	70 Yb	71 Lu
89 Ac	90 Th	91 Pa	92 U	93 Np	94 Pu	95 Am	96 Cm	97 Bk	98 Cf	99 Es	100 Fm	101 Md	102 No	103 Lw

Selecting Ceramics:
Thermodynamic Properties

SELECTING HEAT OF FORMATION OF INORGANIC OXIDES

Reaction	Temperature Range of Validity	ΔH_0
6 V(c) + 13/2 O_2(g) = V_6O_{13}(c)	298.16–1,000K	−1,076,340
3 U(l) + 4 O_2(g) = U_3O_8(c)	1,405–1,500K	−869,460
3 U(α) + 4 O_2(g) = U_3O_8(c)	298.16–935K	−863,370
3 U(γ) + 4 O_2(g) = U_3O_8(c)	1,045–1,405K	−863,230
3 U(β) + 4 O_2(g) = U_3O_8(c)	935–1,045K	−856,720
4W(c) + 11/2 O_2(g) = W_4O_{11}(c)	298.16–1,700K	−745,730
4 P (white) + 5 O_2(g) = P_4H_{10} (hexagonal)	298.16–317.4K	−711,520
2 Ta(c) + 5/2 O_2(g) = Ta_2O_5(c)	298.16–2,000K	−492,790
2 Nb(c) + 5/2 O_2(g) = Nb_2O_5(l)	1,785–2,000K	−463,630
2 Nb(c) + 5/2 O_2(g) = Nb_2O_5(c)	298.16–1,785K	−458,640
2 Ac(c) + 3/2 O_2(g) = Ac_2O_3(c)	298.16–1,000K	−446,090
2 Ce(l) + 3/2 O_2(g) = Ce_2O_3(c)	1,048–1,900K	−440,400
2 Ce(c) + 3/2 O_2(g) = Ce_2O_3(c)	298.16–1,048K	−435,600
2 Y(c) + 3/2 O_2(g) = Y_2O_3(c)	298.16–1,773K	−419,600
2 Al(l) + 3/2 O_2(g) = Al_2O_3 (corundum)	931.7–2,000K	−407,950
2 Al(c) + 3/2 O_2(g) = Al_2O_3 (corundum)	298.16–931.7K	−404,080
2 Nb(c) + 2 O_2(g) = Nb_2O_4(c)	298.16–2,000K	−382,050
2 V(c) + 5/2 O_2(g) = V_2O_5(c)	298.16–943K	−381,960
2 Ti(α) + 3/2 O_2(g) = Ti_2O_3(β)	473–1,150K	−369,710
2 Ti(α) + 3/2 O_2(g) = Ti_2O_3(α)	298.16–473K	−360,660
2 V(c) + 2 O_2(g) = V_2O_4(β)	345–1,818K	−345,330
2 V(c) + 2 O_2(g) = V_2O_4(α)	209.16–345K	−342,890
3 Mn(α) + 2 O_2(g) = Mn_3O_4(α)	298.16–1,000K	−332,400
2 B(c) + 3/2 O_2(g) = B_2O(c)	298.16–723K	−304,690

Selecting Heat of Formation of Inorganic Oxides (Continued)

Reaction	Temperature Range of Validity	ΔH_0
2 Re(c) + 7/2 O_7(g) = Re_2O_7(c)	298.16–569K	–301,470
2 V(c) + 3/2 O_2(g) = V_2O_3(c)	298.16–2,000K	–299,910
2 B(c) + 3/2 O_2(g) = B_2O_3(gl)	298.16–723K	–298,670
2 Re(c) + 7/2 O_7(g) = Re_2O_7(l)	569–635.5K	–295,810
Th(c) + O_2(g) = ThO_2(c)	298.16–2,000K	–294,350
U(α) + 3/2 O_2(g) = UO_3 (hexagonal)	298.16–935K	–294,090
U(γ) + 3/2 O_2(g) = UO_3 (hexagonal)	1,045–1,400K	–294,040
U(β) + 3/2 O_2(g) = UO_3 (hexagonal)	935–1,045K	–291,870
2 Cr(l) + 3/2 O_2(g) = Cr_2O_3(β)	1,823–2,000K	–278,030
3 Fe(γ) + 2 O_2(g) = Fe_3O_4(β)	1,179–1,674K	–276,990
2 Cr(c) + 3/2 O_2(g) = Cr_2O_3(β)	298.16–1,823K	–274,670
3 Fe(α) + 2 O_2(g) = Fe_3O_4(β)	900–1,033K	–272,300
Hf(c) + O_2(g) = HfO_2 (monoclinic)	298.16–2,000K	–268,380
3 Fe(α) + 2 O_2(g) = Fe_3O_4(magnetite)	298.16–900K	–268,310
U(l) + O_2(g) = UO_2(l)	1,405–1,500K	–264,790
Zr(β) + O_2(g) = ZrO_2(α)	1,135–1,478K	–264,190
3 Fe(β) + 2 O_2(g) = Fe_3O_4(β)	1,033–1,179K	–262,990
Zr(α) + O_2(g) = ZrO_2(α)	298.16–1,135K	–262,980
U(α) + O_2(g) = UO_2(c)	298.16–935K	–262,880
U(γ) + O_2(g) = UO_2(c)	1,045–1,405K	–262,830
Zr(β) + O_2(g) = ZrO_2(β)	1.478–2,000K	–262,290
U(β) + O_2(g) = UO_2(c)	935–1,045K	–260,660
Ce(l) + O_2(g) = CeO_2(c)	1,048–2,000K	–247,930
Ce(c) + O_2(g) = CeO_2(c)	298.16–1,048K	–245,490

Selecting Heat of Formation of Inorganic Oxides (Continued)

Reaction	Temperature Range of Validity	ΔH_0
2 $Mn(\alpha)$ + 3/2 $O_2(g)$ = $Mn_2O_3(c)$	298.16–1,000K	−230,610
$Si(l)$ + $O_2(g)$ = $SiO_2(l)$	1,883–2,000K	−228,590
$Ti(\alpha)$ + $O_2(g)$ = TiO_2 (rutile)	1,150–2,000K	−228,380
$Ti(\alpha)$ + $O_2(g)$ = TiO_2 (rutile)	298.16–1,150K	−228,360
2 $As(c)$ + 5/2 $O_2(g)$ = $As_2O_5(c)$	298.16–883K	−217,080
$Si(c)$ + $O_2(g)$ = $SiO_2(\alpha$–quartz$)$	298.16–848K	−210,070
$Si(c)$ + $O_2(g)$ = $SiO_2(\beta$–quartz$)$	848–1,683K	−209,920
$Si(c)$ + $O_2(g)$ = $SiO_2(\beta$–cristobalite$)$	523–1,683K	−209,820
$Si(c)$ + $O_2(g)$ = $SiO_2(\beta$–tridymite$)$	390–1,683K	−209,350
$Si(c)$ + $O_2(g)$ = $SiO_2(\alpha$–cristobalite$)$	298.16–523K	−207,330
$Si(c)$ + $O_2(g$) = $SiO_2(\alpha$–tridymite$)$	298.16–390K	−207,030
$W(c)$ + 3/2 $O_2(g)$ = $WO_3(l)$	1,743–2,000K	−203,140
2 $Fe(\alpha)$ + 3/2 $O_2(g)$ = $Fe_2O_3(\beta)$	950–1,033K	−202,960
2 $Fe(\gamma)$ + 3/2 $O_2(g)$ = $Fe_2O_3(\gamma)$	1,179–1,674K	−202,540
$W(c)$ + 3/2 $O_2(g)$ = $WO_3(c)$	298.16–1,743K	−201,180
2 $Fe(\alpha)$ + 3/2 $O_2(g)$ = Fe_2O_3(hematite)	298.16–950K	−200,000
2 $Fe(\beta)$ + 3/2 $O_2(g)$ = $Fe_2O_3(\beta)$	1,033–1,050K	−196,740
2 $Fe(\beta)$ + 3/2 $O_2(g)$ = $Fe_2O_3(\gamma)$	1,050–1,179K	−193,200
2 $Fe(\alpha)$ + 3/2 $O_2(g)$ = $Fe_2O_3(\gamma)$	1,674–1,800K	−192,920
$Mo(c)$ + 3/2 $O_2(g)$ = $MoO_3(c)$	298.16–1,068K	−182,650
$Mg(g)$ + 1/2 $O_2(g)$ = MgO (periclase)	1,393–2,000K	−180,700
3 $Pb(c)$ + 2 $O_2(g)$ = $Pb_3O_4(c)$	298.16–600.5K	−174,920
2 $Sb(c)$ + 3/2 $O_2(g)$ = Sb_2O_3 (cubic)	298.16–842K	−169,450
2 $Sb(c)$ + 3/2 $O_2(g)$ = Sb_2O_3 (orthorhombic)	298.16–903K	−168,060

Selecting Heat of Formation of Inorganic Oxides (Continued)

Reaction	Temperature Range of Validity	ΔH_0
2 As(c) + 3/2 O_2(g) = As_2O_3 (orthorhombic)	298.16–542K	−154,870
Ca(α) + 1/2 O_2(g) = CaO(c)	298.16–673K	−151,850
Ca(β) + 1/2 O_2(g) = CaO(c)	673–1,124K	−151,730
2 As(c) + 3/2 O_2(g) = As_2O_3 (monoclinic)	298.16–586K	−150,760
Re(c) + 3/2 O_2(g) = ReO_3(c)	298.16–433K	−149,090
2 Cs(g) + 3/2 O_2(g) = Cs_2O_3(l)	963–1,500K	−148,680
Re(c) + 3/2 O_2(g) = ReO_3(l)	433–1,000K	−146,750
Mg(l) + 1/2 O_2(g) = MgO (periclase)	923–1,393K	−145,810
Be(c) + 1/2 O_2(g) = BeO(c)	298.16–1,556K	−144,220
Mg(c) + 1/2 O_2(g) = MgO (periclase)	298.16–923K	−144,090
Cr(c) + O_2(g) = CrO_2 (c)	298.16–1,000K	−142,500
Sr(c) + 1/2 O_2(g) = SrO(c)	298.16–1,043K	−142,410
2 Bi(l) + 3/2 O_2(g) = Bi_2O_3(c)	544–1,090K	−142,270
2 Li(c) + 1/2 O_2(g) = Li_2O(c)	298.16–452K	−142,220
Cr(c) + 3/2 O_2(g) = CrO_3(c)	298.16–471K	−141,590
Cr(c) + 3/2 O_2(g) = Cr_2O_3(l)	471–600K	−141,580
2 Bi(c) + 3/2 O_2(g) = Bi_2O_3(c)	298.16–544K	−139,000
W(c) + O_2(g) = WO_2(c)	298.16–1,500K	−137,180
Ba(α) + 1/2 O_2(g) = BaO(c)	298.16–648K	−134,590
Ba(β) + 1/2 O_2(g) = BaO(c)	648–977K	−134,140
2 K(g) + 1/2 O_2(g) = K_2O(c)	1,049–1,500K	−133,090
Mo(c) + O_2(g) = MoO_2(c)	298.16–2,000K	−132,910
Ra(c) + 1/2 O_2(g) = RaO(c)	298.16–1,000K	−130,000
Mn(α) + O_2(g) = MnO_2(c)	298.16–1,000K	−126,400

Selecting Heat of Formation of Inorganic Oxides (Continued)

Reaction	Temperature Range of Validity	ΔH_0
$Ti(\alpha) + 1/2\ O_2(g) = TiO(\alpha)$	1,150–1,264K	−125,040
$Ti(\alpha) + 1/2\ O_2(g) = TiO(\alpha)$	298.16–1,150K	−125,010
$2\ Na(c) + O_2(g) = Na_2O_2(c)$	298.16–371K	−122,500
$2\ Cs(l) + 3/2\ O_2(g) = Cs_2O_3(c)$	301.5–775K	−113,840
$2\ Cs(g) + 1/2\ O_2(g) = Cs_2O(l)$	963–1,500K	−113,790
$2\ Cs(c) + 3/2\ O_2(g) = Cs_2O_3(c)$	298.16–301.5K	−112,690
$S(rhombohedral) + 3/2\ O_2(g) = SO_3(c–I)$	298.16–335.4K	−111,370
$2\ Cs(l) + 3/2\ O_2(g) = Cs_2O_3(l)$	775–963K	−110,740
$1/2\ S_2(g) + 3/2\ O_2(g) = SO_3(g)$	298.16–1,500K	−110,420
$S(rhombohedral) + 3/2\ O_2(g) = SO_3(c–II)$	298.16–305.7K	−108,680
$S(rhombohedral) + 3/2\ O_2(g) = SO_3(l)$	298.16–335.4K	−107,430
$V(c) + 1/2\ O_2(g) = VO(c)$	298.16–2,000K	−101,090
$2\ Na(l) + 1/2\ O_2(g) = Na_2O(c)$	371–1,187K	−100,150
$2\ Na(c) + 1/2\ O_2(g) = Na_2O(c)$	298.16–371K	−99,820
$2\ Tl(\alpha) + 3/2\ O_2(g) = Tl_2O_3(c)$	298.16–505.5K	−99,410
$S(monoclinic) + 3/2\ O_2(g) = SO_3(g)$	368.6–392K	−95,120
$S(rhombohedral) + 3/2\ O_2(g) = SO_3(g)$	298.16–368.6K	−95,070
$S(1\lambda,\mu) + 3/2\ O_2(g) = SO_3(g)$	392–718K	−94,010
$C(graphite) + O_2(g) = CO_2(g)$	298.16–2,000K	−93,690
$Mn(l) + 1/2\ O_2(g) = MnO(c)$	1,517–2,000K	−93,350
$Mn(\alpha) + 1/2\ O_2(g) = MnO(c)$	298.16–1,000K	−92,600
$Mn(\beta) + 1/2\ O_2(g) = MnO(c)$	1,000–1,374K	−91,900
$Mn(\gamma) + 1/2\ O_2(g) = Mno(c)$	1,374–1,410K	−89,810
$Mn(\delta) + 1/2\ O_2(g) = MnO(c)$	1,410–1,517K	−89,390

Selecting Heat of Formation of Inorganic Oxides (Continued)

Reaction	Temperature Range of Validity	ΔH_0
2 K(l) + 1/2 O_2(g) = K_2O(c)	336.4–1,049K	–87,380
2 K(c) + 1/2 O_2(g) = K_2O(c)	298.16–336.4K	–86,400
1/2 S_2(g) + O_2 (g) = SO_2(g)	298.16–2,000K	–86,330
Zn(c) + 1/2 O_2(g) = ZnO(c)	298.16–692.7K	–84,670
2 Rb(l) + 1/2 O_2(g) = Rb_2O(c)	312.2–750K	–79,950
2 Rb(c) + 1/2 O_2(g) = Rb_2O(c)	298.16–312.2K	–78,900
2 Cs(l) + 1/2 O_2(g) = Cs_2O(c)	301.5–763K	–76,900
2 Cs(c) + 1/2 O_2(g) = Cs_2O(c)	298.16–301.5K	–75,900
2 Cs(l) + 1/2 O_2(g) = Cs_2O(l)	763–963K	–75,370
D_2(g) + 1/2 O_2(g) = D_2O(l)	298.16–374.5K	–72,760
S(monoclinic) + O_2(g) = SO_2(g)	368.6–392K	–71,020
S(rhombohedral) + O_2(g) = SO_2(g)	298.16–368.6K	–70,980
H_2(g) + 1/2 O_2(g) = H_2O(l)	298.16–373.16K	–70,600
S($1\lambda,\mu$) + O_2(g) = SO_2(g)	392–718K	–69,900
Sn(l) + 1/2 O_2(g) = SnO(c)	505–1,300K	–69,670
Sn(c) + 1/2 O_2(g) = SnO(c)	298.16–505K	–68,600
0.947 Fc(β) + 1/2 O_2(g) = $Fe_{0.947}O$(c)	1,179–1,650K	–66,750
Pb(c) + O_2(g) = PbO_2(c)	298.16–600.5K	–66,120
0.947 Fe(α) + 1/2 O_2(g) = $Fe_{0.947}O$(c)	298.16–1,033K	–65,320
0.947 Fe(γ) + 1/2 O_2(g) = $Fe_{0.947}O$(l)	1,650–1,674K	–64,200
0.947 Fe(δ) + 1/2 O_2(g) = $Fe_{0.947}O$(l)	1,803–2,000K	–63,660
Cd(l) + 1/2 O_2(g) = CdO(c)	594–1,038K	–63,240
0.947 Fe(α) + 1/2 O_2(g) = $Fe_{0.947}O$(c)	1,033–1,179K	–62,380
Cd(c) + 1/2 O_2(g) = CdO(c)	298.16–594K	–62,330

Selecting Heat of Formation of Inorganic Oxides (Continued)

Reaction	Temperature Range of Validity	ΔH_0
$0.947\ Fe(\gamma) + 1/2\ O_2(g) = Fe_{0.947}O(l)$	1,647–1,803K	–59,650
$D_2(g) + 1/2\ O_2(g) = D_2O(g)$	298.16–2,000K	–58,970
$Co(\gamma) + 1/2\ O_2(g) = CoO(c)$	1,400–1,763K	–58,160
$I_2(g) + 5/2\ O_2(g) = I_2O_5(c)$	456–500K	–58,020
$Ni(\alpha) + 1/2\ O_2(g) = NiO(c)$	298.16–633K	–57,640
$Ni(\beta) + 1/2\ O_2(g) = NiO(c)$	633–1,725K	–57,460
$H_2(g) + 1/2\ O_2(g) = H_2O(g)$	298.16–2,000K	–56,930
$Co(\alpha,\beta) + 1/2\ O_2(g) = CoO(c)$	298.16–1,400K	–56,910
$Pb(l) + 1/2\ O_2(g) = PbO\ (red)$	600.5–762K	–53,780
$Pb(l) + 1/2\ O_2(g) = PbO\ (yellow)$	600.5–1,159K	–53,020
$Bi(l) + 1/2\ O_2(g) = BiO(c)$	544–1,600K	–52,920
$Pb(c) + 1/2\ O_2(g) = PbO\ (red)$	298.16–600.5K	–52,800
$Pb(c) + 1/2\ O_2(g) = PbO\ (yellow)$	298.16–600.5K	–52,040
$Bi(c) + 1/2\ O_2(g) = BiO(c)$	298.16–544K	–50,450
$2\ Tl(\beta) + O_2(g) = Tl_2O(c)$	505.5–573K	–44,260
$2\ Tl(\alpha) + O_2(g) = Tl_2O(c)$	298.16–505.5K	–44,110
$2\ Cu(l) + 1/2\ O_2(g) = Cu_2O(c)$	1,357–1,502K	–43,880
$I_2(l) + 5/2\ O_2(g) = I_2O_5(c)$	386.8–456K	–43,490
$I_2(c) + 5/2\ O_2(g) = I_2O_5(c)$	298.16–386.8K	–42,040
$Cu(l) + 1/2\ O_2(g) = CuO(l)$	1,720–2,000K	–41,060
$Ir(c) + O_2(g) = IrO_2(c)$	298.16–1,300K	–39,480
$Cu(l) + 1/2\ O_2(g) = CuO(c)$	1,357–1,720K	–39,410
$2\ Al(l) + 1/2\ O_2(g) = Al_2O(g)$	931.7–2,000K	–38,670
$Cu(c) + 1/2\ O_2(g) = CuO(c)$	298.16–1,357K	–37,740

Selecting Heat of Formation of Inorganic Oxides (Continued)

Reaction	Temperature Range of Validity	ΔH_0
2 Cu(l) + 1/2 O_2(g) = Cu_2O(l)	1,502–2,000K	–37,710
2 Al(c) + 1/2 O_2(g) = Al_2O(g)	298.16–931.7K	–31,660
Si(l) + 1/2 O_2(g) = SiO(g)	1,683–2,000K	–30,170
C(graphite) + 1/2 O_2(g) = CO(g)	298.16–2,000K	–25,400
2 Hg(l) + 1/2 O_2(g) = Hg_2O(c)	298.16–629.88K	–22,400
Hg(l) + 1/2 O_2(g) = HgO (red)	298.16–629.88K	–21,760
Si(c) + 1/2 O_2(g) = SiO(g)	298.16–1,683K	–21,090
P(l) + 1/2 O_2(g) = PO(g)	317.4–553K	–9,390
P (white) + 1/2 O_2(g) = PO(g)	298.16–317.4K	–9,370
2 Ag(c) + 1/2 O_2(g) = Ag_2O_2(c)	298.16–1,000K	–7,740
1/2 Se_2(g) + 1/2 O_2(g) = SeO(g)	1,027–2,000K	–7,400
2 Ag(c) + O_2(g) = Ag_2O_2(c)	298.16–500K	–6,620
2 Au(c) + 3/2 O_2(g) = Au_2O_3(c)	298.16–500K	–2,160
1/2 S_2 (g) + 1/2 O_2(g) = SO(g)	298.16–2,000K	+3,890
Al(l) + 1/2 O_2(g) = AlO(g)	931.7–2,000K	+8,170
Se(c) + 1/2 O_2(g) = SeO(g)	298.16–490K	+9,280
Se(l) + 1/2 O_2(g) = SeO(g)	490–1,027K	+9,420
2 Cu(c) + 1/2 O_2(g) = Cu_2O(c)	298.16–1,357K	+10,550
Al(c) + 1/2 O_2(g) = AlO(g)	298.16–931.7K	+10,740
Cl_2(g) + 1/2 O_2(g) = Cl_2O(g)	298.16–2,000K	+17,770
S(monoclmic) + 1/2 O_2(g) = SO(g)	368.6–392K	+19,200
S(rhombohedral) + 1/2 O_2(g) = SO(g)	298.16–368.6K	+19,250
S($1\lambda,\mu$) + 1/2 O_2(g) = SO(g)	392–718K	+20,320
1/2 Cl_2(g) + 1/2 O_2(g) = ClO(g)	298.16–1,000K	+33,000

Selecting Heat of Formation of Inorganic Oxides (Continued)

Reaction	Temperature Range of Validity	ΔH_0
$3/2\ O_2(g)\ =\ O_3(g)$	298.16–2,000K	+33,980
$2\ Cl_2(g)\ +\ 3/2\ O_2(g)\ =\ ClO(g)$	298.16–500K	+37,740
$Te(l)\ +\ 1/2\ O_2(g)\ =\ TeO(g)$	723–1,360K	+39,750
$Te(c)\ +\ 1/2\ O_2(g)\ =\ TeO(g)$	298.16–723K	+43,110
$V(c)\ +\ 1/2\ O_2(g)\ =\ VO(g)$	298.16–2,000K	+52,090

The ΔH_o values are given in **gram calories per mole**.

Source: data from *CRC Handbook of Materials Science, Vol II*, Charles T. Lynch, Ed., CRC Press, Cleveland, (1974).

SELECTING MELTING POINTS OF ELEMENTS AND INORGANIC COMPOUNDS

Compound	Formula	Melting Point °C
Hydrogen	H_2	−259.25
Neon	Ne	−248.6
Fluorine	F_2	−219.6
Oxygen	O_2	−218.8
Nitrogen	N_2	−210
Carbon monoxide	CO	−205
Nitric oxide	NO	−163.7
Boron trifluoride	BF_3	−128.0
Hydrogen chloride	HCl	−114.3
Xenon	Xe	−111.6
Boron trichloride	BCl_3	−107.8
Chlorine	Cl_2	−103±5
Nitrous oxide	N_2O	−90.9
Hydrogen sulfide, di–	H_2S_2	−89.7
Hydrogen bromide	HBr	−86.96
Hydrogen sulfide	H_2S	−85.6
Sulfur dioxide	SO_2	−73.2
Silicon tetrachloride	$SiCl_4$	−67.7
Bromine pentafluoride	BrF_5	−61.4
Carbon dioxide	CO_2	−57.6
Hydrogen iodide	HI	−50.91
Hydrogen telluride	H_2Te	−49.0
Boron tribromide	BBr_3	−48.8
Hydrogen nitrate	HNO_3	−47.2

Selecting Melting Points of Elements and Inorganic Compounds (Continued)

Compound	Formula	Melting Point °C
Mercury	Hg	–39
Tin chloride,tetra–	$SnCl_4$	–33.3
Silane, hexafluoro–	Si_2F_6	–28.6
Cyanogen	C_2N_2	–27.2
Titanium chloride, tetra–	$TiCl_4$	–23.2
Iron pentacarbonyl	$Fe(CO)_5$	–21.2
Arsenic trichloride	$AsCl_3$	–16.0
Nitrogen tetroxide	N_2O_4	–13.2
Bromine	Br_2	–7.2
Arsenic trifluoride	AsF_3	–6.0
Cyanogen chloride	$CNCl$	–5.2
Hydrogen peroxide	H_2O_2	–0.7
Tungsten hexafluoride	WF_6	–0.5
Hydrogen oxide (water)	H_2O	0
Phosphorus oxychloride	$POCl_3$	1.0
Deuterium oxide	D_2O	3.78
Antimony pentachloride	$SbCl_5$	4.0
Seleniumoxychloride	$SeOCl_3$	9.8
Hydrogen sulfate	H_2SO_4	10.4
Iodine chloride (β)	ICl	13.8
Sulfur trioxide (α)	SO_3	16.8
Molybdenum hexafluoride	MoF_6	17
Iodine chloride (α)	ICl	17.1
Phosphorus acid, hypo–	H_3PO_2	17.3

Selecting Melting Points of Elements and Inorganic Compounds (Continued)

Compound	Formula	Melting Point °C
Rhenium hexafluoride	ReF_6	19.0
Niobium pentachloride	$NbCl_5$	21.1
Phosphorus trioxide	P_4O_6	23.7
Cesium	Cs	28.3
Gallium	Ga	29
Tin bromide, tetra–	$SnBr_4$	29.8
Arsenic tribromide	$AsBr_3$	30.0
Sulfur trioxide (β)	SO_3	32.3
Titanium bromide, tetra–	$TiBr_4$	38
Cesium chloride	$CsCl$	38.5
Rubidium	Rb	38.9
Osmium tetroxide (white)	OsO_4	41.8
Phosphoric acid	H_3PO_4	42.3
Phosphorus, yellow	P_4	44.1
Phosphoric acid. hypo–	$H_4P_2O_6$	54.8
Osmium tetroxide (yellow)	OsO_4	55.8
Hydrogen selenate	H_2SeO_4	57.8
Sulfur trioxide (γ)	SO_3	62.1
Potassium	K	63.4
Antimony trichloride	$SbCl_3$	73.3
Phosphorus acid, ortho–	H_3PO_3	73.8
Arsenic pentafluoride	AsF_5	80.8
Hydrogen fluoride	HF	83.11
Aluminum bromide	Al_2Br_6	87.4

Selecting Melting Points of Elements and Inorganic Compounds (Continued)

Compound	Formula	Melting Point °C
Antimony tribromide	$SbBr_3$	96.8
Sodium	Na	97.8
Iodine	I_2	112.9
Sulfur (monatomic)	S	119
Tin iodide, tetra–	SnI_4	143.4
Indium	In	156.3
Lithium	Li	178.8
Potassium thiocyanate	KSCN	179
Argon	Ar	190.2
Aluminum iodide	Al_2I_6	190.9
Aluminum chloride	Al_2Cl_6	192.4
Chromium trioxide	CrO_3	197
Tantalum pentachloride	$TaCl_5$	206.8
Thallium nitrate	$TINO_3$	207
Silver nitrate	$AgNO_3$	209
Selenium	Se	217
Bismuth trichloride	$BiCl_3$	223.8
Tin	Sn	231.7
Tin bromide, di–	$SnBr_2$	231.8
Mercury bromide	$HgBr_2$	241
Tin chloride, di–	$SnCl_2$	247
Lithium nitrate	$LiNO_3$	250
Mercury iodide	HgI_2	250
Sodium chlorate	$NaClO_3$	255

Selecting Melting Points of Elements and
Inorganic Compounds (Continued)

Compound	Formula	Melting Point °C
Bismuth	Bi	271
Thallium carbonate	Tl_2CO_3	273
Mercury chloride	$HgCl_2$	276.8
Zincchloride	$ZnCl_2$	283
Rhenium heptoxide	Re_2O_7	296
Thallium	Tl	302.4
Iron (III) chloride	Fe_2Cl_6	303.8
Rubidium nitrate	$RbNO_3$	305
Sodium nitrate	$NaNO_3$	310
Arsenic trioxide	As_4O_6	312.8
Cadmium	Cd	320.8
Sodium hydroxide	NaOH	322
Sodium thiocyanate	NaSCN	323
Tungsten tetrachloride	WCl_4	327
Lead	Pb	327.3
Potassium nitrate	KNO_3	338
Silver cyanide	AgCN	350
Potassium hydroxide	KOH	360
Cadmium iodide	CdI_2	386.8
Potassium dichromate	$K_2Cr_2O_7$	398
Beryllium chloride	$BeCl_2$	404.8
Cesium nitrate	$CsNO_3$	406.8
Lead iodide	PbI_2	412
Zinc	Zn	419.4

Selecting Melting Points of Elements and Inorganic Compounds (Continued)

Compound	Formula	Melting Point °C
Thallium chloride, mono–	TlCl	427
Copper (I) chloride	CuCl	429
Copper (II) chloride	$CuCl_2$	430
Silver bromide	AgBr	430
Lithium iodide	LiI	440
Thallium iodide, mono–	TlI	440
Boron trioxide	B_2O_3	448.8
Thallium sulfide	Tl_2S	449
Tellurium	Te	453
Silver chloride	AgCl	455
Sodium peroxide	Na_2O_2	460
Thallium bromide, mono–	TlBr	460
Lithium hydroxide	LiOH	462
Copper(l) cyanide	$Cu_2(CN)_2$	473
Beryllium bromide	$BeBr_2$	487.8
Lead bromide	$PbBr_2$	487.8
Potassium peroxide	K_2O_2	490
Antimony trisulfide	Sb_4S_6	546.0
Lithium bromide	LiBr	552
Silver iodide	AgI	557
Calcium nitrate	$Ca(NO_3)_2$	560.8
Sodium cyanide	NaCN	562
Cadmium bromide	$CdBr_2$	567.8
Cadmium chloride	$CdCl_2$	567.8

Selecting Melting Points of Elements and Inorganic Compounds (Continued)

Compound	Formula	Melting Point °C
Phosphorus pentoxide	P_4O_{10}	569.0
Copper (I) iodide	CuI	587
Uranium tetrachloride	UCl_4	590
Barium nitrate	$Ba(NO_3)_2$	594.8
Lithium chloride	LiCl	614
Europium trichloride	$EuCl_3$	622
Potassium cyanide	KCN	623
Antimony	Sb	630
Thallium sulfate	Tl_2SO_4	632
Rubidium iodide	RbI	638
Strontium bromide	$SrBr_2$	643
Magnesium	Mg	650
Manganese dichloride	$MnCl_2$	650
Antimony trioxide	Sb_4O_6	655.0
Silver sulfate	Ag_2SO_4	657
Aluminum	Al	658.5
Sodium iodide	NaI	662
Vanadium pentoxide	V_2O_5	670
Iron (II) chloride	$FeCl_2$	677
Rubidium bromide	RbBr	677
Potassium iodide	KI	682
Sodium molybdate	Na_2MoO_4	687
Sodium tungstate	Na_2WO_4	702
Lithium molybdate	Li_2MoO_4	705

Selecting Melting Points of Elements and Inorganic Compounds (Continued)

Compound	Formula	Melting Point °C
Barium iodide	BaI_2	710.8
Magnesium bromide	$MgBr_2$	711
Magnesium chloride	$MgCl_2$	712
Rubidium chloride	RbCl	717
Barium	Ba	725
Bismuth trifluoride	BiF_3	726.0
Molybdenum dichloride	$MoCl_2$	726.8
Cobalt (II) chloride	$CoCl_2$	727
Zirconium dichloride	$ZrCl_2$	727
Calcium bromide	$CaBr_2$	729.8
Lithium tungstate	Li_2WO_4	742
Potassium bromide	KBr	742
Sodium bromide	NaBr	747
Strontium	Sr	757
Thorium chloride	$ThCl_4$	765
Potassium chloride	KCl	770
Cerium	Ce	775
Calcium chloride	$CaCl_2$	782
Nickel subsulfide	Ni_3S_2	790
Molybdenum trioxide	MoO_3	795
Sodium chloride	NaCl	800
Chromium (II) chloride	$CrCl_2$	814
Bismuth trioxide	Bi_2O_3	815.8
Arsenic	As	816.8

Selecting Melting Points of Elements and Inorganic Compounds (Continued)

Compound	Formula	Melting Point °C
Lead fluoride	PbF_2	823
Ytterbium	Yb	823
Europium	Eu	826
Rubidium fluoride	RbF	833
Silver sulfide	Ag_2S	841
Barium bromide	$BaBr_2$	846.8
Mercury sulfate	$HgSO_4$	850
Calcium	Ca	851
Sodium carbonate	Na_2CO_3	854
Lithium sulfate	Li_2SO_4	857
Strontium chloride	$SrCl_2$	872
Potassium fluoride	KF	875
Sodium silicate, di–	$Na_2Si_2O_5$	884
Sodium sulfate	Na_2SO_4	884
Lead oxide	PbO	890
Lithium fluoride	LiF	896
Potassium carbonate	K_2CO_3	897
Lanthanum	La	920
Sodium sulfide	Na_2S	920
Praseodymium	Pr	931
Potassium borate, meta–	KBO_2	947
Germanium	Ge	959
Barium chloride	$BaCl_2$	959.8
Silver	Ag	961

Selecting Melting Points of Elements and
Inorganic Compounds (Continued)

Compound	Formula	Melting Point °C
Sodium borate, meta–	$NaBO_2$	966
Sodium pyrophosphate	$Na_4P_2O_7$	970
Potassium chromate	K_2CrO_4	984
Sodium phosphate, meta–	$NaPO_3$	988
Titanium oxide	TiO	991
Sodium fluoride	NaF	992
Cadmium sulfate	$CdSO_4$	1000
Neodymium	Nd	1020
Vanadium dichloride	VCl_2	1027
Nickel chloride	$NiCl_2$	1030
Tin oxide	SnO	1042
Actinium227	Ac	1050 ± 50
Gold	Au	1063
Lead molybdate	$PbMoO_4$	1065
Samarium	Sm	1072
Potassium sulfate	K_2SO_4	1074
Copper	Cu	1083
Lead sulfate	$PbSO_4$	1087
Sodium silicate, meta–	Na_2SiO_3	1087
Potassium pyro–phosphate	$K_4P_2O_7$	1092
Sodiumsilicate,aluminum–	$NaAlSi_3O_8$	1107
Cadmium fluoride	CdF_2	1110
Lead sulfide	PbS	1114
Copper (I) sulfide	Cu_2S	1129

Selecting Melting Points of Elements and Inorganic Compounds (Continued)

Compound	Formula	Melting Point °C
Uranium^{235}	U	~1133
Lithium metasilicate	Li_2SiO_3	1177
Iron (II) sulfide	FeS	1195
Manganese	Mn	1220
Magnesium fluoride	MgF_2	1221
Iron carbide	Fe_3C	1226.8
Copper (I) oxide	Cu_2O	1230
Lithium orthosilicate	Li_4SiO_4	1249
Tungsten dioxide	WO_2	1270
Manganese metasilicate	$MnSiO_3$	1274
Beryllium	Be	1278
Calcium carbonate	$CaCO_3$	1282
Barium fluoride	BaF_2	1286.8
Calcium sulfate	$CaSO_4$	1297
Gadolinium	Gd	1312
Magnesium sulfate	$MgSO_4$	1327
Potassium phosphate	K_3PO_4	1340
Barium sulfate	$BaSO_4$	1350
Terbium	Tb	1356
Iron (II) oxide	FeO	1380
Calcium fluoride	CaF_2	1382
Strontium fluoride	SrF_2	1400
Dysprosium	Dy	1407
Silicon	Si	1427

Selecting Melting Points of Elements and Inorganic Compounds (Continued)

Compound	Formula	Melting Point °C
Copper (II) oxide	CuO	1446
Nickel	Ni	1452
Holmium	Ho	1461
Tungsten trioxide	WO_3	1470
Cobalt	Co	1490
Erbium	Er	1496
Yttrium	Y	1504
Niobium pentoxide	Nb_2O_5	1511
Calcium metasilicate	$CaSiO_3$	1512
Magnesium silicate	$MgSiO_3$	1524
Iron	Fe	1530.0
Scandium	Sc	1538
Thulium	Tm	1545
Palladium	Pd	1555
Manganese oxide	Mn_3O_4	1590
Iron oxide	Fe_3O_4	1596
Lutetium	Lu	1651
Barium phosphate	$Ba_3(PO_4)_2$	1727
Zinc sulfide	ZnS	1745
Platinum	Pt	1770
Manganese (II) oxide	MnO	1784
Titanium	Ti	1800
Titanium dioxide	TiO_2	1825

Selecting Melting Points of Elements and Inorganic Compounds (Continued)

Compound	Formula	Melting Point °C
Thorium	Th	1845
Zirconium	Zr	1857
Tantalum pentoxide	Ta_2O_5	1877
Chromium	Cr	1890
Vanadium	V	1917
Barium oxide	BaO	1922.8
Zinc oxide	ZnO	1975
Aluminum oxide	Al_2O_3	2045.0
Vanadium oxide	VO	2077
Hafnium	Hf	2214
Yttrium oxide	Y_2O_3	2227
Chromium (III) sequioxide	Cr_2O_3	2279
Boron	B	2300
Strontium oxide	SrO	2430
Niobium	Nb	2496
Beryllium oxide	BeO	2550.0
Molybdenum	Mo	2622
Magnesium oxide	MgO	2642
Osmium	Os	2700

Selecting Melting Points of Elements and
Inorganic Compounds (Continued)

Compound	Formula	Melting Point °C
Calcium oxide	CaO	2707
Zirconium oxide	ZrO_2	2715
Thorium dioxide	ThO_2	2952
Tantalum	Ta	2996 $\pm$ 50
Rhenium	Re	3167$\pm$60
Tungsten	W	3387

Source: data from: Weast, R C., Ed., *Handbook of Chemistry and Physics, 55th ed.*, CRC Press, Cleveland, (1974); and Bolz, R. E. and Tuve, G. L., Eds., *Handbook of Tables for Applied Engineering Science, 2nd ed.*, CRC Press, Cleveland, (1973), p. 479.

SELECTING MELTING POINTS OF CERAMICS

Compound	(K)		Compound	(K)
TaC	3813		CeO_2	>2873
NbC	3770		UC	2863
VC	3600		BeO	2725
ZrC	3533		B_4C	2720
ThO_2	3493		Si_3N_4	2715
TiC	3433		$TaSi_2$	2670
Ta_2N	3360		MoB	2625
ZrB_2	3313		Cr_2O_3	>2603
TiB_2	3253		VN	2593
ZrN	3250		$MoSi_2$	2553
TiN	3200		BaB_4	2543
CaO	3183		Be_3N_2	2513
UO_2	3151		SrB_6	2508
WB	3133		AlN	>2475
ZrO_2	3123		CeB_6	2463
UN	3123		CeS	2400
MgO	3098		Be_2C	>2375
BN	3000		VB_2	2373
SiC	2970		NbN	2323
Mo_2C	2963		Al_2O_3	2322
SrO	2933		WSi_2	2320
ThN	2903		BaO	2283
WC	2900		SrS	>2275
ThC	2898		MgS	>2275

Selecting Melting Points of Ceramics (Continued)

Compound	(K)	Compound	(K)
ThB_4	>2270	$CrSi_2$	1843
TaB	>2270	MnO	1840
NbB	>2270	ZrS_2	1823
NiO	2257	$TiSi_2$	1813
ZnO	2248	CdO	1773
BeB_2	>2243	UB_2	>1770
$NbSi_2$	2203	CrN	1770
ThS_2	2198	Nb_2O_5	1764
In_2O_3	2183	WO_3	1744
Cr_3C_2	2168	SrF_2	1736
CrB_2	2123	$CaSO_4$	1723
TiO_2	2113	CeF_2	1710
Fe_3C	2110	CaF_2	1675
Ta_2O_5	2100	BaF_2	1627
VSi_2	2023	TaS_4	>1575
CdS	2023	AlF_3	1564
Al_4C_3	2000	MgF_2	1535
SiO_2	1978	WS_2	1523
Li_2O	>1975	Cu_2O	1508
USi_2	1970	TiF_3	1475
SrC_2	>1970	BaS	1473
$SrSO_4$	1878	FeS	1468
Fe_2O_3	1864	Ca_3N_2	1468
$BaSO_4$	1853	MoS_2	1458

Selecting Melting Points of Ceramics (Continued)

Compound	(K)	Compound	(K)
Na_2S	1453	Na_2SO_4	1157
$PbSO_4$	1443	SnS	1153
InF_3	1443	$SrCl_2$	1148
Cu_2S	1400	ZnF_2	1145
$MgSO_4$	1397	Li_2SO_4	1132
PbS	1387	KF	1131
US_2	>1375	MnF_2	1129
ThF_4	1375	CuF_2	1129
Mg_2Si	1375	Cu_4Si	1123
CdF_2	1373	$BaBr_2$	1123
Al_2S_3	1373	$NiSO_4$	1121
SnO	1353	LiF	1119
K_2SO_4	1342	Li_3N	1118
In_2S_3	1323	K_2S	1113
FeF_3	>1275	B_2O_3	1098
$NiCl_3$	1274	Ag_2S	1098
NiF_2	1273	PbF_2	1095
$CdSO_4$	1273	$CeCl_3$	1095
NaF	1267	VF_3	>1075
$NiBr_2$	1236	NaCl	1073
$BaCl_2$	1235	NiS	1070
UF_4	1233	NiI_2	1070
Li_2S	1198	MoO_3	1068
PbO	1159	$CaCl_2$	1055

Selecting Melting Points of Ceramics (Continued)

Compound	(K)	Compound	(K)
FI_2	1048	$SrBr_2$	916
$ThCl_4$	1043	MgI_2	<910
KCl	1043	$ThBr_4$	883
$Al_2(SO_4)_3$	1043	$LiCl$	883
CeI_3	1025	CuI	878
$NaBr$	1023	V_2S_3	>875
Bi_2S_3	1020	ZrF_4	873
BaI_2	1013	$ZnSO_4$	873
KBr	1008	TiI_2	873
TeO_2	1006	$Ba(NO_3)_2$	865
$CaBr_2$	1003	$PtCl_2$	854
BiF_3	1000	CaI_2	848
$MgCl_2$	987	$BeSO_4$	848
$MgBr_2$	984	UCl_4	843
SnF_4	978	$CdCl_2$	841
NaC_2	973	$CdBr_2$	841
KI	958	$Cd(NO_3)_2$	834
$FeBr_2$	955	AgI	831
V_2O_5	947	$LiBr$	823
$FeCl_2$	945	SbS_3	820
NaI	935	BeF_2	813
Ag_2SO_4	933	$BeBr_2$	793
Sb_2O_3	928	UBr_4	789
$MnCl_2$	923	SnI_2	788

Selecting Melting Points of Ceramics (Continued)

Compound	(K)	Compound	(K)
BeI_2	783	$PbBr_2$	643
UI_4	779	$SnSO_4$	>635
$CuBr$	777	PtI_2	633
ZrI_4	772	$ZrBr_2$	>625
$PbCl_2$	771	$ZrCl_2$	623
$Fe_2(SO_4)_3$	753	$Ca(NO_3)_2$	623
$Pb(NO_3)_2$	743	$TeBr_2$	612
$AgCl$	728	KNO_3	610
B_2O_3	723	SrI_2	593
LiI	722	$NaNO_3$	583
ZnI_2	719	$SnCl_2$	581
$BeCl_2$	713	Na_2N	573
$InBr_3$	709	Cu_3N	573
AgF	708	Ag_2O	573
K_2O_3	703	SbF_3	565
$AgBr$	703	$ZnCl_2$	548
$CuCl$	695	WCl_6	548
$Zr(SO_4)_2$	683	$TaBr_5$	538
BiI_3	681	$LiNO_3$	527
$Bi(SO_4)_3$	678	$PtBr_2$	523
PbI_2	675	PtS_2	508
$ZnBr_2$	667	$BiCl_3$	507
BS_4	663	$InCl$	498
$Sr(NO_3)_2$	643	$BiBr_3$	491

Selecting Melting Points of Ceramics (Continued)

Compound	(K)	Compound	(K)
$TaCl_5$	489	TaF_5	370
$SnBr_2$	488	$SbBr_3$	370
InI_3	483	$SbCl_3$	346
$AgNO_3$	483	$TiBr_4$	312
$Ce(SO_4)_2$	468	MoF_6	290
$AlCl_3$	465	$TiCl_4$	250
AlI	464	VCl_4	245
$TeCl_2$	448	BBr_3	227
SbI_3	443	SiF_4	183
CdI_2	423	BCl_3	166
MoI_4	373	BF_3	146
$AlBr_3$	371		

Source: data from Lynch, Charles T., Ed., *CRC Handbook of Materials Science, Vol. 1*, CRC Press, Boca Raton, 1974, 348.

Selecting Ceramics and Glasses: Thermal Properties

SELECTING THERMAL CONDUCTIVITY OF CERAMICS

Ceramic	Thermal Conductivity $(\text{cal} \cdot \text{cm}^{-1} \cdot \text{sec}^{-1} \cdot \text{K}^{-1})$
Zirconium Oxide (ZrO_2) (plasma sprayed)	0.0019-0.0022 at 800°C
Zirconium Oxide (ZrO_2) (plasma sprayed)	0.0019-0.0031 at room temp.
Silicon Dioxide (SiO_2)	0.0025 at 200°C
Cerium Dioxide (CeO_2)	0.00287 at 1400K
Silicon Dioxide (SiO_2)	0.003 at 400°C
Sillimanite (Al_2O_3 SiO_2) (0% porosity)	0.003 at 1500°C
Silicon Carbide (SiC) (cubic, CVD)	0.0032 at 1530°C
Zirconium Oxide (ZrO_2) (plasma sprayed and coated with Cr_2O_3)	0.0033 at 800°C
Zirconium Oxide (ZrO_2) (plasma sprayed and coated with Cr_2O_3)	0.0033 at room temp.
Sillimanite (Al_2O_3 SiO_2) (0% porosity)	0.0035 at 800°C
Sillimanite (Al_2O_3 SiO_2) (0% porosity)	0.0035 at 1200°C
Cordierite (2MgO 2Al_2O_3 5SiO_2) (ρ=2.1g/cm^3)	0.0038 at 800°C
Zirconium Oxide (ZrO_2) (stabilized)	0.004 at 100°C
Sillimanite (Al_2O_3 SiO_2) (0% porosity)	0.004 at 400°C
Silicon Dioxide (SiO_2)	0.004 at 800°C
Cordierite (2MgO 2Al_2O_3 5SiO_2) (ρ=2.1g/cm^3)	0.0040 at 500°C
Cordierite (2MgO 2Al_2O_3 5SiO_2) (ρ=2.1g/cm^3)	0.0041 at 300°C
Sillimanite (Al_2O_3 SiO_2) (0% porosity)	0.0042 at 100°C
Cordierite (2MgO 2Al_2O_3 5SiO_2) (ρ=2.1g/cm^3)	0.0043 at 20°C
Zirconium Oxide (ZrO_2) (stabilized)	0.0044 at 500°C
Zirconium Oxide (ZrO_2) (5-10% CaO stabilized)	0.0045 at 400°C

Selecting Thermal Conductivity of Ceramics (Continued)

Ceramic	Thermal Conductivity $(cal \cdot cm^{-1} \cdot sec^{-1} \cdot K^{-1})$
Zirconium Oxide (ZrO$_2$) (stabilized)	0.0048-0.0055 at 1000°C
Zirconium Oxide (ZrO$_2$) (stabilized)	0.0049-0.0050 at 1200°C
Zirconium Oxide (ZrO$_2$) (5-10% CaO stabilized)	0.0049 at 800°C
Zirconium Oxide (ZrO$_2$) (stabilized, 0% porosity)	0.005 at 100°C
Zirconium Oxide (ZrO$_2$) (stabilized, 0% porosity)	0.005 at 200°C
Zirconium Oxide (ZrO$_2$) (stabilized, 0% porosity)	0.005 at 400°C
Silicon Dioxide (SiO$_2$)	0.005 at 1200°C
Zirconium Oxide (ZrO$_2$) (Y$_2$O$_3$ stabilized)	0.0053 at 800°C
Cordierite (2MgO 2Al$_2$O$_3$ 5SiO$_2$) (ρ=2.3g/cm^3)	0.0055 at 500°C
Zirconium Oxide (ZrO$_2$) (stabilized, 0% porosity)	0.0055 at 800°C
Cordierite (2MgO 2Al$_2$O$_3$ 5SiO$_2$) (ρ=2.3g/cm^3)	0.0055 at 800°C
Zirconium Oxide (ZrO$_2$) (Y$_2$O$_3$ stabilized)	0.0055 at room temp.
Zirconium Oxide (ZrO$_2$) (MgO stabilized)	0.0057 at 800°C
Zirconium Oxide (ZrO$_2$) (5-10% CaO stabilized)	0.0057 at 1200°C
Silicon Carbide (SiC) (cubic, CVD)	0.0059 at 1250°C
Thorium Dioxide (ThO$_2$) (0% porosity)	0.006-0.0076 at 1200°C
Uranium Dioxide (UO$_2$)	0.006 at 1000°C
Uranium Dioxide (UO$_2$)	0.006 at 1200°C
Zirconium Oxide (ZrO$_2$) (stabilized, 0% porosity)	0.006 at 1200°C
Thorium Dioxide (ThO$_2$) (0% porosity)	0.006 at 1400°C
Silicon Dioxide (SiO$_2$)	0.006 at 1600°C
Cordierite (2MgO 2Al$_2$O$_3$ 5SiO$_2$) (ρ=2.3g/cm^3)	0.0062 at 300°C
Zirconium Oxide (ZrO$_2$) (stabilized, 0% porosity)	0.0065 at 1400°C
Thorium Dioxide (ThO$_2$) (0% porosity)	0.007-0.0074 at 1000°C

Selecting Thermal Conductivity of Ceramics (Continued)

Ceramic	Thermal Conductivity $(cal \cdot cm^{-1} \cdot sec^{-1} \cdot K^{-1})$
Zirconium Oxide (ZrO_2) (MgO stabilized)	0.0076 at room temp.
Cordierite ($2MgO\ 2Al_2O_3\ 5SiO_2$) ($\rho=2.3g/cm^3$)	0.0077 at 20°C
Titanium Oxide (TiO_2) (0% porosity)	0.008 at 600°C
Uranium Dioxide (UO_2)	0.008 at 600°C
Uranium Dioxide (UO_2)	0.008 at 700°C
Thorium Dioxide (ThO_2) (0% porosity)	0.008 at 800°C
Titanium Oxide (TiO_2) (0% porosity)	0.008 at 800°C
Titanium Oxide (TiO_2) (0% porosity)	0.008 at 1000°C
Uranium Dioxide (UO_2) (0% porosity)	0.008 at 1000°C
Titanium Oxide (TiO_2) (0% porosity)	0.008 at 1200°C
Titanium Oxide (TiO_2) (0% porosity)	0.009 at 400°C
Uranium Dioxide (UO_2) (0% porosity)	0.009 at 800°C
Mullite ($3Al_2O_3\ 2SiO_2$) (0% porosity)	0.009 at 1000°C
Mullite ($3Al_2O_3\ 2SiO_2$) (0% porosity)	0.009 at 1200°C
Mullite ($3Al_2O_3\ 2SiO_2$) (0% porosity)	0.009 at 1400°C
Mullite ($3Al_2O_3\ 2SiO_2$) (0% porosity)	0.0095 at 800°C
Zircon ($SiO_2\ ZrO_2$) (0% porosity)	0.0095 at 1200°C
Zircon ($SiO_2\ ZrO_2$) (0% porosity)	0.0095 at 1400°C
Magnesium Oxide (MgO)	0.0096-0.0191 at 1800°C
Thorium Dioxide (ThO_2) (0% porosity)	0.010 at 600°C
Uranium Dioxide (UO_2) (0% porosity)	0.010 at 600°C
Mullite ($3Al_2O_3\ 2SiO_2$) (0% porosity)	0.010 at 600°C
Zircon ($SiO_2\ ZrO_2$) (0% porosity)	0.010 at 800°C
Magnesium Oxide (MgO)	0.0108-0.016 at 1600°C

Selecting Thermal Conductivity of Ceramics (Continued)

Ceramic	Thermal Conductivity $(cal \cdot cm^{-1} \cdot sec^{-1} \cdot K^{-1})$
Mullite ($3Al_2O_3$ $2SiO_2$) (0% porosity)	0.011 at 400°C
Nickel monoxide (NiO) (0% porosity)	0.011 at 1000°C
Magnesium Oxide (MgO)	0.012-0.014 at 1400°C
Titanium Oxide (TiO_2) (0% porosity)	0.012 at 200°C
Uranium Dioxide (UO_2)	0.012 at 400°C
Zircon (SiO_2 ZrO_2) (0% porosity)	0.012 at 400°C
Nickel monoxide (NiO) (0% porosity)	0.012 at 800°C
Spinel (Al_2O_3 MgO) (0% porosity)	0.013-0.0138 at 1000°C
Aluminum Oxide (Al_2O_3)	0.013-0.015 at 1200°C
Mullite ($3Al_2O_3$ $2SiO_2$) (0% porosity)	0.013 at 200°C
Spinel (Al_2O_3 MgO) (0% porosity)	0.013 at 1200°C
Aluminum Oxide (Al_2O_3)	0.013 at 1400°C
Zircon (SiO_2 ZrO_2) (0% porosity)	0.0135 at 200°C
Titanium Monocarbide (TiC)	0.0135 at 1000 °C
Magnesium Oxide (MgO)	0.0139-0.0148 at 1200°C
Aluminum Oxide (Al_2O_3)	0.014-0.016 at 1000°C
Thorium Dioxide (ThO_2) (0% porosity)	0.014 at 400°C
Aluminum Oxide (Al_2O_3)	0.014 at 1600°C
Mullite ($3Al_2O_3$ $2SiO_2$) (0% porosity)	0.0145 at 100°C
Zircon (SiO_2 ZrO_2) (0% porosity)	0.0145 at 100°C
Aluminum Oxide (Al_2O_3)	0.015-0.017 at 800°C
Uranium Dioxide (UO_2) (0% porosity)	0.015 at 400°C
Spinel (Al_2O_3 MgO) (0% porosity)	0.015 at 800°C
Zirconium Mononitride (ZrN)	0.015 at 1100 °C

Selecting Thermal Conductivity of Ceramics (Continued)

Ceramic	Thermal Conductivity $(cal \cdot cm^{-1} \cdot sec^{-1} \cdot K^{-1})$
Hafnium Diboride (HfB$_2$)	0.015 at room temp.
Magnesium Oxide (MgO)	0.016-0.020 at 1000°C
Titanium Oxide (TiO$_2$) (0% porosity)	0.016 at 100°C
Zirconium Mononitride (ZrN)	0.016 at 875 °C
Nickel monoxide (NiO) (0% porosity)	0.017 at 400°C
Aluminum Oxide (Al$_2$O$_3$)	0.017 at 1800°C
Uranium Dioxide (UO$_2$)	0.018 at 100°C
Zirconium Mononitride (ZrN)	0.018 at 650 °C
Calcium Oxide (CaO)	0.0186-0.019 at 1000°C
Thorium Dioxide (ThO$_2$) (0% porosity)	0.019 at 200°C
Spinel (Al$_2$O$_3$ MgO) (0% porosity)	0.019 at 600°C
Calcium Oxide (CaO)	0.019 at 800°C
Magnesium Oxide (MgO)	0.0198-0.026 at 800°C
Aluminum Oxide (Al$_2$O$_3$)	0.02-0.031 at 400°C
Thorium Dioxide (ThO$_2$) (0% porosity)	0.020 at 100°C
Uranium Dioxide (UO$_2$) (0% porosity)	0.020 at 200°C
Calcium Oxide (CaO)	0.020 at 600°C
Titanium Mononitride (TiN)	0.020 at 1000 °C
Aluminum Oxide (Al$_2$O$_3$)	0.021-0.022 at 600°C
Trisilicon tetranitride (Si$_3$N$_4$) (pressureless sintered)	0.022-0.072 at 127 °C
Calcium Oxide (CaO)	0.022 at 400°C
Cerium Dioxide (CeO$_2$)	0.0229 at 400K
Dichromium Trioxide (Cr$_2$O$_3$)	0.0239-0.0788
Nickel monoxide (NiO) (0% porosity)	0.024 at 200°C

Selecting Thermal Conductivity of Ceramics (Continued)

Ceramic	Thermal Conductivity $(cal \cdot cm^{-1} \cdot sec^{-1} \cdot K^{-1})$
Spinel (Al_2O_3 MgO) (0% porosity)	0.024 at 400°C
Thorium Dioxide (ThO_2) (0% porosity)	0.024 at room temp.
Uranium Dioxide (UO_2) (0% porosity)	0.025 at 100°C
Zirconium Mononitride (ZrN)	0.025 at 425 °C
Tantalum Diboride (TaB_2)	0.026 at room temp.
Calcium Oxide (CaO)	0.027 at 200°C
Titanium Mononitride (TiN)	0.027 at 650 °C
Hafnium Dioxide (HfO_2)	0.0273 at 25-425°C
Nickel monoxide (NiO) (0% porosity)	0.029 at 100°C
Aluminum Oxide (Al_2O_3) (single crystal)	0.029 at 800°C
Boron Nitride (BN) parallel to a axis	0.0295 at 1000°C
Aluminum Oxide (Al_2O_3)	0.03-0.064 at 200°C
Spinel (Al_2O_3 MgO) (0% porosity)	0.031 at 200°C
Boron Nitride (BN) parallel to a axis	0.0318 at 700°C
Beryllium Oxide (BeO)	0.032-0.34 at 100°C
Trisilicon tetranitride (Si_3N_4) (pressureless sintered)	0.033-0.034 at 1200 °C
Beryllium Oxide (BeO)	0.033-0.039 at 1600°C
Tantalum Diboride (TaB_2)	0.033 at 200 °C.
Beryllium Oxide (BeO)	0.033 at 1700°C
Beryllium Oxide (BeO)	0.034 at 1500°C
Spinel (Al_2O_3 MgO) (0% porosity)	0.035 at 100°C
Aluminum Oxide (Al_2O_3)	0.035 at 500°C
Trisilicon tetranitride (Si_3N_4) (pressureless sintered)	0.036-0.042 at 500 °C
Beryllium Oxide (BeO)	0.036 at 1400°C

Selecting Thermal Conductivity of Ceramics (Continued)

Ceramic	Thermal Conductivity $(cal \cdot cm^{-1} \cdot sec^{-1} \cdot K^{-1})$
Beryllium Oxide (BeO)	0.036 at 1800°C
Beryllium Oxide (BeO)	0.036 at 1900°C
Beryllium Oxide (BeO)	0.036 at 2000°C
Boron Nitride (BN) parallel to a axis	0.0362 at 300°C
Calcium Oxide (CaO)	0.037 at 100°C
Aluminum Oxide (Al_2O_3)	0.037 at 315°C
Magnesium Oxide (MgO)	0.038-0.045 at 400°C
Beryllium Oxide (BeO)	0.038-0.47 at 20°C
Trisilicon tetranitride (Si_3N_4) (pressureless sintered)	0.038 at 1000 °C
Beryllium Oxide (BeO)	0.038 at 1300°C
Aluminum Oxide (Al_2O_3)	0.04-0.069 at 100°C
Titanium Mononitride (TiN)	0.040 at 200 °C
Zirconium Mononitride (ZrN)	0.040 at 200 °C
Beryllium Oxide (BeO)	0.041-0.054 at 1200°C
Titanium Monocarbide (TiC)	0.041-0.074 at room temp.
Trisilicon tetranitride (Si_3N_4) (pressureless sintered)	0.041 at 200-750 °C
Molybdenum Disilicide ($MoSi_2$)	0.041 at 1100°C
Aluminum Nitride (AlN)	0.042 at 800°C
Beryllium Oxide (BeO)	0.043 at 1100°C
Molybdenum Disilicide ($MoSi_2$)	0.046 at 875°C
Aluminum Oxide (Al_2O_3) (single crystal)	0.047 at 300°C
Aluminum Nitride (AlN)	0.048 at 600°C
Chromium Diboride (CrB_2)	0.049-0.076 at room temp.
Silicon Carbide (SiC) (cubic, CVD)	0.049-0.080 at 600°C

Selecting Thermal Conductivity of Ceramics (Continued)

Ceramic	Thermal Conductivity $(cal \cdot cm^{-1} \cdot sec^{-1} \cdot K^{-1})$
Zirconium Monocarbide (ZrC)	0.049 at room temp.
Silicon Carbide (SiC) (cubic, CVD)	0.051 at 1000°C
Aluminum Nitride (AlN)	0.053 at 400°C
Molybdenum Disilicide (MoSi$_2$)	0.053 at 540°C
Hafnium Monocarbide (HfC)	0.053 at room temp.
Tantalum Monocarbide (TaC)	0.053 at room temp.
Zirconium Diboride (ZrB$_2$)	0.055-0.058 at room temp.
Zirconium Diboride (ZrB$_2$)	0.055-0.060 at 200 °C
Titanium Mononitride (TiN)	0.057 at 127 °C
Molybdenum Disilicide (MoSi$_2$)	0.057 at 650°C
Titanium Diboride (TiB$_2$)	0.058-0.062 at room temp.
Aluminum Oxide (Al$_2$O$_3$)	0.06 at room temp.
Beryllium Oxide (BeO)	0.060-0.093 at 800°C
Aluminum Nitride (AlN)	0.060 at 200°C
Zirconium Monocarbide (ZrC)	0.061 at 288°C
Silicon Carbide (SiC) (cubic, CVD)	0.061 at 800°C
Titanium Diboride (TiB$_2$)	0.063 at 200 °C
Boron Nitride (BN) parallel to c axis	0.0637 at 1000°C
Magnesium Oxide (MgO)	0.064-0.065 at 200°C
Boron Nitride (BN) parallel to c axis	0.0646 at 700°C
Boron Carbide (B$_4$C)	0.065-0.069 at room temp.
Zirconium Monocarbide (ZrC)	0.065 at 188°C
Boron Nitride (BN) parallel to c axis	0.0687 at 300°C
Titanium Mononitride (TiN)	0.069 at 25 °C

Selecting Thermal Conductivity of Ceramics (Continued)

Ceramic	Thermal Conductivity $(\text{cal} \cdot \text{cm}^{-1} \cdot \text{sec}^{-1} \cdot \text{K}^{-1})$
Zirconium Monocarbide (ZrC)	0.069 at 150°C
Aluminum Nitride (AlN)	0.072 at 25°C
Trisilicon tetranitride (Si_3N_4) (pressureless sintered)	0.072 at room temp.
Molybdenum Disilicide ($MoSi_2$)	0.074 at 425°C
Magnesium Oxide (MgO)	0.078-0.082 at 100°C
Zirconium Monocarbide (ZrC)	0.080 at 600°C
Silicon Carbide (SiC) (cubic, CVD)	0.0827 at 1327°C
Zirconium Monocarbide (ZrC)	0.083 at 800°C
Zirconium Monocarbide (ZrC)	0.086 at 1000°C
Beryllium Oxide (BeO)	0.089-0.1137 at 600°C
Zirconium Monocarbide (ZrC)	0.089 at 1200°C
Zirconium Monocarbide (ZrC)	0.092 at 1400°C
Zirconium Monocarbide (ZrC)	0.096 at 1600°C
Magnesium Oxide (MgO)	0.097 at room temp.
Silicon Carbide (SiC)	0.098-0.10 at 20°C
Zirconium Monocarbide (ZrC)	0.098 at 50°C
Zirconium Monocarbide (ZrC)	0.099 at 1800°C
Aluminum Oxide (Al_2O_3) (single crystal)	0.103 at 20°C
Zirconium Monocarbide (ZrC)	0.103 at 2000°C
Zirconium Monocarbide (ZrC)	0.105 at 2200°C
Molybdenum Disilicide ($MoSi_2$)	0.129 at 150°C
Titanium Mononitride (TiN)	0.136 at 2300 °C
Beryllium Oxide (BeO)	0.14-0.16 at 400°C
Silicon Carbide (SiC) (with 1 wt% Al additive)	0.143

Selecting Thermal Conductivity of Ceramics (Continued)

Ceramic	Thermal Conductivity $(cal \cdot cm^{-1} \cdot sec^{-1} \cdot K^{-1})$
Titanium Mononitride (TiN)	0.162 at 1500 °C
Boron Carbide (B_4C)	0.198 at 425 °C
Tungsten Monocarbide (WC)	0.201 at 20 °C
Tungsten Monocarbide (WC) (6% Co, 1-3μm grain size)	0.239
Tungsten Monocarbide (WC) (24% Co, 1-3μm grain size)	0.239
Tungsten Monocarbide (WC) (12% Co, 1-3μm grain size)	0.251
Tungsten Monocarbide (WC) (6% Co, 2-4μm grain size)	0.251
Tungsten Monocarbide (WC) (6% Co, 3-6μm grain size)	0.256
Silicon Carbide (SiC) (with 2 wt% BN additive)	0.263
Silicon Carbide (SiC) (cubic, CVD)	0.289 at 127°C
Silicon Carbide (SiC) (with 1 wt% B additive)	0.406
Trichromium Dicarbide (Cr_3C_2)	0.454
Silicon Carbide (SiC) (with 1 wt% Be additive)	0.621
Silicon Carbide (SiC) (with 1.6 wt% BeO additive)	0.645 at room temp.
Silicon Carbide (SiC) (with 3.2 wt% BeO additive)	0.645 at room temp.

Source: *data compiled by* J.S. Park from *No. 1 Materials Index*, Peter T.B. Shaffer, Plenum Press, New York, (1964); *Smithells Metals Reference Book*, Eric A. Brandes, ed., in association with Fulmer Research Institute Ltd. 6th ed. London, Butterworths, Boston, (1983); and *Ceramic Source*, American Ceramic Society (1986-1991)

SELECTING THERMAL CONDUCTIVITY OF CERAMICS AT TEMPERATURE

Temperature (°C)	Ceramic	Thermal Conductivity $(cal \cdot cm^{-1} \cdot sec^{-1} \cdot K^{-1})$
20	Zirconium Oxide (ZrO_2) (plasma sprayed)	0.0019-0.0031
20	Zirconium Oxide (ZrO_2) (plasma sprayed and coated with Cr_2O_3)	0.0033
20	Cordierite ($2MgO\ 2Al_2O_3\ 5SiO_2$) ($\rho=2.1g/cm^3$)	0.0043
20	Zirconium Oxide (ZrO_2) (Y_2O_3 stabilized)	0.0055
20	Zirconium Oxide (ZrO_2) (MgO stabilized)	0.0076
20	Cordierite ($2MgO\ 2Al_2O_3\ 5SiO_2$) ($\rho=2.3g/cm^3$)	0.0077
20	Hafnium Diboride (HfB_2)	0.015
20	Thorium Dioxide (ThO_2) (0% porosity)	0.024
20	Tantalum Diboride (TaB_2)	0.026
20	Beryllium Oxide (BeO)	0.038-0.47
20	Titanium Monocarbide (TiC)	0.041-0.074
20	Zirconium Monocarbide (ZrC)	0.049
20	Chromium Diboride (CrB_2)	0.049-0.076
20	Hafnium Monocarbide (HfC)	0.053
20	Tantalum Monocarbide (TaC)	0.053
20	Zirconium Diboride (ZrB_2)	0.055-0.058
20	Titanium Diboride (TiB_2)	0.058-0.062
20	Aluminum Oxide (Al_2O_3)	0.06
20	Boron Carbide (B_4C)	0.065-0.069
20	Trisilicon tetranitride (Si_3N_4) (pressureless sintered)	0.072
20	Magnesium Oxide (MgO)	0.097
20	Silicon Carbide (SiC)	0.098-0.10
20	Aluminum Oxide (Al_2O_3) (single crystal)	0.103

Selecting Thermal Conductivity of Ceramics at Temperature (Continued)

Temperature (°C)	Ceramic	Thermal Conductivity (cal $\cdot$ cm^{-1} $\cdot$ sec^{-1} $\cdot$ K^{-1})
20	Tungsten Monocarbide (WC)	0.201
20	Silicon Carbide (SiC) (with 1.6 wt% BeO additive)	0.645
20	Silicon Carbide (SiC) (with 3.2 wt% BeO additive)	0.645
25-425	Hafnium Dioxide (HfO$_2$)	0.0273
25	Titanium Mononitride (TiN)	0.069
25	Aluminum Nitride (AlN)	0.072
50	Zirconium Monocarbide (ZrC)	0.098
100	Zirconium Oxide (ZrO$_2$) (stabilized)	0.004
100	Sillimanite (Al$_2$O$_3$ SiO$_2$) (0% porosity)	0.0042
100	Zirconium Oxide (ZrO$_2$) (stabilized, 0% porosity)	0.005
100	Mullite (3Al$_2$O$_3$ 2SiO$_2$) (0% porosity)	0.0145
100	Zircon (SiO$_2$ ZrO$_2$) (0% porosity)	0.0145
100	Titanium Oxide (TiO$_2$) (0% porosity)	0.016
100	Uranium Dioxide (UO$_2$)	0.018
100	Thorium Dioxide (ThO$_2$) (0% porosity)	0.020
100	Uranium Dioxide (UO$_2$) (0% porosity)	0.025
100	Nickel monoxide (NiO) (0% porosity)	0.029
100	Beryllium Oxide (BeO)	0.032-0.34
100	Spinel (Al$_2$O$_3$ MgO) (0% porosity)	0.035
100	Calcium Oxide (CaO)	0.037
100	Aluminum Oxide (Al$_2$O$_3$)	0.04-0.069
100	Magnesium Oxide (MgO)	0.078-0.082

Selecting Thermal Conductivity of Ceramics at Temperature (Continued)

Temperature (°C)	Ceramic	Thermal Conductivity (cal $\cdot$ cm^{-1} $\cdot$ sec^{-1} $\cdot$ K^{-1})
127	Trisilicon tetranitride (Si$_3$N$_4$) (pressureless sintered)	0.022-0.072
127	Titanium Mononitride (TiN)	0.057
127	Silicon Carbide (SiC) (cubic, CVD)	0.289
150	Zirconium Monocarbide (ZrC)	0.069
150	Molybdenum Disilicide (MoSi$_2$)	0.129
188	Zirconium Monocarbide (ZrC)	0.065
200	Silicon Dioxide (SiO$_2$)	0.0025
200	Zirconium Oxide (ZrO$_2$) (stabilized, 0% porosity)	0.005
200	Titanium Oxide (TiO$_2$) (0% porosity)	0.012
200	Mullite (3Al$_2$O$_3$ 2SiO$_2$) (0% porosity)	0.013
200	Zircon (SiO$_2$ ZrO$_2$) (0% porosity)	0.0135
200	Thorium Dioxide (ThO$_2$) (0% porosity)	0.019
200	Uranium Dioxide (UO$_2$) (0% porosity)	0.020
200	Nickel monoxide (NiO) (0% porosity)	0.024
200	Calcium Oxide (CaO)	0.027
200	Aluminum Oxide (Al$_2$O$_3$)	0.03-0.064
200	Spinel (Al$_2$O$_3$ MgO) (0% porosity)	0.031
200.	Tantalum Diboride (TaB$_2$)	0.033
200	Titanium Mononitride (TiN)	0.040
200	Zirconium Mononitride (ZrN)	0.040
200-750	Trisilicon tetranitride (Si$_3$N$_4$) (pressureless sintered)	0.041
200	Zirconium Diboride (ZrB$_2$)	0.055-0.060

Selecting Thermal Conductivity of Ceramics at Temperature (Continued)

Temperature (°C)	Ceramic	Thermal Conductivity $(cal \cdot cm^{-1} \cdot sec^{-1} \cdot K^{-1})$
200	Aluminum Nitride (AlN)	0.060
200	Titanium Diboride (TiB$_2$)	0.063
200	Magnesium Oxide (MgO)	0.064-0.065
288	Zirconium Monocarbide (ZrC)	0.061
300	Cordierite (2MgO 2Al$_2$O$_3$ 5SiO$_2$) (ρ=2.1g/cm^3)	0.0041
300	Cordierite (2MgO 2Al$_2$O$_3$ 5SiO$_2$) (ρ=2.3g/cm^3)	0.0062
300	Boron Nitride (BN) parallel to a axis	0.0362
300	Aluminum Oxide (Al$_2$O$_3$) (single crystal)	0.047
300	Boron Nitride (BN) parallel to c axis	0.0687
315	Aluminum Oxide (Al$_2$O$_3$)	0.037
400	Silicon Dioxide (SiO$_2$)	0.003
400	Sillimanite (Al$_2$O$_3$ SiO$_2$) (0% porosity)	0.004
400	Zirconium Oxide (ZrO$_2$) (5-10% CaO stabilized)	0.0045
400	Zirconium Oxide (ZrO$_2$) (stabilized, 0% porosity)	0.005
400	Titanium Oxide (TiO$_2$) (0% porosity)	0.009
400	Mullite (3Al$_2$O$_3$ 2SiO$_2$) (0% porosity)	0.011
400	Uranium Dioxide (UO$_2$)	0.012
400	Zircon (SiO$_2$ ZrO$_2$) (0% porosity)	0.012
400	Thorium Dioxide (ThO$_2$) (0% porosity)	0.014
400	Uranium Dioxide (UO$_2$) (0% porosity)	0.015
400	Nickel monoxide (NiO) (0% porosity)	0.017
400	Aluminum Oxide (Al$_2$O$_3$)	0.02-0.031
400	Calcium Oxide (CaO)	0.022
400	Spinel (Al$_2$O$_3$ MgO) (0% porosity)	0.024

Selecting Thermal Conductivity of Ceramics at Temperature (Continued)

Temperature (°C)	Ceramic	Thermal Conductivity $(cal \cdot cm^{-1} \cdot sec^{-1} \cdot K^{-1})$
400	Magnesium Oxide (MgO)	0.038-0.045
400	Aluminum Nitride (AlN)	0.053
400	Beryllium Oxide (BeO)	0.14-0.16
425	Zirconium Mononitride (ZrN)	0.025
425	Boron Carbide (B$_4$C)	0.198
425	Molybdenum Disilicide (MoSi$_2$)	0.074
500	Cordierite (2MgO 2Al$_2$O$_3$ 5SiO$_2$) (ρ=2.1g/cm^3)	0.0040
500	Zirconium Oxide (ZrO$_2$) (stabilized)	0.0044
500	Cordierite (2MgO 2Al$_2$O$_3$ 5SiO$_2$) (ρ=2.3g/cm^3)	0.0055
500	Aluminum Oxide (Al$_2$O$_3$)	0.035
500	Trisilicon tetranitride (Si$_3$N$_4$) (pressureless sintered)	0.036-0.042
540	Molybdenum Disilicide (MoSi$_2$)	0.053
600	Titanium Oxide (TiO$_2$) (0% porosity)	0.008
600	Uranium Dioxide (UO$_2$)	0.008
600	Thorium Dioxide (ThO$_2$) (0% porosity)	0.010
600	Uranium Dioxide (UO$_2$) (0% porosity)	0.010
600	Mullite (3Al$_2$O$_3$ 2SiO$_2$) (0% porosity)	0.010
600	Spinel (Al$_2$O$_3$ MgO) (0% porosity)	0.019
600	Calcium Oxide (CaO)	0.020
600	Aluminum Oxide (Al$_2$O$_3$)	0.021-0.022
600	Aluminum Nitride (AlN)	0.048
600	Silicon Carbide (SiC) (cubic, CVD)	0.049-0.080
600	Zirconium Monocarbide (ZrC)	0.080
600	Beryllium Oxide (BeO)	0.089-0.1137

Selecting Thermal Conductivity of Ceramics at Temperature (Continued)

Temperature (°C)	Ceramic	Thermal Conductivity $(cal \cdot cm^{-1} \cdot sec^{-1} \cdot K^{-1})$
650	Zirconium Mononitride (ZrN)	0.018
650	Titanium Mononitride (TiN)	0.027
650	Molybdenum Disilicide (MoSi$_2$)	0.057
700	Uranium Dioxide (UO$_2$)	0.008
700	Boron Nitride (BN) parallel to a axis	0.0318
700	Boron Nitride (BN) parallel to c axis	0.0646
800	Zirconium Oxide (ZrO$_2$) (plasma sprayed)	0.0019-0.0022
800	Zirconium Oxide (ZrO$_2$) (plasma sprayed and coated with Cr$_2$O$_3$)	0.0033
800	Sillimanite (Al$_2$O$_3$ SiO$_2$) (0% porosity)	0.0035
800	Cordierite (2MgO 2Al$_2$O$_3$ 5SiO$_2$) (ρ=2.1g/cm^3)	0.0038
800	Silicon Dioxide (SiO$_2$)	0.004
800	Zirconium Oxide (ZrO$_2$) (5-10% CaO stabilized)	0.0049
800	Zirconium Oxide (ZrO$_2$) (Y$_2$O$_3$ stabilized)	0.0053
800	Zirconium Oxide (ZrO$_2$) (stabilized, 0% porosity)	0.0055
800	Cordierite (2MgO 2Al$_2$O$_3$ 5SiO$_2$) (ρ=2.3g/cm^3)	0.0055
800	Zirconium Oxide (ZrO$_2$) (MgO stabilized)	0.0057
800	Thorium Dioxide (ThO$_2$) (0% porosity)	0.008
800	Titanium Oxide (TiO$_2$) (0% porosity)	0.008
800	Uranium Dioxide (UO$_2$) (0% porosity)	0.009
800	Mullite (3Al$_2$O$_3$ 2SiO$_2$) (0% porosity)	0.0095
800	Zircon (SiO$_2$ ZrO$_2$) (0% porosity)	0.010

Selecting Thermal Conductivity of Ceramics at Temperature (Continued)

Temperature (°C)	Ceramic	Thermal Conductivity $(cal \cdot cm^{-1} \cdot sec^{-1} \cdot K^{-1})$
800	Nickel monoxide (NiO) (0% porosity)	0.012
800	Spinel (Al_2O_3 MgO) (0% porosity)	0.015
800	Aluminum Oxide (Al_2O_3)	0.015-0.017
800	Calcium Oxide (CaO)	0.019
800	Magnesium Oxide (MgO)	0.0198-0.026
800	Aluminum Oxide (Al_2O_3) (single crystal)	0.029
800	Aluminum Nitride (AlN)	0.042
800	Beryllium Oxide (BeO)	0.060-0.093
800	Silicon Carbide (SiC) (cubic, CVD)	0.061
800	Zirconium Monocarbide (ZrC)	0.083
875	Zirconium Mononitride (ZrN)	0.016
875	Molybdenum Disilicide ($MoSi_2$)	0.046
1000	Zirconium Oxide (ZrO_2) (stabilized)	0.0048-0.0055
1000	Uranium Dioxide (UO_2)	0.006
1000	Thorium Dioxide (ThO_2) (0% porosity)	0.007-0.0074
1000	Titanium Oxide (TiO_2) (0% porosity)	0.008
1000	Uranium Dioxide (UO_2) (0% porosity)	0.008
1000	Mullite ($3Al_2O_3$ $2SiO_2$) (0% porosity)	0.009
1000	Nickel monoxide (NiO) (0% porosity)	0.011
1000	Spinel (Al_2O_3 MgO) (0% porosity)	0.013-0.0138
1000	Titanium Monocarbide (TiC)	0.0135
1000	Aluminum Oxide (Al_2O_3)	0.014-0.016
1000	Magnesium Oxide (MgO)	0.016-0.020
1000	Calcium Oxide (CaO)	0.0186-0.019

Selecting Thermal Conductivity of Ceramics at Temperature (Continued)

Temperature (°C)	Ceramic	Thermal Conductivity $(\text{cal} \cdot \text{cm}^{-1} \cdot \text{sec}^{-1} \cdot \text{K}^{-1})$
1000	Titanium Mononitride (TiN)	0.020
1000	Boron Nitride (BN) parallel to a axis	0.0295
1000	Trisilicon tetranitride (Si_3N_4) (pressureless sintered)	0.038
1000	Silicon Carbide (SiC) (cubic, CVD)	0.051
1000	Boron Nitride (BN) parallel to c axis	0.0637
1000	Zirconium Monocarbide (ZrC)	0.086
1100	Zirconium Mononitride (ZrN)	0.015
1100	Molybdenum Disilicide ($MoSi_2$)	0.041
1100	Beryllium Oxide (BeO)	0.043
1200	Sillimanite (Al_2O_3 SiO_2) (0% porosity)	0.0035
1200	Zirconium Oxide (ZrO_2) (stabilized)	0.0049-0.0050
1200	Silicon Dioxide (SiO_2)	0.005
1200	Zirconium Oxide (ZrO_2) (5-10% CaO stabilized)	0.0057
1200	Uranium Dioxide (UO_2)	0.006
1200	Zirconium Oxide (ZrO_2) (stabilized, 0% porosity)	0.006
1200	Thorium Dioxide (ThO_2) (0% porosity)	0.006-0.0076
1200	Titanium Oxide (TiO_2) (0% porosity)	0.008
1200	Mullite ($3Al_2O_3$ $2SiO_2$) (0% porosity)	0.009
1200	Zircon (SiO_2 ZrO_2) (0% porosity)	0.0095
1200	Spinel (Al_2O_3 MgO) (0% porosity)	0.013
1200	Aluminum Oxide (Al_2O_3)	0.013-0.015
1200	Magnesium Oxide (MgO)	0.0139-0.0148
1200	Trisilicon tetranitride (Si_3N_4) (pressureless sintered)	0.033-0.034
1200	Beryllium Oxide (BeO)	0.041-0.054
1200	Zirconium Monocarbide (ZrC)	0.089

Selecting Thermal Conductivity of Ceramics at Temperature (Continued)

Temperature (°C)	Ceramic	Thermal Conductivity (cal $\cdot$ cm^{-1} $\cdot$ sec^{-1} $\cdot$ K^{-1})
1250	Silicon Carbide (SiC) (cubic, CVD)	0.0059
1300	Beryllium Oxide (BeO)	0.038
1327	Silicon Carbide (SiC) (cubic, CVD)	0.0827
1400	Thorium Dioxide (ThO$_2$) (0% porosity)	0.006
1400	Zirconium Oxide (ZrO$_2$) (stabilized, 0% porosity)	0.0065
1400	Mullite (3Al$_2$O$_3$ 2SiO$_2$) (0% porosity)	0.009
1400	Zircon (SiO$_2$ ZrO$_2$) (0% porosity)	0.0095
1400	Magnesium Oxide (MgO)	0.012-0.014
1400	Aluminum Oxide (Al$_2$O$_3$)	0.013
1400	Beryllium Oxide (BeO)	0.036
1400	Zirconium Monocarbide (ZrC)	0.092
1500	Sillimanite (Al$_2$O$_3$ SiO$_2$) (0% porosity)	0.003
1500	Beryllium Oxide (BeO)	0.034
1500	Titanium Mononitride (TiN)	0.162
1530	Silicon Carbide (SiC) (cubic, CVD)	0.0032
1600	Silicon Dioxide (SiO$_2$)	0.006
1600	Magnesium Oxide (MgO)	0.0108-0.016
1600	Aluminum Oxide (Al$_2$O$_3$)	0.014
1600	Beryllium Oxide (BeO)	0.033-0.039
1600	Zirconium Monocarbide (ZrC)	0.096
1700	Beryllium Oxide (BeO)	0.033

Selecting Thermal Conductivity of Ceramics at Temperature (Continued)

Temperature (°C)	Ceramic	Thermal Conductivity $(\text{cal} \cdot \text{cm}^{-1} \cdot \text{sec}^{-1} \cdot \text{K}^{-1})$
1800	Magnesium Oxide (MgO)	0.0096-0.0191
1800	Aluminum Oxide (Al_2O_3)	0.017
1800	Beryllium Oxide (BeO)	0.036
1800	Zirconium Monocarbide (ZrC)	0.099
1900	Beryllium Oxide (BeO)	0.036
2000	Beryllium Oxide (BeO)	0.036
2000	Zirconium Monocarbide (ZrC)	0.103
2200	Zirconium Monocarbide (ZrC)	0.105
2300	Titanium Mononitride (TiN)	0.136

Source: *data compiled by* J.S. Park from *No. 1 Materials Index*, Peter T.B. Shaffer, Plenum Press, New York, (1964); *Smithells Metals Reference Book*, Eric A. Brandes, ed., in association with Fulmer Research Institute Ltd. 6th ed. London, Butterworths, Boston, (1983); and *Ceramic Source*, American Ceramic Society (1986-1991)

SELECTING THERMAL EXPANSION OF CERAMICS

Ceramic	Thermal Expansion ($^{\circ}C^{-1}$)
Hafnium Dioxide (HfO_2) monoclinic, parallel to b axis	0 for 28–262°C
Silicon Dioxide (SiO_2) Vitreous	0.5×10^{-6} for 20–1250°C
Silicon Dioxide (SiO_2) Vitreous	0.527×10^{-6} for 25–500°C
Silicon Dioxide (SiO_2) Vitreous	0.564×10^{-6} for 25–1000°C
Boron Nitride (BN) parallel to a axis	0.59×10^{-6} for 25 to 350°C
Cordierite ($2MgO\ 2Al_2O_3\ 5SiO_2$) ($\rho=1.8g/cm^3$)	0.6×10^{-6} for 25 to 400°C
Boron Nitride (BN) parallel to a axis	0.77×10^{-6} for 25 to 1000°C
Boron Nitride (BN) parallel to a axis	0.89×10^{-6} for 25 to 700°C
Hafnium Dioxide (HfO_2) monoclinic, parallel to a axis	0.9×10^{-6} for 28–494°C
Zirconium Oxide (ZrO_2) tetragonal, parallel to b axis	1.1×10^{-6} for 27 to 759°C
Hafnium Dioxide (HfO_2) monoclinic, parallel to a axis	1.3×10^{-6} for 28–697°C
Hafnium Dioxide (HfO_2) — tetragonal polycrystalline	1.31×10^{-6} for 25–1700°C
Hafnium Dioxide (HfO_2) monoclinic, parallel to a axis	1.4×10^{-6} for 28–903°C
Cordierite ($2MgO\ 2Al_2O_3\ 5SiO_2$) ($\rho=1.8g/cm^3$)	1.5×10^{-6} for 25 to 700°C
Zirconium Oxide (ZrO_2) tetragonal, parallel to b axis	1.5×10^{-6} for 27 to 964°C
Aluminum Oxide (Al_2O_3) perpendicular to c axis	1.65×10^{-6} for 0 to –273°C
Cordierite ($2MgO\ 2Al_2O_3\ 5SiO_2$) ($\rho=1.8g/cm^3$)	1.7×10^{-6} for 25 to 900°C
Aluminum Oxide (Al_2O_3) — polycrystalline	1.89×10^{-6} for 0 to –273°C
Zirconium Oxide (ZrO_2) tetragonal, parallel to b axis	1.9×10^{-6} for 27 to 1110°C
Aluminum Oxide (Al_2O_3) parallel to c axis	1.95×10^{-6} for 0 to –273°C
Zirconium Oxide (ZrO_2) tetragonal, parallel to b axis	2×10^{-6} for 27 to 504°C

Selecting Thermal Expansion of Ceramics (Continued)

Ceramic	Thermal Expansion $(^{\circ}C^{-1})$
Hafnium Dioxide (HfO_2) monoclinic, parallel to a axis	2.1×10^{-6} for 28–1098°C
Trisilicon Tetranitride (Si_3N_4)	2.11×10^{-6} for 25 to 500°C
Cordierite ($2MgO\ 2Al_2O_3\ 5SiO_2$) (ρ=2.1g/cm^3)	2.2×10^{-6} for 25 to 400°C
Cordierite ($2MgO\ 2Al_2O_3\ 5SiO_2$) (ρ=2.3g/cm^3)	2.3×10^{-6} for 25 to 400°C
Beryllium Oxide (BeO) — polycrystalline	2.4×10^{-6} for 25–200°C
Aluminum Oxide (Al_2O_3) perpendicular to c axis	2.55×10^{-6} for 0 to –173°C
Cordierite ($2MgO\ 2Al_2O_3\ 5SiO_2$) (ρ=2.51g/cm^3)	2.7×10^{-6} for 25 to 1100°C
Cordierite ($2MgO\ 2Al_2O_3\ 5SiO_2$) (ρ=2.1g/cm^3)	2.8×10^{-6} for 25 to 700°C
Cordierite ($2MgO\ 2Al_2O_3\ 5SiO_2$) (ρ=2.1g/cm^3)	2.8×10^{-6} for 25 to 900°C
Trisilicon Tetranitride (Si_3N_4)	2.87×10^{-6} for 25 to 1000°C
Trisilicon Tetranitride (Si_3N_4) (reaction sintered)	2.9×10^{-6} for 20 to 1000°C
Aluminum Oxide (Al_2O_3) — polycrystalline	2.91×10^{-6} for 0 to –173°C
Zirconium Oxide (ZrO_2) tetragonal, parallel to b axis	3×10^{-6} for 27 to 264°C
Trisilicon Tetranitride (Si_3N_4) (hot pressed)	$3–3.9 \times 10^{-6}$ for 20 to 1000°C
Aluminum Oxide (Al_2O_3) parallel to c axis	3.01×10^{-6} for 0 to –173°C
Hafnium Dioxide (HfO_2) — tetragonal polycrystalline	3.03×10^{-6} for 25–2000°C
Cordierite ($2MgO\ 2Al_2O_3\ 5SiO_2$) (ρ=2.3g/cm^3)	3.3×10^{-6} for 25 to 700°C
Trisilicon Tetranitride (Si_3N_4) (sintered)	3.5×10^{-6} for 20 to 1000°C
Trisilicon Tetranitride (Si_3N_4)	3.66×10^{-6} for 25 to 1500°C
Thorium Dioxide (ThO_2)	3.67×10^{-6} for 0 to –273°C
Cordierite ($2MgO\ 2Al_2O_3\ 5SiO_2$) (ρ=2.3g/cm^3)	3.7×10^{-6} for 25 to 900°C
Trisilicon Tetranitride (Si_3N_4) (pressureless sintered)	3.7×10^{-6} for 40 to 1000°C
Cordierite ($2MgO\ 2Al_2O_3\ 5SiO_2$) (glass)	$3.7–3.8 \times 10^{-6}$ for 25 to 900°C
Aluminum Oxide (Al_2O_3) perpendicular to c axis	3.75×10^{-6} for 0 to –73°C

Selecting Thermal Expansion of Ceramics (Continued)

Ceramic	Thermal Expansion $(°C^{-1})$
Zircon (SiO$_2$ ZrO$_2$)	3.79×10^{-6} for 25 to 500°C
Zirconium Oxide (ZrO$_2$) — tetragonal	4.0×10^{-6} for 0 to 500°C
Aluminum Nitride (AlN)	4.03×10^{-6} for 25 to 200°C
Aluminum Oxide (Al$_2$O$_3$) — polycrystalline	4.10×10^{-6} for 0 to –73°C
Aluminum Oxide (Al$_2$O$_3$) parallel to c axis	4.39×10^{-6} for 0 to –73°C
Tungsten Monocarbide (WC)	4.42×10^{-6} for 25–500°C
Mullite (3Al$_2$O$_3$ 2SiO$_2$)	4.5×10^{-6} for 20 to 1325°C
Boron Carbide (B$_4$C)	4.5×10^{-6} for room temp.–800°C
Chromium Diboride (CrB$_2$)	$4.6–11.1 \times 10^{-6}$ for 20–1000°C
Titanium Diboride (TiB$_2$)	$4.6–8.1 \times 10^{-6}$
Zircon (SiO$_2$ ZrO$_2$)	4.62×10^{-6} for 25 to 1000°C
Mullite (3Al$_2$O$_3$ 2SiO$_2$)	4.63×10^{-6} for 25 to 500°C
Silicon Carbide (SiC)	4.63×10^{-6} for 25–500°C
Silicon Carbide (SiC)	4.70×10^{-6} for 0–1700°C
Silicon Carbide (SiC)	4.70×10^{-6} for 20–1500°C
Aluminum Oxide (Al$_2$O$_3$) perpendicular to c axis	4.78×10^{-6} for 0 to 27°C
Boron Carbide (B$_4$C)	4.78×10^{-6} for 25–500°C
Aluminum Nitride (AlN)	4.83×10^{-6} for 25 to 600°C
Aluminum Nitride (AlN)	4.84×10^{-6} for 25 to 500°C
Tungsten Monocarbide (WC)	$4.84–4.92 \times 10^{-6}$ for 25–1000°C
Zirconium Oxide (ZrO$_2$) — tetragonal	5.0×10^{-6} for 0 to 1400°C
Mullite (3Al$_2$O$_3$ 2SiO$_2$)	5.0×10^{-6} for 25 to 800°C
Tantalum Diboride (TaB$_2$)	5.1×10^{-6} at room temp.
Silicon Carbide (SiC)	5.12×10^{-6} for 25–1000°C

Selecting Thermal Expansion of Ceramics (Continued)

Ceramic	Thermal Expansion ($°C^{-1}$)
Mullite ($3Al_2O_3\ 2SiO_2$)	5.13×10^{-6} for 25 to 1000°C
Aluminum Oxide (Al_2O_3) parallel to c axis	5.31×10^{-6} for 0 to 27°C
Thorium Dioxide (ThO_2)	5.32×10^{-6} for 0 to −173°C
Tungsten Monocarbide (WC)	$5.35–5.8 \times 10^{-6}$ for 25–1500°C
Hafnium Dioxide (HfO_2) — monoclinic polycrystalline	5.47×10^{-6} for 25–500°C
Silicon Carbide (SiC)	5.48×10^{-6} for 25–1500°C
Zircon ($SiO_2\ ZrO_2$)	5.5×10^{-6} for 20 to 1200°C
Hafnium Diboride (HfB_2)	$5.5 –5.54 \times 10^{-6}$ for 20 to1000°C
Zirconium Oxide (ZrO_2) — tetragonal	$5.5–5.58 \times 10^{-6}$ for 20 to 1200°C
Zirconium Diboride (ZrB_2)	$5.5–6.57 \times 10^{-6}\ °C$ for 25–1000°C
Aluminum Oxide (Al_2O_3) perpendicular to c axis	5.51×10^{-6} for 0 to 127°C
Boron Carbide (B_4C)	5.54×10^{-6} for 25–1000°C
Aluminum Nitride (AlN)	$5.54–5.64 \times 10^{-6}$ for 25 to 1000°C
Aluminum Oxide (Al_2O_3) — polycrystalline	5.60×10^{-6} for 0 to 27°C
Mullite ($3Al_2O_3\ 2SiO_2$)	5.62×10^{-6} for 20 to 1500°C
Zircon ($SiO_2\ ZrO_2$)	5.63×10^{-6} for 20 to 1500°C
Zirconium Diboride (ZrB_2)	5.69×10^{-6} for 25–500°C
Silicon Carbide (SiC)	5.77×10^{-6} for 25–2000°C
Hafnium Dioxide (HfO_2) — monoclinic polycrystalline	5.8×10^{-6} for 25–1300°C
Tungsten Monocarbide (WC)	$5.82–7.4 \times 10^{-6}$ for 25–2000°C
Hafnium Dioxide (HfO_2) — monoclinic polycrystalline	5.85×10^{-6} for 25–1000°C
Silicon Carbide (SiC)	5.94×10^{-6} for 25–2500°C
Boron Carbide (B_4C)	6.02×10^{-6} for 25–1500°C
Aluminum Oxide (Al_2O_3) — polycrystalline	6.03×10^{-6} for 0 to 127°C

Selecting Thermal Expansion of Ceramics (Continued)

Ceramic	Thermal Expansion $(°C^{-1})$
Aluminum Nitride (AlN)	6.09×10^{-6} for 25 to 1350°C
Aluminum Oxide (Al$_2$O$_3$) perpendicular to c axis	6.10×10^{-6} for 0 to 227°C
Zirconium Monocarbide (ZrC)	$6.10x \times 10^{-6}$ for 25–500°C
Zirconium Monocarbide (ZrC)	$6.10–6.73 \times 10^{-6}$ for 25–650°C
Zirconium Mononitride (ZrN)	6.13×10^{-6} for 20–450°C
Hafnium Dioxide (HfO$_2$) monoclinic, parallel to a axis	$6.2x10^{-6}$ for 28–494°C
Hafnium Monocarbide (HfC)	6.25×10^{-6} for 25–1000°C
Aluminum Oxide (Al$_2$O$_3$) parallel to c axis	6.26×10^{-6} for 0 to 127°C
Hafnium Monocarbide (HfC)	$6.27–6.59 \times 10^{-6}$ for 25–650°C
Tantalum Monocarbide (TaC)	$6.29–6.32 \times 10^{-6}$ for 25–500°C
Beryllium Oxide (BeO) parallel to c axis	6.3×10^{-6} for 28 to 252°C
Hafnium Dioxide (HfO$_2$) — monoclinic polycrystalline	6.30×10^{-6} for 25–1500°C
Beryllium Oxide (BeO) — polycrystalline	$6.3–6.4 \times 10^{-6}$ for 25–300°C
Zirconium Monocarbide (ZrC)	$6.32x \times 10^{-6}$ for 0–750°C
Hafnium Dioxide (HfO$_2$) — monoclinic polycrystalline	6.45×10^{-6} for 20–1700°C
Zirconium Monocarbide (ZrC)	$6.46–6.66x \times 10^{-6}$ for 0–1000°C
Thorium Dioxide (ThO$_2$)	6.47×10^{-6} for 0 to –73°C
Tantalum Monocarbide (TaC)	6.50×10^{-6} for 0–1000°C
Aluminum Oxide (Al$_2$O$_3$) perpendicular to c axis	6.52×10^{-6} for 0 to 327°C
Titanium Monocarbide (TiC)	$6.52–7.15 \times 10^{-6}$ for 25–500°C
Zirconium Oxide (ZrO$_2$) — monoclinic	6.53×10^{-6} for 25 to 500°C
Boron Carbide (B$_4$C)	6.53×10^{-6} for 25–2000°C
Aluminum Oxide (Al$_2$O$_3$) — polycrystalline	6.55×10^{-6} for 0 to 227°C
Zirconium Monocarbide (ZrC)	$6.56x \times 10^{-6}$ for 25–1000°C

Selecting Thermal Expansion of Ceramics (Continued)

Ceramic	Thermal Expansion $(^{\circ}C^{-1})$
Sillimanite (Al_2O_3 SiO_2)	6.58×10^{-6} at 20°C
Tantalum Monocarbide (TaC)	6.64×10^{-6} for 0–1200°C
Zirconium Monocarbide (ZrC)	6.65×10^{-6} for 25–800°C
Tantalum Monocarbide (TaC)	6.67×10^{-6} for 25–1000°C
Zirconium Monocarbide (ZrC)	6.68×10^{-6} for 0–1275°C
Beryllium Oxide (BeO) parallel to c axis	6.7×10^{-6} for 28 to 474°C
Hafnium Dioxide (HfO_2) monoclinic, parallel to a axis	6.7×10^{-6} for 28–697°C
Zirconium Oxide (ZrO_2) tetragonal, parallel to a axis	6.8×10^{-6} for 27 to 759°C
Hafnium Dioxide (HfO_2) monoclinic, parallel to a axis	6.8×10^{-6} for 28–262°C
Beryllium Oxide (BeO) average for (2a+c)/3	6.83×10^{-6} for 28 to 252°C
Zirconium Monocarbide (ZrC)	6.83×10^{-6} for 0–1525°C
Aluminum Oxide (Al_2O_3) parallel to c axis	6.86×10^{-6} for 0 to 227°C
Aluminum Oxide (Al_2O_3) perpendicular to c axis	6.88×10^{-6} for 0 to 427°C
Aluminum Oxide (Al_2O_3) — polycrystalline	6.93×10^{-6} for 0 to 327°C
Zirconium Diboride (ZrB_2)	6.98×10^{-6} for 20–1500°C
Zirconium Monocarbide (ZrC)	6.98×10^{-6} for 0–1775°C
Zirconium Mononitride (ZrN)	7.03×10^{-6} for 20–680°C
Zirconium Monocarbide (ZrC)	7.06×10^{-6} for 25–1500°C
Titanium Monocarbide (TiC)	7.08×10^{-6} for 0–750°C
Boron Carbide (B_4C)	7.08×10^{-6} for 25–2500°C
Beryllium Oxide (BeO) perpendicular to c axis	7.1×10^{-6} for 28 to 252°C
Tantalum Monocarbide (TaC)	7.12×10^{-6} for 25–1500°C
Aluminum Oxide (Al_2O_3) perpendicular to c axis	7.15×10^{-6} for 0 to 527°C
Boron Nitride (BN) parallel to c axis	7.15×10^{-6} for 25 to 1000°C
Titanium Monocarbide (TiC)	$7.18–7.45 \times 10^{-6}$ for 25–750°C

Selecting Thermal Expansion of Ceramics (Continued)

Ceramic	Thermal Expansion ($°C^{-1}$)
Zirconium Oxide (ZrO_2) — tetragonal	7.2×10^{-6} for -10 to $1000°C$
Aluminum Oxide (Al_2O_3) — polycrystalline	7.24×10^{-6} for 0 to $427°C$
Aluminum Oxide (Al_2O_3) parallel to c axis	7.31×10^{-6} for 0 to $327°C$
Aluminum Oxide (Al_2O_3) perpendicular to c axis	7.35×10^{-6} for 0 to $627°C$
Titanium Monocarbide (TiC)	$7.40–8.82 \times 10^{-6}$ for $25–1000°C$
Beryllium Oxide (BeO) average for $(2a+c)/3$	7.43×10^{-6} for 28 to $474°C$
Zirconium Oxide (ZrO_2) tetragonal, parallel to a axis	7.5×10^{-6} for 27 to $504°C$
Aluminum Oxide (Al_2O_3) — polycrystalline	7.50×10^{-6} for 0 to $527°C$
Hafnium Dioxide (HfO_2) monoclinic, parallel to a axis	7.5×10^{-6} for $28–903°C$
Aluminum Oxide (Al_2O_3) perpendicular to c axis	7.53×10^{-6} for 0 to $727°C$
Zirconium Oxide (ZrO_2) — monoclinic	7.59×10^{-6} for 25 to $1000°C$
Beryllium Oxide (BeO) — polycrystalline	7.59×10^{-6} for $25–500°C$
Tantalum Monocarbide (TaC)	7.64×10^{-6} for $25–2000°C$
Zirconium Monocarbide (ZrC)	7.65×10^{-6} for $25–650°C$
Aluminum Oxide (Al_2O_3) perpendicular to c axis	7.67×10^{-6} for 0 to $827°C$
Aluminum Oxide (Al_2O_3) parallel to c axis	7.68×10^{-6} for 0 to $427°C$
Aluminum Oxide (Al_2O_3) — polycrystalline	7.69×10^{-6} for 0 to $627°C$
Zirconium Oxide (ZrO_2) — monoclinic	7.72×10^{-6} for 25 to $1050°C$
Spinel (Al_2O_3 MgO)	7.79×10^{-6} for 25 to $500°C$
Molybdenum Disilicide ($MoSi_2$)	7.79×10^{-6} for $25–500°C$
Tungsten Disilicide (WSi_2)	7.79×10^{-6} for $25–500°C$
Titanium Oxide (TiO_2) — polycrystalline	7.8×10^{-6} for $20–600°C$
Thorium Dioxide (ThO_2)	7.8×10^{-6} for 27 to $223°C$
Zirconium Oxide (ZrO_2) tetragonal, parallel to a axis	7.8×10^{-6} for 27 to $964°C$

Selecting Thermal Expansion of Ceramics (Continued)

Ceramic	Thermal Expansion $(^{\circ}C^{-1})$
Beryllium Oxide (BeO) perpendicular to c axis	7.8×10^{-6} for 28 to 474°C
Beryllium Oxide (BeO) parallel to c axis	7.8×10^{-6} for 28 to 749°C
Aluminum Oxide (Al$_2$O$_3$) perpendicular to c axis	7.80×10^{-6} for 0 to 927°C
Aluminum Oxide (Al$_2$O$_3$) — polycrystalline	7.83×10^{-6} for 0 to 727°C
Titanium Monocarbide (TiC)	$7.85–7.86 \times 10^{-6}$ for 0–1000°C
Aluminum Oxide (Al$_2$O$_3$) perpendicular to c axis	7.88×10^{-6} for 0 to 1027°C
Titanium Oxide (TiO$_2$) perpendicular to a axis	7.9×10^{-6} for 26 to 240°C
Titanium Monocarbide (TiC)	7.90×10^{-6} for 0–2500°C
Hafnium Dioxide (HfO$_2$) monoclinic, parallel to a axis	7.9×10^{-6} for 28–1098°C
Aluminum Oxide (Al$_2$O$_3$) parallel to c axis	7.96×10^{-6} for 0 to 527°C
Aluminum Oxide (Al$_2$O$_3$) perpendicular to c axis	7.96×10^{-6} for 0 to 1127°C
Aluminum Oxide (Al$_2$O$_3$) — polycrystalline	7.97×10^{-6} for 0 to 827°C
Zirconium Oxide (ZrO$_2$) — monoclinic	8.0×10^{-6} for 25 to 1080°C
Trichromium Dicarbide (Cr$_3$C$_2$)	8.00×10^{-6} for 25–500°C
Titanium Monocarbide (TiC)	8.02×10^{-6} for 0–1275°C
Aluminum Oxide (Al$_2$O$_3$) perpendicular to c axis	8.05×10^{-6} for 0 to 1227°C
Thorium Dioxide (ThO$_2$)	8.06×10^{-6} for 0 to 127°C
Boron Nitride (BN) parallel to c axis	8.06×10^{-6} for 25 to 700°C
Aluminum Oxide (Al$_2$O$_3$) — polycrystalline	8.08×10^{-6} for 0 to 927°C
Titanium Oxide (TiO$_2$) perpendicular to a axis	8.1×10^{-6} for 26 to 670°C
Thorium Dioxide (ThO$_2$)	8.10×10^{-6} for 0 to 27°C
Aluminum Oxide (Al$_2$O$_3$) perpendicular to c axis	8.12×10^{-6} for 0 to 1327°C
Titanium Monocarbide (TiC)	$8.15–9.45 \times 10^{-6}$ for 25–1500°C
Aluminum Oxide (Al$_2$O$_3$) perpendicular to c axis	8.16×10^{-6} for 0 to 1427°C

Selecting Thermal Expansion of Ceramics (Continued)

Ceramic	Thermal Expansion ($°C^{-1}$)
Aluminum Oxide (Al_2O_3) — polycrystalline	8.18×10^{-6} for 0 to 1027°C
Aluminum Oxide (Al_2O_3) parallel to c axis	8.19×10^{-6} for 0 to 627°C
Titanium Oxide (TiO_2) perpendicular to a axis	8.2×10^{-6} for 26 to 455°C
Titanium Oxide (TiO_2) perpendicular to a axis	8.2×10^{-6} for 26 to 940°C
Beryllium Oxide (BeO) parallel to c axis	8.2×10^{-6} for 28 to 872°C
Aluminum Oxide (Al_2O_3) perpendicular to c axis	8.20×10^{-6} for 0 to 1527°C
Tungsten Disilicide (WSi_2)	8.21×10^{-6} for 0–1000°C
Cerium Dioxide (CeO_2)	8.22×10^{-6} for 25–500°C
Titanium Oxide (TiO_2) — polycrystalline	8.22×10^{-6} for 25–500°C
Aluminum Oxide (Al_2O_3) — polycrystalline	8.25×10^{-6} for 0 to 1127°C
Aluminum Oxide (Al_2O_3) perpendicular to c axis	8.26×10^{-6} for 0 to 1627°C
Titanium Monocarbide (TiC)	8.26×10^{-6} for 0–1525°C
Beryllium Oxide (BeO) average for (2a+c)/3	8.27×10^{-6} for 28 to 749°C
Titanium Monocarbide (TiC)	8.29×10^{-6} for 0–1400°C
Titanium Oxide (TiO_2) perpendicular to a axis	8.3×10^{-6} for 26 to 1110°C
Aluminum Oxide (Al_2O_3) perpendicular to c axis	8.30×10^{-6} for 0 to 1727°C
Thorium Dioxide (ThO_2)	8.31×10^{-6} for 0 to 227°C
Tungsten Disilicide (WSi_2)	8.31×10^{-6} for 25–1000°C
Aluminum Oxide (Al_2O_3) — polycrystalline	8.32×10^{-6} for 0 to 1227°C
Aluminum Oxide (Al_2O_3) parallel to c axis	8.38×10^{-6} for 0 to 727°C
Aluminum Oxide (Al_2O_3) — polycrystalline	8.39×10^{-6} for 0 to 1327°C
Zirconium Oxide (ZrO_2) tetragonal, parallel to a axis	8.4×10^{-6} for 27 to 264°C
Titanium Monocarbide (TiC)	8.40×10^{-6} for 0–1775°C
Tantalum Monocarbide (TaC)	8.40×10^{-6} for 25–2500°C

Selecting Thermal Expansion of Ceramics (Continued)

Ceramic	Thermal Expansion $(°C^{-1})$
Beryllium Oxide (BeO) — polycrystalline	$8.4–8.5 \times 10^{-6}$ for 25–800°C
Molybdenum Disilicide (MoSi$_2$)	8.41×10^{-6} for 0–1000°C
Spinel (Al$_2$O$_3$ MgO)	8.41×10^{-6} for 25 to 1000°C
Dichromium Trioxide (Cr$_2$O$_3$)	8.43×10^{-6} for 25–500°C
Aluminum Oxide (Al$_2$O$_3$) — polycrystalline	8.45×10^{-6} for 0 to 1427°C
Aluminum Oxide (Al$_2$O$_3$) — polycrystalline	8.49×10^{-6} for 0 to 1527°C
Beryllium Oxide (BeO) perpendicular to c axis	8.5×10^{-6} for 28 to 749°C
Molybdenum Disilicide (MoSi$_2$)	8.51×10^{-6} for 25–1000°C
Aluminum Oxide (Al$_2$O$_3$) parallel to c axis	8.52×10^{-6} for 0 to 827°C
Thorium Dioxide (ThO$_2$)	8.53×10^{-6} for 0 to 327°C
Aluminum Oxide (Al$_2$O$_3$) — polycrystalline	8.53×10^{-6} for 0 to 1627°C
Titanium Oxide (TiO$_2$) average for (2a+c)/3	8.53×10^{-6} for 26 to 240°C
Molybdenum Disilicide (MoSi$_2$)	8.56×10^{-6} for 0–1400°C
Aluminum Oxide (Al$_2$O$_3$) — polycrystalline	8.58×10^{-6} for 0 to 1727°C
Dichromium Trioxide (Cr$_2$O$_3$)	8.62×10^{-6} for 25–1000°C
Thorium Dioxide (ThO$_2$)	8.63×10^{-6} for 25 to 500°C
Zirconium Oxide (ZrO$_2$) — tetragonal	8.64×10^{-6} for –20 to 600°C
Aluminum Oxide (Al$_2$O$_3$) parallel to c axis	8.65×10^{-6} for 0 to 927°C
Thorium Dioxide (ThO$_2$)	8.7×10^{-6} for 27 to 498°C
Zirconium Oxide (ZrO$_2$) tetragonal, parallel to a axis	8.7×10^{-6} for 27 to 1110°C
Thorium Dioxide (ThO$_2$)	8.71×10^{-6} for 0 to 427°C
Aluminum Oxide (Al$_2$O$_3$) parallel to c axis	8.75×10^{-6} for 0 to 1027°C
Trichromium Dicarbide (Cr$_3$C$_2$)	8.8×10^{-6} for 25–120°C
Tungsten Disilicide (WSi$_2$)	8.81×10^{-6} for 0–1400°C

Selecting Thermal Expansion of Ceramics (Continued)

Ceramic	Thermal Expansion ($°C^{-1}$)
Titanium Monocarbide (TiC)	8.81×10^{-6} for 25–2000°C
Dichromium Trioxide (Cr_2O_3)	8.82×10^{-6} for 25–1500°C
Titanium Oxide (TiO_2) — polycrystalline	8.83×10^{-6} for 25–1000°C
Aluminum Oxide (Al_2O_3) parallel to c axis	8.84×10^{-6} for 0 to 1127°C
Thorium Dioxide (ThO_2)	8.87×10^{-6} for 0 to 527°C
Beryllium Oxide (BeO) average for (2a+c)/3	8.87×10^{-6} for 28 to 872°C
Thorium Dioxide (ThO_2)	8.9×10^{-6} for 27 to 755°C
Beryllium Oxide (BeO) parallel to c axis	8.9×10^{-6} for 28 to 1132°C
Aluminum Oxide (Al_2O_3) parallel to c axis	8.92×10^{-6} for 0 to 1227°C
Cerium Dioxide (CeO_2)	8.92×10^{-6} for 25–1000°C
Titanium Oxide (TiO_2) average for (2a+c)/3	8.93×10^{-6} for 26 to 670°C
Thorium Dioxide (ThO_2)	8.96×10^{-6} for 0 to 1000°C
Titanium Oxide (TiO_2) average for (2a+c)/3	8.97×10^{-6} for 26 to 455°C
Titanium Oxide (TiO_2) average for (2a+c)/3	8.97×10^{-6} for 26 to 940°C
Aluminum Oxide (Al_2O_3) parallel to c axis	8.98×10^{-6} for 0 to 1327°C
Titanium Oxide (TiO_2) — polycrystalline	8.98×10^{-6} for 0–1000°C
Spinel (Al_2O_3 MgO)	9.0×10^{-6} for 20 to 1250°C
Thorium Dioxide (ThO_2)	9.00×10^{-6} for 0 to 627°C
Molybdenum Disilicide ($MoSi_2$)	$9.00–9.18 \times 10^{-6}$ for 25–1500°C
Zirconium Monocarbide (ZrC)	$9.0x \times 10^{-6}$ for 1000–2000°C
Aluminum Oxide (Al_2O_3) parallel to c axis	9.02×10^{-6} for 0 to 1427°C
Beryllium Oxide (BeO) — polycrystalline	9.03×10^{-6} for 25–1000°C
Uranium Dioxide (UO_2) (heating)	9.07×10^{-6} for 27 to 400°C
Aluminum Oxide (Al_2O_3) parallel to c axis	9.08×10^{-6} for 0 to 1527°C

Selecting Thermal Expansion of Ceramics (Continued)

Ceramic	Thermal Expansion $(°C^{-1})$
Thorium Dioxide (ThO$_2$)	9.1×10^{-6} for 27 to 1087°C
Aluminum Oxide (Al$_2$O$_3$) parallel to c axis	9.13×10^{-6} for 0 to 1627°C
Titanium Oxide (TiO$_2$) average for (2a+c)/3	9.13×10^{-6} for 26 to 1110°C
Thorium Dioxide (ThO$_2$)	9.14×10^{-6} for 0 to 727°C
Spinel (Al$_2$O$_3$ MgO)	9.17×10^{-6} for 25 to 1500°C
Aluminum Oxide (Al$_2$O$_3$) parallel to c axis	9.18×10^{-6} for 0 to 1727°C
Beryllium Oxide (BeO) — polycrystalline	9.18×10^{-6} for 25–1250°C
Uranium Dioxide (UO$_2$)	9.18×10^{-6} for 27 to 400°C
Thorium Dioxide (ThO$_2$)	9.2×10^{-6} for 27 to 994°C
Beryllium Oxide (BeO) perpendicular to c axis	9.2×10^{-6} for 28 to 872°C
Thorium Dioxide (ThO$_2$)	9.24×10^{-6} for 0 to 827°C
Uranium Dioxide (UO$_2$) (cooling)	9.28×10^{-6} for 27 to 400°C
Titanium Monocarbide (TiC)	9.32×10^{-6} for 25–1250°C
Thorium Dioxide (ThO$_2$)	9.34×10^{-6} for 0 to 927°C
Titanium Mononitride (TiN)	9.35×10^{-6}
Thorium Dioxide (ThO$_2$)	9.35×10^{-6} for 0 to 1200°C
Beryllium Oxide (BeO) — polycrystalline	9.40×10^{-6} for 500–1200°C
Thorium Dioxide (ThO$_2$)	9.42×10^{-6} for 0 to 1027°C
Thorium Dioxide (ThO$_2$)	9.44×10^{-6} for 25 to 1000°C
Uranium Dioxide (UO$_2$)	9.47×10^{-6} for 25 to 500°C
Titanium Oxide (TiO$_2$) — polycrystalline	9.50×10^{-6} for 25–1500°C
Thorium Dioxide (ThO$_2$)	9.53×10^{-6} for 0 to 1127°C
Thorium Dioxide (ThO$_2$)	9.55×10^{-6} for 20 to 800°C
Thorium Dioxide (ThO$_2$)	9.55×10^{-6} for 20 to 1400°C

Selecting Thermal Expansion of Ceramics (Continued)

Ceramic	Thermal Expansion $(°C^{-1})$
Dichromium Trioxide (Cr_2O_3)	9.55×10^{-6} for 20–1400°C
Beryllium Oxide (BeO) average for (2a+c)/3	9.57×10^{-6} for 28 to 1132°C
Thorium Dioxide (ThO_2)	9.60×10^{-6} for 0 to 1227°C
Thorium Dioxide (ThO_2)	9.68×10^{-6} for 0 to 1327°C
Thorium Dioxide (ThO_2)	9.76×10^{-6} for 0 to 1427°C
Titanium Oxide (TiO_2) parallel to c axis	9.8×10^{-6} for 26 to 240°C
Thorium Dioxide (ThO_2)	9.83×10^{-6} for 0 to 1527°C
Thorium Dioxide (ThO_2)	9.84×10^{-6} for 0 to 1400°C
Beryllium Oxide (BeO) perpendicular to c axis	9.9×10^{-6} for 28 to 1132°C
Thorium Dioxide (ThO_2)	9.91×10^{-6} for 0 to 1627°C
Trichromium Dicarbide (Cr_3C_2)	9.95×10^{-6} for 25–500°C
Thorium Dioxide (ThO_2)	9.97×10^{-6} for 0 to 1727°C
Boron Nitride (BN) parallel to c axis	10.15×10^{-6} for 25 to 350°C
Thorium Dioxide (ThO_2)	10.17×10^{-6} for 25 to 1500°C
Beryllium Oxide (BeO) — polycrystalline	10.3×10^{-6} for 25–1500°C
Thorium Dioxide (ThO_2)	10.43×10^{-6} for 25 to 1700°C
Silicon Dioxide (SiO_2) β_2 tridymite	10.45×10^{-6} for 25–1000°C
Zirconium Oxide (ZrO_2) — tetragonal	10.5×10^{-6} for 0 to 1000°C
Titanium Oxide (TiO_2) parallel to c axis	10.5×10^{-6} for 26 to 455°C
Titanium Oxide (TiO_2) parallel to c axis	10.5×10^{-6} for 26 to 940°C
Zirconium Oxide (ZrO_2) — tetragonal	10.52×10^{-6} for 0 to 1000°C (MgO)
Zirconium Oxide (ZrO_2) — tetragonal	10.6×10^{-6} for 0 to 1200°C (CaO)
Titanium Oxide (TiO_2) parallel to c axis	10.6×10^{-6} for 26 to 670°C
Titanium Oxide (TiO_2) parallel to c axis	10.8×10^{-6} for 26 to 1110°C

Selecting Thermal Expansion of Ceramics (Continued)

Ceramic	Thermal Expansion $(°C^{-1})$
Uranium Dioxide (UO_2) (cooling)	10.8×10^{-6} for 400 to 800°C
Uranium Dioxide (UO_2) (cooling)	10.8×10^{-6} for 400 to 800°C
Hafnium Dioxide (HfO_2) monoclinic, parallel to c axis	10.8×10^{-6} for 28–697°C
Trichromium Dicarbide (Cr_3C_2)	10.9×10^{-6} for 150–980°C
Zirconium Oxide (ZrO_2) — tetragonal	11.0×10^{-6} for 0 to 1500°C
Hafnium Dioxide (HfO_2) monoclinic, parallel to c axis	11×10^{-6} for 28–262°C
Beryllium Oxide (BeO) — polycrystalline	11.1×10^{-6} for 25–2000°C
Uranium Dioxide (UO_2) (heating)	11.1×10^{-6} for 400 to 800°C
Uranium Dioxide (UO_2)	11.15×10^{-6} for 25 to 1750°C
Uranium Dioxide (UO_2)	11.19×10^{-6} for 25 to 1000°C
Hafnium Dioxide (HfO_2) monoclinic, parallel to c axis	11.4×10^{-6} for 28–494°C
Zirconium Oxide (ZrO_2) tetragonal, parallel to c axis	11.9×10^{-6} for 27 to 759°C
Hafnium Dioxide (HfO_2) monoclinic, parallel to c axis	11.9×10^{-6} for 28–903°C
Hafnium Dioxide (HfO_2) monoclinic, parallel to c axis	12.1×10^{-6} for 28–1098°C
Uranium Dioxide (UO_2)	12.19×10^{-6} for 25 to 1200°C
Boron Nitride (BN)	12.2×10^{-6} for 25 to 500°C
Uranium Dioxide (UO_2) (cooling)	12.6×10^{-6} for 800 to 1250°C
Zirconium Oxide (ZrO_2) tetragonal, parallel to c axis	12.8×10^{-6} for 27 to 964°C
Magnesium Oxide (MgO)	12.83×10^{-6} for 25–500°C
Uranium Dioxide (UO_2) (cooling)	12.9×10^{-6} for 800 to 1200°C
Zirconium Oxide (ZrO_2) tetragonal, parallel to c axis	13×10^{-6} for 27 to 504°C
Uranium Dioxide (UO_2) (heating)	13.0×10^{-6} for 800 to 1200°C
Magnesium Oxide (MgO)	13.3×10^{-6} for 20–1700°C
Boron Nitride (BN)	13.3×10^{-6} for 25 to 1000°C

Selecting Thermal Expansion of Ceramics (Continued)

Ceramic	Thermal Expansion ($°C^{-1}$)
Zirconium Oxide (ZrO_2) tetragonal, parallel to c axis	13.6×10^{-6} for 27 to 1110°C
Magnesium Oxide (MgO)	13.63×10^{-6} for 25–1000°C
Magnesium Oxide (MgO)	13.90×10^{-6} for 0–1000°C
Zirconium Oxide (ZrO_2) tetragonal, parallel to c axis	14×10^{-6} for 27 to 264°C
Magnesium Oxide (MgO)	14.0×10^{-6} for 20–1400°C
Magnesium Oxide (MgO)	$14.2–14.9 \times 10^{-6}$ for 20–1700°C
Magnesium Oxide (MgO)	14.46×10^{-6} for 0–1200°C
Silicon Dioxide (SiO_2) β quartz	14.58×10^{-6} for 25–1000°C
Magnesium Oxide (MgO)	15.06×10^{-6} for 0–1400°C
Magnesium Oxide (MgO)	15.11×10^{-6} for 25–1500°C
Magnesium Oxide (MgO)	15.89×10^{-6} for 25–1800°C
Silicon Dioxide (SiO_2) α tridymite	18.5×10^{-6} for 25–117°C
Silicon Dioxide (SiO_2) α quartz	19.35×10^{-6} for 25–500°C
Silicon Dioxide (SiO_2) $β_2$ tridymite	19.35×10^{-6} for 25–500°C
Silicon Dioxide (SiO_2) α quartz	22.2×10^{-6} for 25–575°C
Silicon Dioxide (SiO_2) $β_1$ tridymite	25.0×10^{-6} for 25–117°C
Silicon Dioxide (SiO_2) $β_1$ tridymite	27.5×10^{-6} for 25–163°C
Silicon Dioxide (SiO_2) β quartz	27.8×10^{-6} for 25–575°C
Silicon Dioxide (SiO_2) $β_2$ tridymite	31.9×10^{-6} for 25–163°C

Source: *data compiled* by J.S. Park from *No. 1 Materials Index*, Peter T.B. Shaffer, Plenum Press, New York, (1964); Smithells Metals Reference Book, Eric A. Brandes, ed., in association with Fulmer Research Institute Ltd. 6th ed. London, Butterworths, Boston, (1983); and *Ceramic Source*, American Ceramic Society (1986-1991)

SELECTING THERMAL EXPANSION OF GLASSES

Glass	Temperature Range of Validity	Thermal Expansion (K^{-1})
SiO_2 glass	–60—20°C	3.50×10^{-7}
SiO_2 glass	–40—20°C	3.80×10^{-7}
SiO_2 glass	–20—20°C	4.00×10^{-7}
SiO_2 glass	0–20°C	4.30×10^{-7}
SiO_2 glass	20–100°C	5.35×10^{-7}
SiO_2 glass	20–150°C	5.75×10^{-7}
SiO_2 glass	20–200°C	5.85×10^{-7}
SiO_2 glass	20–350°C	5.90×10^{-7}
SiO_2 glass	20–250°C	5.92×10^{-7}
SiO_2 glass	20–300°C	5.94×10^{-7}
SiO_2–Al_2O_3 glass (3.1% mol Al_2O_3, 1000°C for 115 hr)	20–980°C	6.2×10^{-7}
SiO_2–Al_2O_3 glass (3.1% mol Al_2O_3, water quenching)	20–980°C	6.2×10^{-7}
SiO_2–Al_2O_3 glass (8.2% mol Al_2O_3, water quenching)	20–800°C	8.8×10^{-7}
SiO_2–Al_2O_3 glass (5.4% mol Al_2O_3, 1130°C for 20 hr)	20–350°C	12.2×10^{-7}
SiO_2–Al_2O_3 glass (8.2% mol Al_2O_3, 1000°C for 115 hr)	20–950°C	14.5×10^{-7}
SiO_2–Al_2O_3 glass (13.9% mol Al_2O_3, water quenching)	20–600°C	17.2×10^{-7}
SiO_2–Al_2O_3 glass (17.4% mol Al_2O_3, water quenching)	20–700°C	20.7×10^{-7}

Selecting Thermal Expansion of Glasses (Continued)

Glass	Temperature Range of Validity	Thermal Expansion (K^{-1})
SiO$_2$–Al$_2$O$_3$ glass (13.9% mol Al$_2$O$_3$, 1000°C for 115 hr)	20–900°C	22.7×10^{-7}
SiO$_2$–Al$_2$O$_3$ glass (17.4% mol Al$_2$O$_3$, 1000°C for 115 hr)	20–800°C	28.3×10^{-7}
SiO$_2$–B$_2$O$_3$ glass (39.2% mol B$_2$O$_3$)	100–200°C	44.9×10^{-7}
SiO$_2$–B$_2$O$_3$ glass (39.2% mol B$_2$O$_3$)	0–100°C	47.5×10^{-7}
SiO$_2$–B$_2$O$_3$ glass (44.2% mol B$_2$O$_3$)	0–100°C	49.8×10^{-7}
SiO$_2$–B$_2$O$_3$ glass (44.2% mol B$_2$O$_3$)	100–200°C	50.8×10^{-7}
SiO$_2$–PbO glass (25.7% mol PbO)	20–170°C	$51.45–52.23 \times 10^{-7}$
SiO$_2$–B$_2$O$_3$ glass (50.8% mol B$_2$O$_3$)	100–200°C	54.8×10^{-7}
B$_2$O$_3$–CaO glass (29.3% mol CaO)	room temp. to 100°C	$54.9–56.4 \times 10^{-7}$
B$_2$O$_3$–CaO glass (31.4% mol CaO)	room temp. to 100°C	$57.3–58.2 \times 10^{-7}$
SiO$_2$–B$_2$O$_3$ glass (50.8% mol B$_2$O$_3$)	0–100°C	57.6×10^{-7}
SiO$_2$–PbO glass (30.0% mol PbO)	20–170°C	$57.68–59.08 \times 10^{-7}$
B$_2$O$_3$–CaO glass (34.9% mol CaO)	room temp. to 100°C	$60.1–66.2 \times 10^{-7}$
B$_2$O$_3$–CaO glass (29.3% mol CaO)	100–200°C	$60.2–60.8 \times 10^{-7}$
SiO$_2$–PbO glass (32.5% mol PbO)	20–170°C	$60.62–62.31 \times 10^{-7}$
SiO$_2$–PbO glass (33.2% mol PbO)	20–170°C	$61.58–63.33 \times 10^{-7}$
B$_2$O$_3$–CaO glass (37.1% mol CaO)	room temp. to 100°C	$63.1–64.0 \times 10^{-7}$
B$_2$O$_3$–CaO glass (31.4% mol CaO)	100–200°C	$63.5–65.1 \times 10^{-7}$
B$_2$O$_3$–CaO glass (29.3% mol CaO)	200–300°C	$63.9–65.4 \times 10^{-7}$
SiO$_2$–PbO glass (35.0% mol PbO)	20–170°C	$63.99–66.17 \times 10^{-7}$
B$_2$O$_3$–Na$_2$O glass (16.2% mol Na$_2$O)	–196—25°C	65.9×10^{-7}
B$_2$O$_3$–Na$_2$O glass (15.8% mol Na$_2$O)	–196—25°C	67.4×10^{-7}

Selecting Thermal Expansion of Glasses (Continued)

Glass	Temperature Range of Validity	Thermal Expansion (K^{-1})
B_2O_3–CaO glass (31.4% mol CaO)	200–300°C	67.4–68.1×10^{-7}
B_2O_3–CaO glass (34.9% mol CaO)	100–200°C	67.5–67.6×10^{-7}
B_2O_3–CaO glass (37.1% mol CaO)	100–200°C	68.4–70.4×10^{-7}
SiO_2–PbO glass (37.5% mol PbO)	20–170°C	68.75–71.44×10^{-7}
B_2O_3–Na_2O glass		
(15% mol Na_2O, $T_g = 407$°C)	below T_g	69×10^{-7}
B_2O_3–Na_2O glass (18.4% mol Na_2O)	-196—25°C	69.1×10^{-7}
B_2O_3–Na_2O glass (13.7% mol Na_2O)	-196—25°C	69.3×10^{-7}
SiO_2–B_2O_3 glass (58.4% mol B_2O_3)	100–200°C	70.1×10^{-7}
B_2O_3–CaO glass (29.3% mol CaO)	300–400°C	71.3–71.6×10^{-7}
B_2O_3–Na_2O glass (11.5% mol Na_2O)	-196—25°C	71.5×10^{-7}
SiO_2–B_2O_3 glass (58.4% mol B_2O_3)	0–100°C	71.9×10^{-7}
B_2O_3–Na_2O glass (22.5% mol Na_2O)	-196—25°C	71.9×10^{-7}
B_2O_3–CaO glass (37.1% mol CaO)	200–300°C	74.6–75.8×10^{-7}
B_2O_3–CaO glass (34.9% mol CaO)	200–300°C	74.7–75.2×10^{-7}
SiO_2–PbO glass (42.6% mol PbO)	20–170°C	75.16–78.58×10^{-7}
B_2O_3–CaO glass (31.4% mol CaO)	300–400°C	76.5–76.7×10^{-7}
B_2O_3–CaO glass (29.3% mol CaO)	400–500°C	76.9–77.1×10^{-7}
B_2O_3–Na_2O glass		
(10% mol Na_2O, $T_g = 354$°C)	below T_g	77×10^{-7}
B_2O_3–CaO glass (34.9% mol CaO)	300–400°C	77.8–78.5×10^{-7}
SiO_2–PbO glass (45.8% mol PbO)	20–170°C	78.85–82.60×10^{-7}
B_2O_3–CaO glass (31.4% mol CaO)	400–500°C	79.2–81.0×10^{-7}
B_2O_3–Na_2O glass (15.8% mol Na_2O)	20–50°C	80.7×10^{-7}

Selecting Thermal Expansion of Glasses (Continued)

Glass	Temperature Range of Validity	Thermal Expansion (K^{-1})
B_2O_3–CaO glass (29.3% mol CaO)	500–600°C	$80.9–86.8 \times 10^{-7}$
B_2O_3–Na_2O glass (28.9% mol Na_2O)	–196—25°C	81.4×10^{-7}
B_2O_3–CaO glass (37.1% mol CaO)	300–400°C	$81.6–82.2 \times 10^{-7}$
SiO_2–PbO glass (47.8% mol PbO)	20–170°C	$83.03–87.03 \times 10^{-7}$
B_2O_3–CaO glass (31.4% mol CaO)	500–600°C	$83.1–88.5 \times 10^{-7}$
B_2O_3–CaO glass (34.9% mol CaO)	400–500°C	$83.8–95.0 \times 10^{-7}$
SiO_2–PbO glass (49.8% mol PbO)	20–170°C	$85.57–89.82 \times 10^{-7}$
B_2O_3–Na_2O glass (17.4% mol Na_2O)	20–50°C	85.6×10^{-7}
B_2O_3–Na_2O glass (20% mol Na_2O, T_g = 456°C)	below T_g	86×10^{-7}
B_2O_3–Na_2O glass (16.2% mol Na_2O)	20–50°C	86.0×10^{-7}
B_2O_3–Na_2O glass (18.4% mol Na_2O)	20–50°C	86.2×10^{-7}
B_2O_3–Na_2O glass (19.6% mol Na_2O)	20–50°C	86.8×10^{-7}
B_2O_3–CaO glass (37.1% mol CaO)	400–500°C	$86.9–87.6 \times 10^{-7}$
SiO_2–B_2O_3 glass (72.7% mol B_2O_3)	0–100°C	87.0×10^{-7}
B_2O_3–Na_2O glass (13.7% mol Na_2O)	20–50°C	87.5×10^{-7}
B_2O_3–Na_2O glass (20.0% mol Na_2O)	20–50°C	87.6×10^{-7}
B_2O_3–Na_2O glass (16.2% mol Na_2O)	20–150°C	87.7×10^{-7}
B_2O_3–Na_2O glass (15.8% mol Na_2O)	20–150°C	87.8×10^{-7}
B_2O_3–Na_2O glass (11.5% mol Na_2O)	20–50°C	88.7×10^{-7}
B_2O_3–Na_2O glass (17.4% mol Na_2O)	20–150°C	89.1×10^{-7}
B_2O_3–Na_2O glass (18.4% mol Na_2O)	20–150°C	89.2×10^{-7}
SiO_2–B_2O_3 glass (72.7% mol B_2O_3)	100–200°C	89.7×10^{-7}
B_2O_3–Na_2O glass (22.5% mol Na_2O)	20–50°C	90.4×10^{-7}

Selecting Thermal Expansion of Glasses (Continued)

Glass	Temperature Range of Validity	Thermal Expansion (K^{-1})
B_2O_3–Na_2O glass (23.6% mol Na_2O)	20–50°C	90.4×10^{-7}
SiO_2–PbO glass (53.8% mol PbO)	20–170°C	90.62–95.25×10^{-7}
B_2O_3–Na_2O glass (13.7% mol Na_2O)	20–250°C	90.9×10^{-7}
B_2O_3–Na_2O glass (16.2% mol Na_2O)	20–250°C	90.9×10^{-7}
B_2O_3–Na_2O glass (19.6% mol Na_2O)	20–150°C	91.2×10^{-7}
B_2O_3–Na_2O glass (20.0% mol Na_2O)	20–150°C	91.6×10^{-7}
B_2O_3–CaO glass (34.9% mol CaO)	500–600°C	91.8–92.1×10^{-7}
B_2O_3–Na_2O glass (13.7% mol Na_2O)	20–150°C	92.3×10^{-7}
B_2O_3–Na_2O glass (17.4% mol Na_2O)	20–250°C	92.4×10^{-7}
B_2O_3–Na_2O glass (15.8% mol Na_2O)	20–250°C	93.3×10^{-7}
B_2O_3–CaO glass (37.1% mol CaO)	500–600°C	93.5–95.5×10^{-7}
B_2O_3–Na_2O glass (18.4% mol Na_2O)	20–250°C	94.1×10^{-7}
B_2O_3–Na_2O glass (4.4% mol Na_2O)	−196—25°C	94.6×10^{-7}
B_2O_3–Na_2O glass (22.5% mol Na_2O)	20–150°C	94.7×10^{-7}
B_2O_3–Na_2O glass (11.5% mol Na_2O)	20–150°C	94.9×10^{-7}
B_2O_3–Na_2O glass (25% mol Na_2O, $T_g = 466$°C)	below T_g	95×10^{-7}
B_2O_3–Na_2O glass (19.6% mol Na_2O)	20–250°C	95.3×10^{-7}
SiO_2–PbO glass (57.5% mol PbO)	20–170°C	95.64–100.45×10^{-7}
B_2O_3–Na_2O glass (18.4% mol Na_2O)	20–350°C	96.2×10^{-7}
B_2O_3–Na_2O glass (17.4% mol Na_2O)	20–350°C	96.3×10^{-7}
B_2O_3–Na_2O glass (23.6% mol Na_2O)	20–150°C	96.7×10^{-7}
B_2O_3–Na_2O glass (16.2% mol Na_2O)	20–350°C	96.9×10^{-7}

Selecting Thermal Expansion of Glasses (Continued)

Glass	Temperature Range of Validity	Thermal Expansion (K^{-1})
SiO$_2$–PbO glass (59.0% mol PbO)	20–170°C	97.00–101.90x10^{-7}
SiO$_2$–Na$_2$O glass (20.3% mol Na$_2$O)	room temp–100°C	97.5x10^{-7}
B$_2$O$_3$–Na$_2$O glass (20.0% mol Na$_2$O)	20–250°C	97.6x10^{-7}
B$_2$O$_3$–Na$_2$O glass (11.5% mol Na$_2$O)	20–250°C	97.9x10^{-7}
B$_2$O$_3$–Na$_2$O glass (15.8% mol Na$_2$O)	20–350°C	97.9x10^{-7}
B$_2$O$_3$–Na$_2$O glass (22.5% mol Na$_2$O)	20–250°C	98.7x10^{-7}
B$_2$O$_3$–Na$_2$O glass (8.7% mol Na$_2$O)	20–50°C	98.8x10^{-7}
SiO$_2$–Na$_2$O glass (20.3% mol Na$_2$O)	100–200°C	99.3x10^{-7}
B$_2$O$_3$–Na$_2$O glass (19.6% mol Na$_2$O)	20–350°C	99.6x10^{-7}
B$_2$O$_3$–Na$_2$O glass (8.7% mol Na$_2$O)	20–150°C	100.5x10^{-7}
SiO$_2$–Na$_2$O glass (20.3% mol Na$_2$O)	200–300°C	100.6x10^{-7}
SiO$_2$–PbO glass (61.0% mol PbO)	20–170°C	100.66–105.58x10^{-7}
B$_2$O$_3$–Na$_2$O glass (23.6% mol Na$_2$O)	20–250°C	101.2x10^{-7}
B$_2$O$_3$–Na$_2$O glass (20.0% mol Na$_2$O)	20–350°C	101.3x10^{-7}
SiO$_2$–PbO glass (61.75% mol PbO)	20–170°C	101.36–106.30x10^{-7}
B$_2$O$_3$–Na$_2$O glass (28.9% mol Na$_2$O)	20–50°C	102.1x10^{-7}
B$_2$O$_3$–Na$_2$O glass (4.4% mol Na$_2$O)	20–50°C	103.0x10^{-7}
B$_2$O$_3$–Na$_2$O glass (22.5% mol Na$_2$O)	20–350°C	104.0x10^{-7}
B$_2$O$_3$–Na$_2$O glass (8.7% mol Na$_2$O)	20–250°C	105.3x10^{-7}
B$_2$O$_3$–Na$_2$O glass (23.6% mol Na$_2$O)	20–350°C	106.5x10^{-7}
SiO$_2$–Na$_2$O glass (20.3% mol Na$_2$O)	300–400°C	106.9x10^{-7}
B$_2$O$_3$–Na$_2$O glass (28.9% mol Na$_2$O)	20–150°C	107.4x10^{-7}
SiO$_2$–Na$_2$O glass (24.0% mol Na$_2$O)	room temp–100°C	109.7x10^{-7}
B$_2$O$_3$–Na$_2$O glass (4.4% mol Na$_2$O)	20–150°C	109.9x10^{-7}

Selecting Thermal Expansion of Glasses (Continued)

Glass	Temperature Range of Validity	Thermal Expansion (K^{-1})
SiO_2–PbO glass (67.7% mol PbO)	20–170°C	110.38–115.48×10^{-7}
SiO_2–B_2O_3 glass (83.2% mol B_2O_3)	0–100°C	111.4×10^{-7}
B_2O_3–Na_2O glass (28.9% mol Na_2O)	20–250°C	112.8×10^{-7}
SiO_2–Na_2O glass (24.0% mol Na_2O)	100–200°C	114.3×10^{-7}
B_2O_3–Na_2O glass		
(5% mol Na_2O, $T_g = 318$°C)	below T_g	115×10^{-7}
B_2O_3–Na_2O glass (4.4% mol Na_2O)	20–250°C	116.0×10^{-7}
SiO_2–B_2O_3 glass (83.2% mol B_2O_3)	100–200°C	116.6×10^{-7}
SiO_2–Na_2O glass (24.0% mol Na_2O)	200–300°C	116.6×10^{-7}
B_2O_3–Na_2O glass (28.9% mol Na_2O)	20–350°C	117.1×10^{-7}
SiO_2–B_2O_3 glass (88.6% mol B_2O_3)	0–100°C	118.1×10^{-7}
SiO_2–Na_2O glass		
(20% mol Na_2O, $T_g = 478$°C)	below T_g	120×10^{-7}
SiO_2–Na_2O glass (24.0% mol Na_2O)	300–400°C	121.7×10^{-7}
SiO_2–B_2O_3 glass (88.6% mol B_2O_3)	100–200°C	126.0×10^{-7}
B_2O_3–Na_2O glass		
(30% mol Na_2O, $T_g = 468$°C)	below T_g	128×10^{-7}
SiO_2–B_2O_3 glass (94.0% mol B_2O_3)	0–100°C	131.7×10^{-7}
SiO_2–Na_2O glass (31.1% mol Na_2O)	room temp–100°C	136.0×10^{-7}
B_2O_3–Na_2O glass (0.01% mol Na_2O)	–196—25°C	140×10^{-7}
SiO_2–B_2O_3 glass (94.0% mol B_2O_3)	100–200°C	141.9×10^{-7}
SiO_2–Na_2O glass (31.1% mol Na_2O)	100–200°C	142.5×10^{-7}
SiO_2–Na_2O glass (33.8% mol Na_2O)	room temp–100°C	143.9×10^{-7}

Selecting Thermal Expansion of Glasses (Continued)

Glass	Temperature Range of Validity	Thermal Expansion (K^{-1})
SiO_2–Na_2O glass (31.1% mol Na_2O)	200–300°C	148.3×10^{-7}
B_2O_3–Na_2O glass (0.01% mol Na_2O)	20–150°C	149.0×10^{-7}
B_2O_3–Na_2O glass (0.01% mol Na_2O)	20–50°C	149.3×10^{-7}
B_2O_3 glass	20–200°C	150 ± 3–$158 \pm 3 \times 10^{-7}$
SiO_2–Na_2O glass (30% mol Na_2O, $T_g = 455$°C)	below T_g	152×10^{-7}
SiO_2–Na_2O glass (37.2% mol Na_2O)	room temp–100°C	152.1×10^{-7}
SiO_2–Na_2O glass (33.8% mol Na_2O)	100–200°C	153.6×10^{-7}
B_2O_3 glass	100–200°C	154.5–169×10^{-7}
B_2O_3 glass	0–100°C	154.5–183×10^{-7}
SiO_2–Na_2O glass (33.8% mol Na_2O)	200–300°C	159.1×10^{-7}
SiO_2–Na_2O glass (31.1% mol Na_2O)	300–400°C	160.0×10^{-7}
SiO_2–Na_2O glass (37.2% mol Na_2O)	100–200°C	160.9×10^{-7}
SiO_2–Na_2O glass (33% mol Na_2O, $T_g = 445$°C)	below T_g	165×10^{-7}
SiO_2–Na_2O glass (37.2% mol Na_2O)	200–300°C	171.6×10^{-7}
SiO_2–Na_2O glass (33.8% mol Na_2O)	300–400°C	173.6×10^{-7}
SiO_2–Na_2O glass (40% mol Na_2O, $T_g = 421$°C)	below T_g	179×10^{-7}
SiO_2–Na_2O glass (37.2% mol Na_2O)	300–400°C	187.7×10^{-7}
SiO_2–Na_2O glass (45% mol Na_2O, $T_g = 417$°C)	below T_g	219×10^{-7}
SiO_2–B_2O_3 glass (39.2% mol B_2O_3)	390–410°C	301×10^{-7}

Selecting Thermal Expansion of Glasses (Continued)

Glass	Temperature Range of Validity	Thermal Expansion (K^{-1})
SiO_2–Na_2O glass (20% mol Na_2O, $T_g = 478°C$)	above T_g	315×10^{-7}
SiO_2–Na_2O glass (30% mol Na_2O, $T_g = 455°C$)	above T_g	402×10^{-7}
SiO_2–B_2O_3 glass (44.2% mol B_2O_3)	380–400°C	450×10^{-7}
SiO_2–Na_2O glass (33% mol Na_2O, $T_g = 445°C$)	above T_g	465×10^{-7}
SiO_2–Na_2O glass (40% mol Na_2O, $T_g = 421°C$)	above T_g	500×10^{-7}
SiO_2–CaO glass (35% mol CaO)	1700°C	$53 \pm 5 \times 10^{-6}$
SiO_2–Na_2O glass (45% mol Na_2O, $T_g = 417°C$)	above T_g	574×10^{-7}
SiO_2–B_2O_3 glass (50.8% mol B_2O_3)	350–370°C	579×10^{-7}
B_2O_3–Na_2O glass (20% mol Na_2O, $T_g = 456°C$)	above T_g	586×10^{-7}
SiO_2–CaO glass (40% mol CaO)	1700°C	$64 \pm 4 \times 10^{-6}$
SiO_2–CaO glass (30% mol CaO)	1700°C	$66 \pm 5 \times 10^{-6}$
SiO_2–Na_2O glass (20% mol Na_2O)	liquidus temp. to 1400°C	6.7×10^{-5}
SiO_2–B_2O_3 glass (58.4% mol B_2O_3)	320–340°C	694×10^{-7}
SiO_2–PbO glass (50% mol PbO)	1100°C	723×10^{-7}
SiO_2–CaO glass (42.5% mol CaO)	1700°C	$76 \pm 4 \times 10^{-6}$
SiO_2–CaO glass (47.5% mol CaO)	1700°C	$76 \pm 4 \times 10^{-6}$
SiO_2–CaO glass (52.5% mol CaO)	1700°C	$76–107 \pm 4 \times 10^{-6}$
B_2O_3–Na_2O glass (15% mol Na_2O, $T_g = 407°C$)	above T_g	761×10^{-7}

Selecting Thermal Expansion of Glasses (Continued)

Glass	Temperature Range of Validity	Thermal Expansion (K^{-1})
B_2O_3–Na_2O glass		
(25% mol Na_2O, T_g = 466°C)	above T_g	834×10^{-7}
SiO_2–CaO glass (50% mol CaO)	1700°C	84–$85 \pm 4 \times 10^{-6}$
SiO_2–CaO glass (45% mol CaO)	1700°C	85–$100 \pm 4 \times 10^{-6}$
SiO_2–PbO glass (66.7% mol PbO)	1100°C	867×10^{-7}
SiO_2–B_2O_3 glass (72.7% mol B_2O_3)	300–320°C	899×10^{-7}
SiO_2–CaO glass (55% mol CaO)	1700°C	94–$95 \pm 4 \times 10^{-6}$
SiO_2–CaO glass (57.5% mol CaO)	1700°C	$95 \pm 4 \times 10^{-6}$
SiO_2–B_2O_3 glass (83.2% mol B_2O_3)	280–300°C	970×10^{-7}
SiO_2–B_2O_3 glass (88.6% mol B_2O_3)	280–300°C	1023×10^{-7}
SiO_2–CaO glass (60% mol CaO)	1700°C	$103 \pm 4 \times 10^{-6}$
B_2O_3–Na_2O glass		
(30% mol Na_2O, T_g = 468°C)	above T_g	1150×10^{-7}
SiO_2–B_2O_3 glass (94.0% mol B_2O_3)	270–290°C	1200×10^{-7}
B_2O_3–Na_2O glass		
(10% mol Na_2O, T_g = 354°C)	above T_g	1230×10^{-7}
B_2O_3–Na_2O glass		
(5% mol Na_2O, T_g = 318°C)	above T_g	1400×10^{-7}
SiO_2–Na_2O glass (33.3% mol Na_2O)	liquidus temp.to 1400°C	17.2×10^{-5}
SiO_2–Na_2O glass (40% mol Na_2O)	liquidus temp. to 1400°C	20.0×10^{-5}
SiO_2–Na_2O glass (50% mol Na_2O)	liquidus temp. to 1400°C	23.7×10^{-5}

Source: *data compiled by* Jun S. Park *from* O. V. Mazurin, M. V. Streltsina and T. P. Shvaiko–Shvaikovskaya, *Handbook of Glass Data, Part A and Part B*, Elsevier, New York, 1983

Selecting Ceramics and Glasses: Mechanical Properties

SELECTING TENSILE STRENGTHS OF CERAMICS

Ceramic	Temperature	Tensile Strength (psi)
Boron Nitride (BN)	1000°C	0.35×10^3
Boron Nitride (BN)	1500°C	0.35×10^3
Beryllium Oxide (BeO)	1300°C	0.6×10^3
Spinel (Al$_2$O$_3$ MgO)	1300°C	1.1×10^3
Boron Nitride (BN)	1800°C	1.15×10^3
Aluminum Oxide (Al$_2$O$_3$)	1460°C	1.5×10^3
Tantalum Monocarbide (TaC)		$2\text{-}42 \times 10^3$
Beryllium Oxide (BeO)	1140°C	2.0×10^3
Boron Nitride (BN)	2000°C	2.25×10^3
Cordierite (2MgO 2Al$_2$O$_3$ 5SiO$_2$)(ρ=1.8g/cm^3)	1200°C	2.5×10^3
Cordierite (2MgO 2Al$_2$O$_3$ 5SiO$_2$)(ρ=2.1g/cm^3)	800°C	3.5×10^3
Zircon (SiO$_2$ ZrO$_2$)	1200°C	3.6×10^3
Aluminum Oxide (Al$_2$O$_3$)	1400°C	4.3×10^3
Silicon Carbide (SiC)	25°C	$5\text{-}20 \times 10^3$
Beryllium Oxide (BeO)	1000°C	5.0×10^3
Silicon Carbide (SiC) (hot pressed)	1400°C	$5.75\text{-}21.75 \times 10^3$
Magnesium Oxide (MgO)	1300°C	6×10^3
Spinel (Al$_2$O$_3$ MgO)	1150°C	6.1×10^3
Aluminum Oxide (Al$_2$O$_3$)	1300°C	6.4×10^3
Zirconium Oxide (ZrO$_2$)	1000°C	$6.75\text{-}17.0 \times 10^3$
Boron Nitride (BN)	2400°C	6.80×10^3
Beryllium Oxide (BeO)	900°C	7.0×10^3
Cordierite (2MgO 2Al$_2$O$_3$ 5SiO$_2$)(ρ=2.51g/cm^3)	25°C	7.8×10^3
Magnesium Oxide (MgO)	1200°C	8×10^3

Selecting Tensile Strengths of Ceramics (Continued)

Ceramic	Temperature	Tensile Strength (psi)
Zircon (SiO$_2$ ZrO$_2$)	1050°C	8.7x10^3
Magnesium Oxide (MgO)	1100°C	10 x10^3
Zirconium Oxide (ZrO$_2$)	1300°C	10.2x10^3
Chromium Diboride (CrB$_2$)		10.6x10^4
Beryllium Oxide (BeO)	500°C	11.1 x10^3
Silicon Carbide (SiC) (reaction bonded)	20°C	11.17x10^3
Magnesium Oxide (MgO)	1000°C	11.5 x10^3
Zirconium Monocarbide (ZrC)	980°C	11.7-14.45x10^3
Zirconium Oxide (ZrO$_2$)	1200°C	12.1x10^3
Zircon (SiO$_2$ ZrO$_2$)	room temp.	12.7x10^3
Zirconium Monocarbide (ZrC)	1250°C	12.95-15.85x10^3
Zirconium Oxide (ZrO$_2$)	1100°C	13.0-13.5x10^3
Beryllium Oxide (BeO)	room temp.	13.5-20 x10^3
Spinel (Al$_2$O$_3$ MgO)	550°C	13.7x10^3
Magnesium Oxide (MgO)	room temp.	14 x10^3
Magnesium Oxide (MgO)	200°C	14 x10^3
Thorium Dioxide (ThO$_2$)	room temp.	14x10^3
Magnesium Oxide (MgO)	400°C	15.2 x10^3
Magnesium Oxide (MgO)	800°C	16 x10^3
Zirconium Monocarbide (ZrC)	room temp.	16.0x10^3
Zirconium Oxide (ZrO$_2$)	800°C	16.0x10^3
Mullite (3Al$_2$O$_3$ 2SiO$_2$)	25°C	16x10^3
Zirconium Oxide (ZrO$_2$)	200°C	16.8x10^3
Titanium Monocarbide (TiC)	1000°C	17.2x10^3

Selecting Tensile Strengths of Ceramics (Continued)

Ceramic	Temperature	Tensile Strength (psi)
Zirconium Oxide (ZrO_2)	400°C	17.5×10^3
Zirconium Oxide (ZrO_2)	600°C	17.6×10^3
Zirconium Oxide (ZrO_2)	room temp.	$17.9\text{-}20 \times 10^3$
Titanium Diboride (TiB_2)		18.4×10^3
Aluminum Oxide (Al_2O_3)	1200°C	$18.5\text{-}20 \times 10^3$
Spinel (Al_2O_3 MgO)	room temp.	19.2×10^3
Zirconium Oxide (ZrO_2)	500°C	20.0×10^3
Trisilicon tetranitride (Si_3N_4) (reaction bonded)	1400°C	20.3×10^3
Trisilicon tetranitride (Si_3N_4) (hot pressed)	1400°C	21.8×10^3
Boron Carbide (B_4C)	980°C	22.5×10^3
Trisilicon tetranitride (Si_3N_4) (reaction bonded)	20°C	24.7×10^3
Zirconium Diboride (ZrB_2)		28.7×10^3
Silicon Carbide (SiC) (hot pressed)	20°C	29×10^3
Aluminum Oxide (Al_2O_3)	1140°C	31.4×10^3
Aluminum Oxide (Al_2O_3)	300°C	33.6×10^3
Aluminum Oxide (Al_2O_3)	1050°C	33.9×10^3
Aluminum Oxide (Al_2O_3)	800°C	34.6×10^3
Aluminum Oxide (Al_2O_3)	1000°C	35×10^3
Aluminum Oxide (Al_2O_3)	room temp.	$37\text{-}37.8 \times 10^3$

Selecting Tensile Strengths of Ceramics (Continued)

Ceramic	Temperature	Tensile Strength (psi)
Aluminum Oxide (Al$_2$O$_3$)	500°C	40 x10^3
Molybdenum Disilicide (MoSi$_2$)	980°C	40x10^3
Molybdenum Disilicide (MoSi$_2$)	1300°C	41.07x10^3
Molybdenum Disilicide (MoSi$_2$)	1090°C	42.16x10^3
Molybdenum Disilicide (MoSi$_2$)	1200°C	42.8x10^3
Tungsten Monocarbide (WC)		50x10^3
Trisilicon tetranitride (Si$_3$N$_4$) (hot pressed)	20°C	54.4 x10^3
Spinel (Al$_2$O$_3$ MgO)	900°C	110.8x10^3

To convert **psi** to **MPa**, multiply by **145**.

Source: *data compiled by* J.S. Park from *No. 1 Materials Index*, Peter T.B. Shaffer, Plenum Press, New York, (1964); *Smithells Metals Reference Book*, Eric A. Brandes, ed., in association with Fulmer Research Institute Ltd. 6th ed. London, Butterworths, Boston, (1983); and *Ceramic Source*, American Ceramic Society (1986-1991).

SELECTING TENSILE STRENGTHS OF GLASS

Glass	Tensile Strength $(Kg \cdot mm^{-2})$
(Corning 7940 silica glass @ 100°C)	5.6
SiO_2 glass (1.5 mm diameter rod, 0.5 g/mm^2•s stress rate)	5.84–7.08
(Corning 7940 silica glass @ 300°C)	6.2
(Corning 7940 silica glass @ 500°C)	6.6
(Corning 7940 silica glass @ 700°C)	7.1
(Corning 7940 silica glass @ 900°C)	7.6
SiO_2 glass (1.5 mm diameter rod, 54 g/mm^2•s stress rate)	8.52±2.52
SiO_2 glass (1.5 mm diameter rod, 50 g/mm^2•s stress rate)	9.73±2.13
SiO_2–Na_2O glass (5 mm diameter rod, 20% mol Na_2O)	15
SiO_2 glass (112 µm diameter fiber)	28.3
SiO_2 glass (108 µm diameter fiber)	28.8
SiO_2 glass (78 µm diameter fiber)	35.8
SiO_2 glass (74 µm diameter fiber)	36.5
SiO_2 glass (65 µm diameter fiber)	39.7
SiO_2 glass (60 µm diameter fiber)	42.3
SiO_2–PbO glass (17.2 µm diameter fiber, 50% mol PbO)	43–51.6
SiO_2 glass (56 µm diameter fiber)	44.3
SiO_2 glass (48 µm diameter fiber)	49.6
SiO_2–PbO glass (11.4 µm diameter fiber, 50% mol PbO)	51.9–56
B_2O_3 glass (10–30 µm diameter fiber)	60
SiO_2–PbO glass (7.1 µm diameter fiber, 50% mol PbO)	62–71.3
SiO_2–PbO glass (4.3 µm diameter fiber, 50% mol PbO)	64
SiO_2–PbO glass (8.0 µm diameter fiber, 50% mol PbO)	64.5
SiO_2–PbO glass (5.7 µm diameter fiber, 50% mol PbO)	66–67.2

Selecting Tensile Strengths of Glass (Continued)

Glass	Tensile Strength $(Kg \cdot mm^{-2})$
SiO_2–PbO glass (3.0 μm diameter fiber, 50% mol PbO)	70.8
SiO_2–Na_2O glass (11.4μm diameter fiber, 36.3% mol Na_2O)	91.2±1.480
SiO_2–Na_2O glass (25.7μm diameter fiber, 19.5% mol Na_2O)	92.5±10.08
SiO_2–Na_2O glass (8.6μm diameter fiber, 36.3% mol Na_2O)	98.0±0.344
B_2O_3–Na_2O glass (10–30 μm diameter fiber, 10% mol Na_2O)	102
SiO_2–Na_2O glass (12.8μm diameter fiber, 25.5% mol Na_2O)	103±1.020
SiO_2–Na_2O glass (5.4μm diameter fiber, 36.3% mol Na_2O)	107.6±0.308
SiO_2–Na_2O glass (6.3μm diameter fiber, 25.5% mol Na_2O)	127±0.259
SiO_2–Na_2O glass (8.6μm diameter fiber, 19.5% mol Na_2O)	134±1.34
B_2O_3–Na_2O glass (10–30 μm diameter fiber, 20% mol Na_2O)	137
SiO_2–Na_2O glass (3.6μm diameter fiber, 25.5% mol Na_2O)	142±0.189
B_2O_3–Na_2O glass (10–30 μm diameter fiber, 30% mol Na_2O)	152
SiO_2–Na_2O glass (6.0μm diameter fiber, 19.5% mol Na_2O)	173±1.36

Source: *data compiled by* J.S. Park *from* O. V. Mazurin, M. V. Streltsina and T. P. Shvaiko–Shvaikovskaya, *Handbook of Glass Data, Part A and Part B,* Elsevier, New York, 1983

SELECTING COMPRESSIVE STRENGTHS OF CERAMICS

Ceramic	Temperature (°C)	Compressive Strength (psi)
Thorium Dioxide (ThO$_2$)	1500	1.5×10^3
Zirconium Oxide (ZrO$_2$)	1500	2.8×10^3
Thorium Dioxide (ThO$_2$)	1400	5.7×10^3
Aluminum Oxide (Al$_2$O$_3$)	1600	7×10^3
Beryllium Oxide (BeO)	1600	7×10^3
Spinel (Al$_2$O$_3$ MgO)	1600	8.5×10^3
Trisilicon tetranitride (Si$_3$N$_4$)	25	$10\text{-}100 \times 10^3$
Trisilicon tetranitride (Si$_3$N$_4$)	1000	$10\text{-}30 \times 10^3$
Aluminum Oxide (Al$_2$O$_3$)	1500	14×10^3
Beryllium Oxide (BeO)	1500	17×10^3
Zirconium Oxide (ZrO$_2$)	1400	18.5×10^3
Cordierite (2MgO 2Al$_2$O$_3$ 5SiO$_2$) (ρ=1.8g/cm^3)	1200	18.5×10^3
Spinel (Al$_2$O$_3$ MgO)	1400	21.4×10^3
Beryllium Oxide (BeO)	1400	24×10^3
Beryllium Oxide (BeO)	1145	28.5×10^3
Thorium Dioxide (ThO$_2$)	1200	28.5×10^3
Cordierite (2MgO 2Al$_2$O$_3$ 5SiO$_2$) (ρ=2.1g/cm^3)	800	30×10^3
Boron Nitride (BN), parallel to c axis		34.0×10^3
Beryllium Oxide (BeO)	1000	$35.5\text{-}40 \times 10^3$
Aluminum Oxide (Al$_2$O$_3$)	1400	35.6×10^3
Boron Nitride (BN), parallel to a axis		45×10^3
Titanium Diboride (TiB$_2$)		$47\text{-}97 \times 10^3$
Cordierite (2MgO 2Al$_2$O$_3$ 5SiO$_2$) (ρ=2.51g/cm^3)	25	50×10^3
Cordierite (2MgO 2Al$_2$O$_3$ 5SiO$_2$) (ρ=2.3g/cm^3)	400	50×10^3

Selecting Compressive Strengths of Ceramics (Continued)

Ceramic	Temperature (°C)	Compressive Strength (psi)
Thorium Dioxide (ThO$_2$)	1000	51×10^3
Beryllium Oxide (BeO)	800	64×10^3
Aluminum Oxide (Al$_2$O$_3$)	1200	71×10^3
Beryllium Oxide (BeO)	500	71×10^3
Thorium Dioxide (ThO$_2$)	800	71×10^3
Spinel (Al$_2$O$_3$ MgO)	1200	71×10^3
Mullite (3Al$_2$O$_3$ 2SiO$_2$)	25	$80-190 \times 10^3$
Silicon Carbide (SiC)	25	$82-200 \times 10^3$
Aluminum Oxide (Al$_2$O$_3$)	1100	85×10^3
Thorium Dioxide (ThO$_2$)	600	85×10^3
Spinel (Al$_2$O$_3$ MgO)	1100	85.5×10^3
Titanium Monocarbide (TiC)	room temp.	$109-190 \times 10^3$
Magnesium Oxide (MgO)	room temp.	112×10^3
Beryllium Oxide (BeO)	room temp.	$114-310 \times 10^3$
Zirconium Oxide (ZrO$_2$)	1200	114×10^3
Aluminum Oxide (Al$_2$O$_3$)	1000	128×10^3
Titanium Mononitirde (TiN)		141×10^3
Thorium Dioxide (ThO$_2$)	room temp.	$146-214 \times 10^3$
Thorium Dioxide (ThO$_2$)	400	156×10^3
Zirconium Oxide (ZrO$_2$)	1000	171×10^3
Spinel (Al$_2$O$_3$ MgO)	800	171×10^3
Aluminum Oxide (Al$_2$O$_3$)	800	183×10^3
Aluminum Oxide (Al$_2$O$_3$)	600	199×10^3
Spinel (Al$_2$O$_3$ MgO)	500	199×10^3

Selecting Compressive Strengths of Ceramics (Continued)

Ceramic	Temperature (°C)	Compressive Strength (psi)
Zirconium Oxide (ZrO$_2$)	room temp.	205-300x10^3
Aluminum Oxide (Al$_2$O$_3$)	400	214 x10^3
Zirconium Oxide (ZrO$_2$)	500	228x10^3
Zirconium Monocarbide (ZrC)	room temp.	238x10^3
Spinel (Al$_2$O$_3$ MgO)	room temp.	270x10^3
Boron Carbide (B$_4$C)	room temp.	414x10^3
Aluminum Oxide (Al$_2$O$_3$)	room temp.	427x10^3
Trichromium Dicarbide (Cr$_3$C$_2$)		600x10^3

To convert **psi** to **MPa**, multiply by **145**.

Source: *data compiled by* J.S. Park from *No. 1 Materials Index*, Peter T.B. Shaffer, Plenum Press, New York, (1964); *Smithells Metals Reference Book*, Eric A. Brandes, ed., in association with Fulmer Research Institute Ltd. 6th ed. London, Butterworths, Boston, (1983); and *Ceramic Source*, American Ceramic Society (1986-1991).

SELECTING HARDNESS OF CERAMICS

Ceramic	Hardness
Tantalum Monocarbide (TaC)	Brinell: 840
Titanium Oxide (TiO$_2$)	Knoop: 713-1121 kg/mm^2
Trisilicon tetranitride (Si$_3$N$_4$) (α)	Knoop: 815-1936kg/mm^2
Zirconium Oxide (ZrO$_2$) (partially stabilized)	Knoop: 1019-1121 kg/mm^2
Zirconium Oxide (ZrO$_2$)(fully stabilized)	Knoop: 1019-1529 kg/mm^2
Trichromium Dicarbide (Cr$_3$C$_2$)	Knoop: 1019-1834 kg/mm^2
Hafnium Monocarbide (HfC)	Knoop: 1790-1870 kg/mm^2
Zirconium Monocarbide (ZrC)	Knoop: 2138 kg/mm^2
Silicon Carbide (SiC) (cubic, CVD)	Knoop: 2853-4483 kg/mm^2
Dichromium Trioxide (Cr$_2$O$_3$)	Knoop: 2955 kg/mm^2
Zirconium Mononitride (ZrN)	Knoop 30g: 1983 kg/mm^2
Titanium Mononitirde (TiN)	Knoop 30g: 2160 kg/mm^2
Tantalum Diboride (TaB$_2$)	Knoop 30g: 2537 kg/mm^2
Titanium Diboride (TiB$_2$)	Knoop 30g: 3370 kg/mm^2
Tantalum Monocarbide (TaC)	Knoop 50g: 1800-1952 kg/mm^2
Calcium Oxide (CaO)	Knoop 100g: 560 kg/mm^2
Uranium Dioxide (UO$_2$)	Knoop 100g: 600 kg/mm^2
Silicon Dioxide (SiO$_2$) (parallel to optical axis)	Knoop 100g: 710 kg/mm^2
Silicon Dioxide (SiO$_2$) (normal to optical axis)	Knoop 100g: 790 kg/mm^2
Tantalum Monocarbide (TaC)	Knoop 100g: 825 kg/mm^2
Thorium Dioxide (ThO$_2$)	Knoop 100g: 945 kg/mm^2
Tungsten Disilicide (WSi$_2$)	Knoop 100g: 1090 kg/mm^2
Zirconium Oxide (ZrO$_2$)	Knoop 100g: 1200 kg/mm^2

Selecting Hardness of Ceramics (Continued)

Ceramic	Hardness
Aluminum Nitride (AlN)	Knoop 100g: 1225-1230 kg/mm^2
Molybdenum Disilicide (MoSi$_2$)	Knoop 100g: 1257 kg/mm^2
Beryllium Oxide (BeO)	Knoop 100g: 1300 kg/mm^2
Zirconium Mononitride (ZrN)	Knoop 100g: 1510 kg/mm^2
Zirconium Diboride (ZrB$_2$)	Knoop 100g: 1560 kg/mm^2
Chromium Diboride (CrB2)	Knoop 100g: 1700 kg/mm^2
Titanium Mononitirde (TiN)	Knoop 100g: 1770 kg/mm^2
Tungsten Monocarbide (WC)	Knoop 100g: 1870-1880 kg/mm^2
Aluminum Oxide (Al$_2$O$_3$)	Knoop 100g: 2000-2050 kg/mm^2
Titanium Monocarbide (TiC)	Knoop 100g: 2470 kg/mm^2
Silicon Carbide (SiC)	Knoop 100g: 2500-2550 kg/mm^2
Tantalum Diboride (TaB$_2$)	Knoop 100g: 2615 ± 120 kg/mm^2
Titanium Diboride (TiB$_2$)	Knoop 100g: 2710-3000 kg/mm^2
Silicon Carbide (SiC)	Knoop 100g: 2745 kg/mm^2 (green)
Boron Carbide (B$_4$C)	Knoop 100g: 2800 kg/mm^2
Silicon Carbide (SiC)	Knoop 100g: 2960 kg/mm^2 (black)
Titanium Diboride (TiB$_2$) (single crystal)	Knoop 100g: 3250±100 kg/mm^2
Zirconium Diboride (ZrB$_2$) (single crystal)	Knoop 160g: 2000 kg/mm^2
Zirconium Diboride (ZrB$_2$)	Knoop 160g: 2100 kg/mm^2
Hafnium Diboride (HfB2) (polycrystalline)	Knoop 160g: 2400kg/mm at 24 °C
Titanium Diboride (TiB$_2$)	Knoop 160g: 3500 kg/mm^2
Hafnium Diboride (HfB2) (single crystal)	Knoop 160g: 3800kg/mm at 24 °C
Titanium Monocarbide (TiC)	Knoop 1000g: 1905 kg/mm^2
Boron Carbide (B$_4$C)	Knoop 1000g: 2230 kg/mm^2

Selecting Hardness of Ceramics (Continued)

Ceramic	Hardness
Tantalum Diboride (TaB$_2$)	Micro: 1700 kg/mm^2
Zirconium Monocarbide (ZrC)	Micro: 2090 kg/mm^2
Titanium Monocarbide (TiC)	Micro 20g: 3200 kg/mm^2
Molybdenum Disilicide (MoSi$_2$)	Micro 50g: 1200 kg/mm^2
Tungsten Disilicide (WSi$_2$)	Micro 50g: 1260 kg/mm^2
Molybdenum Disilicide (MoSi$_2$)	Micro 100g: 1290 kg/mm^2
Chromium Diboride (CrB2)	Micro 100g: 1800 kg/mm^2
Boron Nitride (BN) (hexagonal)	Mohs: 2
Aluminum Nitride (AlN)	Mohs: 5-5.5
Magnesium Oxide (MgO)	Mohs: 5.5
Uranium Dioxide (UO$_2$)	Mohs: 6-7
Sillimanite (Al$_2$O$_3$ SiO$_2$)	Mohs: 6-7
Thorium Dioxide (ThO$_2$)	Mohs: 6.5
Zirconium Oxide (ZrO$_2$)	Mohs: 6.5
Mullite (3Al$_2$O$_3$ 2SiO$_2$)	Mohs: 7.5
Zircon (SiO$_2$ ZrO$_2$)	Mohs: 7.5
Zirconium Mononitride (ZrN)	Mohs: 8+
Titanium Mononitirde (TiN)	Mohs: 8-10
Aluminum Oxide (Al$_2$O$_3$) (single crystal)	Mohs: 9
Trisilicon tetranitride (Si$_3$N$_4$)	Mohs: 9+
Silicon Carbide (SiC)	Mohs: 9.2
Beryllium Oxide (BeO)	R45N: 64-67
Mullite (3Al$_2$O$_3$ 2SiO$_2$)	R45N: 71
Aluminum Oxide (Al$_2$O$_3$)	R45N: 78-90

Selecting Hardness of Ceramics (Continued)

Ceramic   Hardness

Aluminum Nitride (AlN) (thin film)   Rockwell 15N: 94.0
Aluminum Nitride (AlN) (thick film)   Rockwell 15N: 94.5

Tungsten Monocarbide (WC)
      (6% Co, 1-3μm grain size)   Rockwell A: 81.4 ± 0.4
Tungsten Monocarbide (WC)
      (24% Co, 1-3μm grain size)   Rockwell A: 86.9 ± 0.6

Zirconium Diboride (ZrB$_2$)   Rockwell A: 87-89
Tungsten Monocarbide (WC)
      (6% Co, 3-6μm grain size)   Rockwell A: 87.3 ± 0.5

Titanium Monocarbide (TiC) (98.6% density)   Rockwell A: 88-89
Tungsten Monocarbide (WC)
      (6% Co, 2-4μm grain size)   Rockwell A: 88.6 ± 0.5

Tantalum Diboride (TaB$_2$)   Rockwell A: 89
Tantalum Monocarbide (TaC)   Rockwell A: 89
Tungsten Monocarbide (WC)
      (12% Co, 1-3μm grain size)   Rockwell A: 89.4 ± 0.5

Titanium Monocarbide (TiC) (99.5% density)   Rockwell A: 91-93.5
Titanium Monocarbide (TiC) (100% density)   Rockwell A: 91-93.5
Tungsten Monocarbide (WC)   Rockwell A: 92

Zirconium Monocarbide (ZrC)   Rockwell A: 92.5
Trisilicon tetranitride (Si$_3$N$_4$)   Rockwell A: 99

Cordierite (2MgO 2Al$_2$O$_3$ 5SiO$_2$) (glass)   Vickers: 672.5 kg/mm^2
Titanium Oxide (TiO$_2$)   Vickers: 713-1121 kg/mm^2
Trisilicon tetranitride (Si$_3$N$_4$) (α)   Vickers: 815-1936kg/mm^2
Cordierite (2MgO 2Al$_2$O$_3$ 5SiO$_2$)   Vickers: 835.6 kg/mm^2

Selecting Hardness of Ceramics (Continued)

Ceramic   Hardness

Zirconium Oxide (ZrO_2) (partially stabilized)   Vickers: 1019-1121 kg/mm^2
Zirconium Oxide (ZrO_2)(fully stabilized)   Vickers: 1019-1529 kg/mm^2
Trichromium Dicarbide (Cr_3C_2)   Vickers: 1019-1834 kg/mm^2
Mullite ($3Al_2O_3\ 2SiO_2$)   Vickers: 1120 kg/mm^2

Boron Carbide (B_4C)   Vickers: 2400 kg/mm^2
Silicon Carbide (SiC) (cubic, CVD)   Vickers: 2853-4483 kg/mm^2
Dichromium Trioxide (Cr_2O_3)   Vickers: 2955 kg/mm^2

Tungsten Disilicide (WSi_2)   Vickers 10g: 1632 kg/mm^2
Aluminum Oxide (Al_2O_3)   Vickers 20g: 2600 kg/mm^2
Silicon Carbide (SiC)   Vickers 25g: 3000-3500 kg/mm^2

Chromium Diboride (CrB2)   Vickers 50g: 1800 kg/mm^2
Tantalum Monocarbide (TaC)   Vickers 50g: 1800 kg/mm^2
Zirconium Diboride (ZrB_2)   Vickers 50g: 2200 kg/mm^2

Tungsten Monocarbide (WC)   Vickers 50g: 2400 kg/mm^2
Hafnium Monocarbide (HfC)   Vickers 50g: 2533-3202 kg/mm^2
Zirconium Monocarbide (ZrC)   Vickers 50g: 2600 kg/mm^2

Aluminum Oxide (Al_2O_3)   Vickers 50g: 2720 kg/mm^2
Titanium Monocarbide (TiC)   Vickers 50g: 2900-3200 kg/mm^2
Titanium Diboride (TiB_2)   Vickers 50g: 3400 kg/mm^2

Tungsten Disilicide (WSi_2)   Vickers 100g: 1090 kg/mm^2
Molybdenum Disilicide ($MoSi_2$)   Vickers 100g: 1290-1550 kg/mm^2
Tungsten Monocarbide (WC)   Vickers 100g: 1730 kg/mm^2

Zirconium Monocarbide (ZrC)   Vickers 100g: 2836-3840 kg/mm^2
Titanium Monocarbide (TiC)   Vickers 100g: 2850-3390 kg/mm^2

Selecting Hardness of Ceramics (Continued)

Ceramic   Hardness

Silicon Dioxide (SiO_2)
    (1011 face) 10 μm diagonal   Vickers 500g: 1040-1130 kg/mm^2
Silicon Dioxide (SiO_2) (normal to optical axis)   Vickers 500g: 1103 kg/mm^2
Silicon Dioxide (SiO_2)   Vickers 500g: 1120 kg/mm^2
Silicon Dioxide (SiO_2) (parallel to optical axis)   Vickers 500g: 1260 kg/mm^2

Silicon Dioxide (SiO_2)
    (polished 1010 face) 10 μm diagonal   Vickers 500g: 1300 kg/mm^2
Silicon Dioxide (SiO_2)
    (1010 face) 10 μm diagonal   Vickers 500g:1120-1230 kg/mm^2

Source: data compiled by J.S. Park from *No. 1 Materials Index*, Peter T.B. Shaffer, Plenum Press, New York, (1964); *Smithells Metals Reference Book*, Eric A. Brandes, ed., in association with Fulmer Research Institute Ltd. 6th ed. London, Butterworths, Boston, (1983); and *Ceramic Source*, American Ceramic Society (1986-1991).

SELECTING MICROHARDNESS OF GLASS

SiO_2 glass	Knoop	500–679
SiO_2 glass	Knoop	500–679
B_2O_3 glass	Vickers	194–205
SiO_2–B_2O_3 glass (95% mol B_2O_3)	Vickers	227–253
SiO_2–B_2O_3 glass (90% mol B_2O_3)	Vickers	231–257
SiO_2–B_2O_3 glass (75% mol B_2O_3)	Vickers	237–269–345
SiO_2–B_2O_3 glass (85% mol B_2O_3)	Vickers	239–267
SiO_2–B_2O_3 glass (80% mol B_2O_3)	Vickers	239–271
SiO_2–B_2O_3 glass (70% mol B_2O_3)	Vickers	251–279
B_2O_3–Na_2O glass (5% mol Na_2O)	Vickers	276
B_2O_3–Na_2O glass (10% mol Na_2O)	Vickers	292
SiO_2–B_2O_3 glass (65% mol B_2O_3)	Vickers	293–297
B_2O_3–Na_2O glass (15% mol Na_2O)	Vickers	297
SiO_2–B_2O_3 glass (60% mol B_2O_3)	Vickers	328–345
SiO_2–Na_2O glass (45% mol Na_2O)	Vickers	378±2
B_2O_3–Na_2O glass (20% mol Na_2O)	Vickers	380
SiO_2–Na_2O glass (40% mol Na_2O)	Vickers	394±2
SiO_2–Na_2O glass (30% mol Na_2O)	Vickers	413±3
SiO_2–Na_2O glass (35% mol Na_2O)	Vickers	414±4
SiO_2–Na_2O glass (25% mol Na_2O)	Vickers	423±4
B_2O_3–Na_2O glass (25% mol Na_2O)	Vickers	460
B_2O_3–Na_2O glass (30% mol Na_2O)	Vickers	503

Source: *data compiled by* J.S. Park *from* O. V. Mazurin, M. V. Streltsina and T. P. Shvaiko–Shvaikovskaya, *Handbook of Glass Data, Part A and Part B,* Elsevier, New York, 1983

SELECTING YOUNG'S MODULI OF CERAMICS

Ceramic	Temperature	Young's Modulus (psi)
Boron Nitride (BN), parallel to c axis	700°C	0.51×10^6
Boron Nitride (BN), parallel to a axis	700°C	1.54×10^6
Boron Nitride (BN), parallel to a axis	1000°C	1.65×10^6
Zirconium Oxide (ZrO$_2$) (plasma sprayed)	500°C	2×10^6
Zirconium Oxide (ZrO$_2$) (plasma sprayed)	1100°C	3.05×10^6
Zirconium Diboride (ZrB$_2$) (22.4% density, foam)		3.305×10^6
Boron Nitride (BN), parallel to c axis	300°C	3.47×10^6
Magnesium Oxide (MgO)	1300°C	4×10^6
Mullite (3Al$_2$O$_3$ 2SiO$_2$) (ρ=2.77 g/cm^3)	1200°C	4.00×10^6
Boron Nitride (BN), parallel to c axis	23°C	4.91×10^6
Titanium Diboride (TiB$_2$) (12.0 μm grain size, ρ=4.66g/cm^3, 9.6wt% Ni)		6.29×10^6
Zirconium Oxide (ZrO$_2$) (plasma sprayed)	room temp.	6.96×10^6
Hafnium Dioxide (HfO$_2$)		8.2×10^6
Boron Nitride (BN), parallel to a axis	300°C	8.79×10^6
Magnesium Oxide (MgO)	1200°C	10×10^6
Titanium Mononitride (TiN)		$11.47\text{-}36.3 \times 10^6$
Boron Nitride (BN), parallel to a axis	23°C	12.46×10^6
Thorium Dioxide (ThO$_2$)	1200°C	12.8×10^6
Zirconium Oxide (ZrO$_2$) (plasma sprayed)	1500°C	12.8×10^6
Cordierite (2MgO 2Al$_2$O$_3$ 5SiO$_2$) glass		13.92×10^6
Zirconium Oxide (ZrO$_2$) (fully stabilized)	room temp.	$14.1\text{-}30.0 \times 10^6$
Zirconium Oxide (ZrO$_2$) (plasma sprayed)	1400°C	14.2×10^6
Trisilicon tetranitride (Si$_3$N$_4$) (reaction sintered)	20°C	$14.5\text{-}31.9 \times 10^6$

Selecting Young's Moduli of Ceramics (Continued)

Ceramic	Temperature	Young's Modulus (psi)
Mullite ($3Al_2O_3$ $2SiO_2$) (ρ=2.77 g/cm^3)	800°C	14.79x10^6
Dichromium Trioxide (Cr_2O_3)		>14.9x10^6
Zirconium Oxide (ZrO_2) (plasma sprayed)	1200°C	17.1-18.0x10^6
Thorium Dioxide (ThO_2)	1000°C	17.1x10^6
Trisilicon tetranitride (Si_3N_4) (reaction sintered)	1400°C	17.4-29.0x10^6
Thorium Dioxide (ThO_2)	room temp.	17.9-34.87x10^6
Thorium Dioxide (ThO_2)	800°C	18-18.5x10^6
Mullite ($3Al_2O_3$ $2SiO_2$) (ρ=2.77 g/cm^3)	25°C	18.42x10^6
Zirconium Oxide (ZrO_2) (plasma sprayed)	1000°C	18.5-25x10^6
Mullite ($3Al_2O_3$ $2SiO_2$) (ρ=2.77 g/cm^3)	400°C	18.89x10^6
Zirconium Oxide (ZrO_2) (plasma sprayed)	800°C	18.9x10^6
Zirconium Oxide (ZrO_2) (stabilized, ρ=5.634 g/cm^3)	room temp.	19.96x10^6
Beryllium Oxide (BeO)	1145°C	20 x10^6
Spinel (Al_2O_3 MgO)	1300°C	20.1x10^6
Cordierite (2MgO $2Al_2O_3$ $5SiO_2$)		20.16x10^6
Mullite ($3Al_2O_3$ $2SiO_2$) (ρ=2.779 g/cm^3)	room temp.	20.75x10^6
Magnesium Oxide (MgO)	1000°C	21 x10^6
Uranium Dioxide (UO_2)	0-1000°C	21x10^6
Zircon (SiO_2 ZrO_2)	room temp.	24x10^6
Zirconium Oxide (ZrO_2) (plasma sprayed)	room temp.	24.8-27x10^6
Cerium Dioxide (CeO_2)		24.9x10^6
Spinel (Al_2O_3 MgO)	1200°C	25.0x10^6

Selecting Young's Moduli of Ceramics (Continued

Ceramic	Temperature	Young's Modulus (psi)
Uranium Dioxide (UO_2)	20°C	25×10^6
Trisilicon tetranitride (Si_3N_4) (hot pressed)	1400°C	$25.38\text{-}36.25 \times 10^6$
Aluminum Oxide (Al_2O_3)	1500°C	25.6×10^6
Uranium Dioxide (UO_2) (ρ=10.37 g/cm^3)	room temp.	27.98×10^6
Trisilicon tetranitride (Si_3N_4) (sintered)	20°C	$28.28\text{-}45.68 \times 10^6$
Zirconium Monocarbide (ZrC)	room temp.	$28.3\text{-}69.6 \times 10^6$
Silicon Carbide (SiC) (reaction sintered)	1400°C	$29\text{-}46.4 \times 10^6$
Magnesium Oxide (MgO)	600°C	29.5×10^6
Zirconium Oxide (ZrO_2) (partially stabilized)	room temp.	29.7×10^6
Spinel (Al_2O_3 MgO)	1000°C	30.4×10^6
Magnesium Oxide (MgO)	room temp.	$30.5\text{-}36.3 \times 10^6$
Chromium Diboride (CrB_2)		30.6×10^6
Aluminum Oxide (Al_2O_3)	1250°C	32×10^6
Aluminum Oxide (Al_2O_3)	1400°C	32.7×10^6
Spinel (Al_2O_3 MgO)	800°C	32.9×10^6
Beryllium Oxide (BeO)	1000°C	33×10^6
Mullite ($3Al_2O_3$ $2SiO_2$) (full density)	room temp.	33.35×10^6
Spinel (Al_2O_3 MgO)	600°C	34×10^6
Spinel (Al_2O_3 MgO)	200°C	34.4×10^6
Spinel (Al_2O_3 MgO)	room temp.	34.5×10^6
Spinel (Al_2O_3 MgO)	400°C	34.5×10^6
Zirconium Oxide (ZrO_2) (plasma sprayed)	20°C	36×10^6
Trisilicon tetranitride (Si_3N_4) (hot pressed)	20°C	$36.25\text{-}47.13 \times 10^6$
Tantalum Diboride (TaB_2)		37×10^6

Selecting Young's Moduli of Ceramics (Continued)

Ceramic	Temperature	Young's Modulus (psi)
Spinel (Al_2O_3 MgO) (ρ=3.510 g/cm^3)	room temp.	38.23x10^6
Molybdenum Disilicide (MoSi$_2$)	room temp.	39.3-56.36x10^6
Aluminum Oxide (Al$_2$O$_3$)	1200°C	39.8-53.65 x10^6
Beryllium Oxide (BeO)	800°C	40 x10^6
Aluminum Nitride (AlN)	1400°C	40x10^6
Titanium Oxide (TiO$_2$)		41x10^6
Tantalum Monocarbide (TaC)	room temp.	41.3-91.3x10^6
Boron Carbide (B$_4$C)	room temp.	42-65.2x10^6
Magnesium Oxide (MgO) (ρ = 3.506 g/cm^3)	room temp.	42.74x10^6
Beryllium Oxide (BeO)	room temp.	42.8-45.5x10^6
Silicon Carbide (SiC) (sintered)	1400°C	43.5-58.0x10^6
Silicon Carbide (SiC) (pressureless sintered)	room temp.	43.9x10^6
Titanium Monocarbide (TiC)	1000°C	45-55x10^6
Aluminum Oxide (Al$_2$O$_3$)	1000°C	45.5-50 x10^6
Aluminum Nitride (AlN)	1000°C	46x10^6
Zirconium Diboride (ZrB$_2$)		49.8-63.8x10^6
Aluminum Oxide (Al$_2$O$_3$)	500°C	50-57.275 x10^6
Aluminum Oxide (Al$_2$O$_3$)	room temp.	50-59.3x10^6
Aluminum Nitride (AlN)	25°C	50x10^6
Silicon Carbide (SiC) (reaction sintered)	20°C	50.75-54.38x10^6
Silicon Carbide (SiC) (reaction sintered)	1200°C	51x10^6
Aluminum Oxide (Al$_2$O$_3$)	800°C	51.2 x10^6
Silicon Carbide (SiC) (reaction sintered)	800°C	53x10^6
Titanium Diboride (TiB$_2$)		53.2x10^6

Selecting Young's Moduli of Ceramics (Continued

Ceramic	Temperature	Young's Modulus (psi)
Trichromium Dicarbide (Cr$_3$C$_2$)		54.1x10^6
Silicon Carbide (SiC) (sintered)	20°C	54.38-60.9x10^6
Silicon Carbide (SiC) (reaction sintered)	400°C	55x10^6
Silicon Carbide (SiC) (hot presses)	1400°C	55.1x10^6
Silicon Carbide (SiC) (ρ = 3.128 g/cm^3)	room temp.	58.2x10^6
Silicon Carbide (SiC) (self bonded)	room temp.	59.5x10^6
Silicon Carbide (SiC) (ρ = 3.120 g/cm^3)	room temp.	59.52x10^6
Silicon Carbide (SiC) (cubic, CVD)	room temp.	60.2-63.9x10^6
Hafnium Monocarbide (HfC) (ρ = 11.94 g/cm^3)	room temp.	61.55x10^6
Silicon Carbide (SiC) (hot pressed)	20°C	62.4-65.3x10^6
Titanium Monocarbide (TiC)	room temp.	63.715x10^6
Silicon Carbide (SiC) (hot pressed)	room temp.	63.8x10^6
Titanium Diboride (TiB$_2$) (3.5 µm grain size, ρ=4.37g/cm^3, 0.8wt% Ni)		75.0x10^6
Titanium Diboride (TiB$_2$) (6.0 µm grain size, ρ=4.56g/cm^3, 0.16wt% Ni)		77.9x10^6
Titanium Diboride (TiB$_2$) (6.0 µm grain size, ρ=4.46g/cm^3)		81.6x10^6
Tungsten Monocarbide (WC)	room temp.	96.91-103.5x10^6

To convert from **psi** to **MPa**, multiply by **145**.

Source: *data compiled by* J.S. Park from *No. 1 Materials Index*, Peter T.B. Shaffer, Plenum Press, New York, (1964); *Smithells Metals Reference Book*, Eric A. Brandes, ed., in association with Fulmer Research Institute Ltd. 6th ed. London, Butterworths, Boston, (1983); and *Ceramic Source*, American Ceramic Society (1986-1991)

SELECTING YOUNG'S MODULI OF GLASS

Glass	Temperature	Young's Modulus (GPa)
B_2O_3 glass	room temp.	17.2–17.7
SiO_2–B_2O_3 glass (90% mol B_2O_3)		20.9
SiO_2–B_2O_3 glass (85% mol B_2O_3)		21.2
SiO_2–B_2O_3 glass (95% mol B_2O_3)		21.2
SiO_2–B_2O_3 glass (65% mol B_2O_3)		22.5
SiO_2–B_2O_3 glass (80% mol B_2O_3)		22.8
SiO_2–B_2O_3 glass (60% mol B_2O_3)		23.3
SiO_2–B_2O_3 glass (70% mol B_2O_3)		23.5
SiO_2–B_2O_3 glass (75% mol B_2O_3)		24.1
B_2O_3–Na_2O glass (10% mol Na_2O)	15°C	31.4
SiO_2–PbO glass (65.0% mol PbO)		41.2
B_2O_3–Na_2O glass (20% mol Na_2O)	15°C	43.2
SiO_2–PbO glass (60.0% mol PbO)		43.6
SiO_2–PbO glass (50.0% mol PbO)		44.1
SiO_2–Na_2O glass (40% mol Na_2O)	200–250°C	46.1
SiO_2–PbO glass (35.7% mol PbO)		46.3
SiO_2–PbO glass (24.6% mol PbO)		47.1
SiO_2–PbO glass (55.0% mol PbO)		49.3
SiO_2–PbO glass (30.0% mol PbO)		50.1
SiO_2–Na_2O glass (33% mol Na_2O)	200–250°C	51.0
SiO_2–PbO glass (45.0% mol PbO)		51.7
SiO_2–Na_2O glass (40% mol Na_2O)	–196°C	51.9
SiO_2–PbO glass (38.4% mol PbO)		52.8
B_2O_3–Na_2O glass (25% mol Na_2O)	15°C	53.7

Selecting Young's Moduli of Glass (Continued)

Glass	Temperature	Young's Modulus (GPa)
SiO_2–Na_2O glass (25% mol Na_2O)	200–250°C	53.9
SiO_2–Na_2O glass (33% mol Na_2O)	–196°C	54.9
SiO_2–Na_2O glass (25% mol Na_2O)	–196°C	56.9
B_2O_3–Na_2O glass (37% mol Na_2O)	15°C	57.1
B_2O_3–Na_2O glass (33.3% mol Na_2O)	15°C	59.4
SiO_2–Na_2O glass (35% mol Na_2O)	room temp.	60.2
SiO_2–Na_2O glass (33% mol Na_2O)	room temp.	60.3
SiO_2–Na_2O glass (30% mol Na_2O)	room temp.	60.5
SiO_2–Na_2O glass (25% mol Na_2O)	room temp.	61.4
SiO_2–Na_2O glass (20% mol Na_2O)	room temp.	62.0
SiO_2–Na_2O glass (15% mol Na_2O)	room temp.	64.4
SiO_2 glass	20°C	72.76–74.15
SiO_2 glass	998°C (annealing point)	79.87
SiO_2 glass	1096°C (straining point)	80.80

Source: *data compiled by* J.S. Park *from* O. V. Mazurin, M. V. Streltsina and T. P. Shvaiko–Shvaikovskaya, *Handbook of Glass Data, Part A and Part B,* Elsevier, New York, 1983.

SELECTING BULK MODULI OF GLASS

Glass	Temperature	Bulk Modulus (GPa)
B_2O_3-Na_2O glass (10% mol Na_2O)	15°C	23.2
SiO_2-PbO glass (38.4% mol PbO)		25.1
SiO_2-PbO glass (30.0% mol PbO)		25.6
SiO_2-PbO glass (55.0% mol PbO)		29.5
SiO_2-PbO glass (50.0% mol PbO)		30.5
SiO_2-PbO glass (45.0% mol PbO)		30.6
SiO_2 glass		31.01-37.62
SiO_2-PbO glass (35.7% mol PbO)		31.1
SiO_2-PbO glass (65.0% mol PbO)		31.6
SiO_2-PbO glass (60.0% mol PbO)		33.1
B_2O_3-Na_2O glass (20% mol Na_2O)	15°C	33.6
SiO_2-Na_2O glass (15% mol Na_2O)	room temp.	33.8
SiO_2-PbO glass (24.6% mol PbO)		33.9
SiO_2-Na_2O glass (20% mol Na_2O)	room temp.	34.8
SiO_2-Na_2O glass (25% mol Na_2O)	room temp.	36.5
SiO_2-Na_2O glass (30% mol Na_2O)	room temp.	38.2
B_2O_3-Na_2O glass (25% mol Na_2O)	15°C	39.2
SiO_2-Na_2O glass (35% mol Na_2O)	room temp.	39.8
SiO_2-Na_2O glass (33% mol Na_2O)	room temp.	40.1
B_2O_3-Na_2O glass (37% mol Na_2O)	15°C	42.1
B_2O_3-Na_2O glass (33.3% mol Na_2O)	15°C	44.4

Source: *data compiled by* J.S. Park *from* O. V. Mazurin, M. V. Streltsina and T. P. Shvaiko-Shvaikovskaya, *Handbook of Glass Data, Part A and Part B,* Elsevier, New York, 1983

SELECTING SHEAR MODULI OF GLASS

Glass	Temperature	Shear Modulus (GPa)
B_2O_3 glass	300°C	4.75
B_2O_3 glass	290°C	5.15
B_2O_3 glass	280°C	5.49
B_2O_3 glass	270°C	5.78
B_2O_3 glass	260°C	6.07
B_2O_3 glass	250°C	6.29
B_2O_3 glass	room temp.	6.55
B_2O_3–Na_2O glass (10% mol Na_2O)	15°C	12.3
SiO_2–PbO glass (65.0% mol PbO)		16.1
B_2O_3–Na_2O glass (20% mol Na_2O)	15°C	16.8
SiO_2–PbO glass (60.0% mol PbO)		17.0
SiO_2–PbO glass (50.0% mol PbO)		17.5
SiO_2–PbO glass (35.7% mol PbO)		18.5
SiO_2–PbO glass (55.0% mol PbO)		20.2
SiO_2–PbO glass (24.6% mol PbO)		20.4
B_2O_3–Na_2O glass (25% mol Na_2O)	15°C	21.1
SiO_2–PbO glass (45.0% mol PbO)		21.2
SiO_2–PbO glass (30.0% mol PbO)		21.4
B_2O_3–Na_2O glass (37% mol Na_2O)	15°C	22.4
SiO_2–PbO glass (38.4% mol PbO)		23.0
B_2O_3–Na_2O glass (33.3% mol Na_2O)	15°C	23.2
SiO_2–Na_2O glass (35% mol Na_2O)	room temp.	24.1
SiO_2–Na_2O glass (18% mol Na_2O)	160°C	24.2
SiO_2–Na_2O glass (33% mol Na_2O)	room temp.	24.2

Selecting Shear Moduli of Glass (Continued)

Glass	Temperature	Shear Modulus (GPa)
SiO_2–Na_2O glass (30% mol Na_2O)	room temp.	24.5
SiO_2–Na_2O glass (18% mol Na_2O)	80°C	24.8
SiO_2–Na_2O glass (18% mol Na_2O)	0°C	25.0
SiO_2–Na_2O glass (25% mol Na_2O)	room temp.	25.2
SiO_2–Na_2O glass (18% mol Na_2O)	–100°C	25.8
SiO_2–Na_2O glass (20% mol Na_2O)	room temp.	25.8
SiO_2–Na_2O glass (7.5% mol Na_2O)	–100—160°C	26.9
SiO_2–Na_2O glass (5% mol Na_2O)	–100°C	27.2
SiO_2–Na_2O glass (5% mol Na_2O)	160°C	27.2
SiO_2–Na_2O glass (15% mol Na_2O)	room temp.	27.2
SiO_2–Na_2O glass (5% mol Na_2O)	0°C	27.4
SiO_2–Na_2O glass (5% mol Na_2O)	80°C	27.6
SiO_2 glass	20°C	31.38
SiO_2 glass	998°C (annealing point)	33.57
SiO_2 glass	1096°C (straining point)	34.15

Source: *data compiled by* J.S. Park *from* O. V. Mazurin, M. V. Streltsina and T. P. Shvaiko–Shvaikovskaya, *Handbook of Glass Data, Part A and Part B,* Elsevier, New York, 1983.

SELECTING MODULI OF RUPTURE FOR CERAMICS

Ceramic	Temperature (°C)	Modulus of Rupture (psi)
Boron Nitride (BN) parallel to c axis	1000	1.08×10^3
Boron Nitride (BN) parallel to c axis	1500	1.25×10^3
Boron Nitride (BN) parallel to c axis	1800	1.50×10^3
Boron Nitride (BN) parallel to c axis	700	1.90×10^3
Boron Nitride (BN) parallel to a axis	1000	2.18×10^3
Boron Nitride (BN) parallel to c axis	2000	2.45×10^3
Zirconium Monocarbide (ZrC)	2000	2.5×10^3
Cordierite (2MgO 2Al$_2$O$_3$ 5SiO$_2$) (ρ=1.8g/cm^3)	1200	3.4×10^3
Boron Nitride (BN) parallel to a axis	700	3.84×10^3
Hafnium Monocarbide (HfC) (ρ = 11.9 g/cm^3)	2200	4.78×10^3
Zirconium Monocarbide (ZrC)	1750	5.14×10^3
Titanium Diboride (TiB$_2$) (98% dense)		5.37×10^3
Titanium Diboride (TiB$_2$) (3.5 μm grain size, ρ=4.37g/cm^3, 0.8wt% Ni)		5.7×10^3
Mullite (3Al$_2$O$_3$ 2SiO$_2$)	25	$6\text{-}27 \times 10^3$
Titanium Diboride (TiB$_2$) (6.0 μm grain size, ρ=4.46g/cm^3)		6.2×10^3
Titanium Diboride (TiB$_2$) (12.0 μm grain size, ρ=4.66g/cm^3, 9.6wt% Ni)		6.29×10^3
Boron Nitride (BN) parallel to c axis	300	7.03×10^3
Trisilicon Tetranitride (Si$_3$N$_4$) (reaction sintered)	20	$7.25\text{-}43.5 \times 10^3$
Boron Nitride (BN) parallel to c axis	25	$7.28\text{-}13.2 \times 10^3$
Cordierite (2MgO 2Al$_2$O$_3$ 5SiO$_2$) (ρ=2.1g/cm^3)	800	8×10^3
Zirconium Monocarbide (ZrC)	1250	8.3×10^3
Mullite (3Al$_2$O$_3$ 2SiO$_2$) (ρ=2.77g/cm^3)	25	8.5×10^3

Selecting Moduli of Rupture for Ceramics (Continued)

Ceramic	Temperature (°C)	Modulus of Rupture (psi)
Titanium Oxide (TiO_2)	room temp.	$10\text{-}14.9 \times 10^3$
Hafnium Dioxide (HfO_2)		10×10^3
Titanium Diboride (TiB_2)		
(6.0 μm grain size, ρ=4.56g/cm³, 0.16wt% Ni)		11.0×10^3
Silicon Carbide (SiC)	1400	11×10^3
Mullite ($3Al_2O_3$ $2SiO_2$) (ρ=2.77g/cm³)	1200	11.5×10^3
Hafnium Monocarbide (HfC) (ρ = 11.9 g/cm³)	2000	12.64×10^3
Titanium Mononitirde (TiN) (10wt% AlO & 10wt% AlN)		13.34×10^3
Mullite ($3Al_2O_3$ $2SiO_2$) (ρ=2.77g/cm³)	400	13.5×10^3
Titanium Monocarbide (TiC) (ρ = 4.85 g/cm³)	2000	13.6×10^3
Silicon Carbide (SiC)	1800	15×10^3
Cordierite ($2MgO$ $2Al_2O_3$ $5SiO_2$) (ρ=2.3g/cm³)	400	15×10^3
Boron Nitride (BN) parallel to a axis	300	15.14×10^3
Boron Nitride (BN) parallel to a axis	25	15.88×10^3
Cordierite ($2MgO$ $2Al_2O_3$ $5SiO_2$) (ρ=2.51g/cm³)	25	16×10^3
Zirconium Monocarbide (ZrC)	room temp.	$16.6\text{-}22.5 \times 10^3$
Mullite ($3Al_2O_3$ $2SiO_2$) (ρ=2.77g/cm³)	800	16.7×10^3
Aluminum Nitride (AlN)	1400	18.1×10^3
Molybdenum Disilicide ($MoSi_2$) (ρ = 5.57 g/cm³)	room temp.	18.57×10^3
Titanium Diboride (TiB_2)		19×10^3
Zirconium Oxide (ZrO_2) (5-10 CaO stabilized)	room temp.	$20\text{-}35 \times 10^3$
Titanium Mononitirde (TiN) (30wt% AlO & 10wt% AlN)		23.93×10^3
Beryllium Oxide (BeO)	room temp.	$24\text{-}29 \times 10^3$
Silicon Carbide (SiC)	1300	25×10^3

Selecting Moduli of Rupture for Ceramics (Continued)

Ceramic	Temperature (°C)	Modulus of Rupture (psi)
Silicon Carbide (SiC)	room temp.	27×10^3
Aluminum Nitride (AlN)	1000	27×10^3
Aluminum Oxide (Al$_2$O$_3$) (80% dense, 20μm grain size)	600	28×10^3
Aluminum Oxide (Al$_2$O$_3$) (80% dense, 20μm grain size)	20	30×10^3
Aluminum Oxide (Al$_2$O$_3$) (80% dense, 20μm grain size)	1100	30×10^3
Zirconium Oxide (ZrO$_2$) (MgO stabilized)	room temp.	30×10^3
Aluminum Oxide (Al$_2$O$_3$) (80% dense, 20μm grain size)	900	31×10^3
Titanium Monocarbide (TiC) ($\rho = 4.85$ g/cm^3)	room temp.	32.67×10^3
Titanium Mononitirde (TiN) (30wt% AlO & 30wt% AlN)		33.25×10^3
Titanium Mononitirde (TiN)		34×10^3
Hafnium Monocarbide (HfC) ($\rho = 11.9$ g/cm^3)	room temp.	34.67×10^3
Molybdenum Disilicide (MoSi$_2$) (hot pressed)	room temp.	$36\text{-}57 \times 10^3$
Dichromium Trioxide (Cr$_2$O$_3$)		$>38 \times 10^3$
Aluminum Nitride (AlN) (hot pressed)	25	38.5×10^3
Trisilicon Tetranitride (Si$_3$N$_4$) (sintered)	20	$39.9\text{-}121.8 \times 10^3$
Silicon Carbide (SiC) (with 1wt% B addictive)		42×10^3
Aluminum Oxide (Al$_2$O$_3$) (80% dense, 3μm grain size)	1100	42×10^3
Molybdenum Disilicide (MoSi$_2$) (sintered)	room temp.	50.7×10^3
Molybdenum Disilicide (MoSi$_2$) (hot pressed)	1200	55.00×10^3
Tungsten Monocarbide (WC)	room temp.	$55.65\text{-}84 \times 10^3$
Aluminum Oxide (Al$_2$O$_3$) (80% dense, 3μm grain size)	20	56×10^3
Silicon Carbide (SiC) (with 1 wt% Be addictive)		58×10^3
Aluminum Oxide (Al$_2$O$_3$) (80% dense, 3μm grain size)	900	58×10^3
Aluminum Oxide (Al$_2$O$_3$)	room temp.	60×10^3

Selecting Moduli of Rupture for Ceramics (Continued)

Ceramic	Temperature (°C)	Modulus of Rupture (psi)
Aluminum Oxide (Al$_2$O$_3$) (80% dense, 3µm grain size)	600	62x10^3
Trisilicon Tetranitride (Si$_3$N$_4$) (hot pressed)	20	65.3-159.5x10^3
Molybdenum Disilicide (MoSi$_2$) (sintered)	980	67.25x10^3
Molybdenum Disilicide (MoSi$_2$) (hot pressed)	1090	72.00x10^3
Molybdenum Disilicide (MoSi$_2$) (sintered)	1090	86.00x10^3
Aluminum Oxide (Al$_2$O$_3$) (single crystal)		131 x10^3
Silicon Carbide (SiC) (with 1wt% Al addictive)		136x10^3
Aluminum Oxide (Al$_2$O$_3$) (zirconia toughened alumina, 15 vol% ZrO$_2$)		137x10^3
Aluminum Oxide (Al$_2$O$_3$) (zirconia toughened alumina, 25 vol% ZrO$_2$)		139x10^3
Aluminum Oxide (Al$_2$O$_3$) (zirconia toughened alumina, 50 vol% ZrO$_2$)		145x10^3
Zirconium Oxide (ZrO$_2$) (sintered yittria doped zirconia)		148x10^3
Zirconium Oxide (ZrO$_2$) (hot pressed yittria doped zirconia)		222x10^3

To convert from **psi** to **MPa**, multiply by **145**.

Source: data compiled by J.S. Park from *No. 1 Materials Index*, Peter T.B. Shaffer, Plenum Press, New York, (1964); *Smithells Metals Reference Book*, Eric A. Brandes, ed., in association with Fulmer Research Institute Ltd. 6th ed. London, Butterworths, Boston, (1983); and *Ceramic Source*, American Ceramic Society (1986-1991).

SELECTING POISSON'S RATIOS FOR CERAMICS

Ceramic	Poisson's Ratio
Titanium Diboride (TiB$_2$)	0.09—0.28
Titanium Diboride (TiB$_2$) (6.0 μm grain size, ρ=4.46g/cm^3)	0.10
Titanium Diboride (TiB$_2$) (6.0 μm grain size, ρ=4.56g/cm^3, 0.16wt% Ni)	0.11
Titanium Diboride (TiB$_2$) (3.5 μm grain size, ρ=4.37g/cm^3, 0.8wt% Ni)	0.12
Zirconium Diboride (ZrB$_2$)	0.144
Titanium Diboride (TiB$_2$) (12.0 μm grain size, ρ=4.66g/cm^3, 9.6wt% Ni)	0.15
Molybdenum Disilicide (MoSi$_2$)	0.158—0.172
Magnesium Oxide (MgO) (ρ = 3.506 g/cm^3) (room temp)	0.163
Hafnium Monocarbide (HfC)	0.166
Cordierite (2MgO 2Al$_2$O$_3$ 5SiO$_2$) (ρ=2.1g/cm^3)	0.17
Tantalum Monocarbide (TaC)	0.1719—0.24
Silicon Carbide (SiC) (ρ = 3.128 g/cm^3) (room temp)	0.183—0.192
Titanium Monocarbide (TiC)	0.187—189
Boron Carbide (B$_4$C)	0.207
Cordierite (2MgO 2Al$_2$O$_3$ 5SiO$_2$) (ρ=2.3g/cm^3)	0.21
Aluminum Oxide (Al$_2$O$_3$)	0.21—0.27
Trisilicon tetranitride (Si$_3$N$_4$) (presureless sintered)	0.22—0.27
Zirconium Oxide (ZrO$_2$) (partially stabilized)	0.23
Zirconium Oxide (ZrO$_2$) (fully stabilized)	0.23—0.32

Selecting Poisson's Ratios for Ceramics (Continued)

Ceramic	Poisson's Ratio
Mullite ($3Al_2O_3\ 2SiO_2$) ($\rho=2.779$ g/cm^3)	0.238
Tungsten Monocarbide (WC)	0.24
Trisilicon tetranitride (Si_3N_4)	0.24
Zirconium Oxide (ZrO_2) (plasma sprayed)	0.25
Zirconium Monocarbide (ZrC) ($\rho = 6.118$ g/cm^3)	0.257
Cordierite ($2MgO\ 2Al_2O_3\ 5SiO_2$) (glass)	0.26
Beryllium Oxide (BeO)	0.26—0.34
Cerium Dioxide (CeO_2)	0.27—0.31
Thorium Dioxide (ThO_2) ($\rho=9.722$ g/cm^3)	0.275
Titanium Oxide (TiO_2)	0.28
Spinel ($Al_2O_3\ MgO$) ($\rho=3.510$ g/cm^3)	0.294
Uranium Dioxide (UO_2) ($\rho=10.37$ g/cm^3)	0.302
Zirconium Oxide (ZrO_2) (room temp)	0.324—0.337

Source: data compiled by J.S. Park from *No. 1 Materials Index*, Peter T.B. Shaffer, Plenum Press, New York, (1964); *Smithells Metals Reference Book*, Eric A. Brandes, ed., in association with Fulmer Research Institute Ltd. 6th ed. London, Butterworths, Boston, (1983); and *Ceramic Source*, American Ceramic Society (1986-1991)

SELECTING POISSON'S RATIOS OF GLASS

Glass	Temperature (°C)	Poisson's Ratio
SiO_2–PbO glass (38.4% mol PbO)		0.150
SiO_2 glass	room temp.	0.166–0.177
SiO_2–PbO glass (30.0% mol PbO)		0.174
SiO_2–Na_2O glass (15% mol Na_2O)	room temp.	0.183
SiO_2–Na_2O glass (20% mol Na_2O)	room temp.	0.203
SiO_2–Na_2O glass (25% mol Na_2O)	room temp.	0.219
SiO_2–PbO glass (45.0% mol PbO)		0.219
SiO_2–PbO glass (55.0% mol PbO)		0.222
SiO_2–Na_2O glass (30% mol Na_2O)	room temp.	0.236
SiO_2–Na_2O glass (35% mol Na_2O)	room temp.	0.248
SiO_2–Na_2O glass (33% mol Na_2O)	room temp.	0.249
SiO_2–PbO glass (24.6% mol PbO)		0.249
SiO_2–PbO glass (35.7% mol PbO)		0.252
SiO_2–PbO glass (50.0% mol PbO)		0.259
B_2O_3–Na_2O glass (15.4% mol Na_2O)		0.271
B_2O_3–Na_2O glass (25% mol Na_2O)	15	0.2713
B_2O_3–Na_2O glass (22.8% mol Na_2O)		0.272
B_2O_3–Na_2O glass (37% mol Na_2O)	15	0.2739
B_2O_3–Na_2O glass (29.8% mol Na_2O)		0.274
B_2O_3–Na_2O glass (10% mol Na_2O)	15	0.2740

Selecting Poisson's Ratios of Glass (Continued)

Glass	Temperature (°C)	Poisson's Ratio
B_2O_3–Na_2O glass (33.3% mol Na_2O)	15	0.2771
B_2O_3–Na_2O glass (5.5% mol Na_2O)		0.279
SiO_2–PbO glass (60.0% mol PbO)		0.281
SiO_2–PbO glass (65.0% mol PbO)		0.283
B_2O_3–Na_2O glass (20% mol Na_2O)	15	0.2860
B_2O_3 glass	room temp.	0.288–0.309
B_2O_3–Na_2O glass (37.25% mol Na_2O)		0.292

Source: *data compiled by* J.S. Park *from* O. V. Mazurin, M. V. Streltsina and T. P. Shvaiko–Shvaikovskaya, *Handbook of Glass Data, Part A and Part B,* Elsevier, New York, 1983

Selecting Ceramics and Glasses: Electrical Properties

SELECTING RESISTIVITY OF CERAMICS

Ceramic	Temperature Range of Validity	Resistivity (Ω–cm)
Boron Carbide (B$_4$C)		0.3–0.8
Titanium Monocarbide (TiC)		0.3–0.8
Zircoium Oxide (ZrO$_2$) (stabilized)	2200°C	0.37
Zircoium Oxide (ZrO$_2$) (stabilized)	2000°C	0.59
Silicon Carbide (SiC) (with 1 wt% Al additive)		0.8
Zircoium Oxide (ZrO$_2$) (stabilized)	1700°C	1.6
Zircoium Oxide (ZrO$_2$) (stabilized)	1300°C	9.4
Zircoium Oxide (ZrO$_2$) (stabilized)	1200°C	77
Silicon Carbide (SiC)	20°C	10^2–10^{12}
Magnesium Oxide (MgO)	1727°C	4×10^2
Zircoium Oxide (ZrO$_2$) (stabilized)	700°C	2300
Cordierite (2MgO 2Al$_2$O$_3$ 5SiO$_2$) (ρ=2.3g/cm^3)	900°C	1.9×10^4
Silicon Carbide (SiC) (with 1 wt% B additive)		2×10^4
Boron Nitride (BN)	1000°C	3.1×10^4
Cordierite (2MgO 2Al$_2$O$_3$ 5SiO$_2$) (ρ=2.3g/cm^3)	700°C	8.0×10^4
Cordierite (2MgO 2Al$_2$O$_3$ 5SiO$_2$) (ρ=2.1g/cm^3)	900°C	3.5×10^5
Cordierite (2MgO 2Al$_2$O$_3$ 5SiO$_2$) (ρ=1.8g/cm^3)	900°C	7.0×10^5
Cordierite (2MgO 2Al$_2$O$_3$ 5SiO$_2$) (ρ=2.3g/cm^3)	500°C	7.7×10^5
Zirconium Diboride (ZrB$_2$)	liquid air temperature	1.8×10^6
Aluminum Oxide (Al$_2$O$_3$)	1000°C	2×10^6
Cordierite (2MgO 2Al$_2$O$_3$ 5SiO$_2$) (ρ=2.1g/cm^3)	700°C	3.0×10^6
Titanium Diboride (TiB$_2$) (polycrystalline) (100% dense, extrapolated)	liquid air temp.	3.7×10^6
Zirconium Mononitirde (TiN)	liquid air	3.97×10^6

Selecting Resistivity of Ceramics (Continued)

Ceramic	Temperature Range of Validity	Resistivity (Ω–cm)
Cordierite (2MgO 2Al$_2$O$_3$ 5SiO$_2$) (ρ=1.8g/cm^3)	700°C	4.7×10^6
Titanium Diboride (TiB$_2$) (monocrystalline) (crystal length 5 cm, 39 deg. and 59 deg. orientation with respect to growth axis)	room temp.	$6.6 \pm 0.2 \times 10^6$
Titanium Diboride (TiB$_2$) (monocrystalline) (crystal length 1.5 cm, 16.5 deg. and 90 deg. orientation with respect to growth axis)	room temp.	$6.7 \pm 0.2 \times 10^6$
Tantalum Monocarbide (TaC) (80% dense)	4.2K	8×10^6
Titanium Mononitirde (TiN)	liquid air	8.13×10^6
Titanium Diboride (TiB$_2$) (polycrystalline) (100% dense, extrapolated)	room temp.	$8.7\text{–}14.1 \times 10^6$
Titanium Diboride (TiB$_2$) (polycrystalline) (85% dense)	room temp.	9.0×10^6
Zirconium Diboride (ZrB$_2$)	20 °C	9.2×10^6
Tantalum Monocarbide (TaC) (80% dense)	80K	10×10^6
Hafnium Diboride (HfB$_2$)	room temp.	$10\text{–}12 \times 10^6$
Titanium Mononitirde (TiN)	room temp.	$11.07\text{–}130 \times 10^6$
Zirconium Mononitirde (TiN)	room temp.	$11.52\text{–}160 \times 10^6$
Tantalum Monocarbide (TaC) (80% dense)	160K	15×10^6
Molybdenum Disilicide (MoSi$_2$)	–80°C	18.9×10^6
Magnesium Oxide (MgO)	1000°C	$0.2\text{–}1 \times 10^8$
Tantalum Monocarbide (TaC) (80% dense)	240K	20×10^6
Chromium Diboride (CrB$_2$)		21×10^6

Selecting Resistivity of Ceramics (Continued)

Ceramic	Temperature Range of Validity	Resistivity (Ω–cm)
Molybdenum Disilicide (MoSi$_2$)	22°C	21.5x10^6
Tantalum Monocarbide (TaC) (80% dense)	300K	25x10^6
Titanium Diboride (TiB$_2$)		
(polycrystalline) (85% dense)	room temp.	26.5–28.4x10^6
Cordierite (2MgO 2Al$_2$O$_3$ 5SiO$_2$) (ρ=2.3g/cm^3)	300°C	3.3x10^7
Tungsten Disilicide (WSi$_2$)		33.4–54.9x10^6
Hafnium Monocarbide (HfC)	4.2K	41x10^6
Hafnium Monocarbide (HfC)	80K	41x10^6
Zirconium Monocarbide (ZrC)	4.2K	41x10^6
Hafnium Monocarbide (HfC)	160K	45x10^6
Zirconium Monocarbide (ZrC)	80K	45x10^6
Zirconium Monocarbide (ZrC)	160K	47x10^6
Cordierite (2MgO 2Al$_2$O$_3$ 5SiO$_2$) (ρ=1.8g/cm^3)	500°C	4.9x10^7
Hafnium Monocarbide (HfC)	240K	49x10^6
Zirconium Monocarbide (ZrC)	240K	53x10^6
Hafnium Monocarbide (HfC)	300K	60x10^6
Zirconium Monocarbide (ZrC)	300K	61–64x10^6
Tantalum Diboride (TaB$_2$)		68 x10^6
Molybdenum Disilicide (MoSi$_2$)	1600°C	75–80x10^6
Cordierite (2MgO 2Al$_2$O$_3$ 5SiO$_2$) (ρ=2.1g/cm^3)	500°C	9.0x10^7
Zirconium Monocarbide (ZrC)	773K	97x10^6
Mullite (3Al$_2$O$_3$ 2SiO$_2$)	500°C	10^8
Zirconium Monocarbide (ZrC)	1273K	137x10^6
Zirconium Mononitirde (TiN)	melting temp.	320x10^6
Titanium Mononitirde (TiN)	melting temp.	340x10^6

Selecting Resistivity of Ceramics (Continued)

Ceramic	Temperature Range of Validity	Resistivity (Ω–cm)
Aluminum Oxide (Al_2O_3)	700°C	5.0×10^8
Cordierite ($2MgO\ 2Al_2O_3\ 5SiO_2$) ($\rho=1.8g/cm^3$)	300°C	3.0×10^9
Boron Nitride (BN) (90% humidity)	25°C	5.0×10^9
Mullite ($3Al_2O_3\ 2SiO_2$)	300°C	10^{10}
Cordierite ($2MgO\ 2Al_2O_3\ 5SiO_2$) ($\rho=2.1g/cm^3$)	300°C	2.0×10^{10}
Boron Nitride (BN)	480°C	2.3×10^{10}
Aluminum Oxide (Al_2O_3)	500°C	6.3×10^{10}
Boron Nitride (BN) (50% humidity)	25°C	7.0×10^{10}
Silicon Carbide (SiC) (with 2.0 wt% BN additive)		1×10^{11}
Aluminum Nitride (AlN)	room temp.	$2 \times 10^{11} – 10^{13}$
Cordierite ($2MgO\ 2Al_2O_3\ 5SiO_2$) ($\rho=2.3g/cm^3$)	100°C	2.5×10^{11}
Boron Nitride (BN) (20% humidity)	25°C	1.0×10^{12}
Cordierite ($2MgO\ 2Al_2O_3\ 5SiO_2$) ($\rho=1.8g/cm^3$)	100°C	1.0×10^{13}
Aluminum Oxide (Al_2O_3)	300°C	1×10^{13}
Silicon Carbide (SiC) (with 1.6 wt% BeO additive)		$>10^{13}$
Trisilicon tetranitride (Si_3N_4)		$>10^{13}$
Boron Nitride (BN)	25°C	1.7×10^{13}
Aluminum Oxide (Al_2O_3)	100°C	2×10^{13}
Cordierite ($2MgO\ 2Al_2O_3\ 5SiO_2$) ($\rho=2.1g/cm^3$)	100°C	3.0×10^{13}
Silicon Carbide (SiC) (with 1 wt% Be additive)		3×10^{13}
Silicon Carbide (SiC) (with 3.2 wt% BeO additive)		4×10^{13}
Cordierite ($2MgO\ 2Al_2O_3\ 5SiO_2$) ($\rho=1.8g/cm^3$)	25°C	1.0×10^{14}
Cordierite ($2MgO\ 2Al_2O_3\ 5SiO_2$) ($\rho=2.3g/cm^3$)	25°C	1×10^{14}
Mullite ($3Al_2O_3\ 2SiO_2$)	25°C	$>10^{14}$

Selecting Resistivity of Ceramics (Continued)

Ceramic	Temperature Range of Validity	Resistivity (Ω–cm)
Cordierite (2MgO 2Al$_2$O$_3$ 5SiO$_2$) (ρ=2.1g/cm^3)	25°C	>1x10^{14}
Beryllium Oxide (BeO)	500°C	1–5x10^{15}
Beryllium Oxide (BeO)	300°C	>10^{15}
Aluminum Oxide (Al$_2$O$_3$)	25°C	>10x10^{14}
Magnesium Oxide (MgO)	27°C	1.3x10^{15}
Beryllium Oxide (BeO)	700°C	1.5–2x10^{15}
Beryllium Oxide (BeO)	1000°C	4–7x10^{15}
Beryllium Oxide (BeO)	25°C	>10^{17}
Silicon Dioxide (SiO$_2$)	room temp.	10^{18}

Source: data compiled by J.S. Park from *No. 1 Materials Index*, Peter T.B. Shaffer, Plenum Press, New York, (1964); *Smithells Metals Reference Book*, Eric A. Brandes, ed., in association with Fulmer Research Institute Ltd. 6th ed. London, Butterworths, Boston, (1983); and *Ceramic Source*, American Ceramic Society (1986–1991).

SELECTING VOLUME RESISTIVITY OF GLASS

Glass	Temperature ($^{\circ}$C)	Resistivity (log Ω cm)
SiO_2–Na_2O glass (57.5% mol Na_2O)	1300	–0.67
SiO_2–Na_2O glass (49.3% mol Na_2O)	1300	–0.61
SiO_2–Na_2O glass (57.5% mol Na_2O)	1200	–0.61
SiO_2–Na_2O glass (49.3% mol Na_2O)	1200	–0.56
SiO_2–Na_2O glass (44.5% mol Na_2O)	1300	–0.52
SiO_2–Na_2O glass (57.5% mol Na_2O)	1100	–0.52
SiO_2–Na_2O glass (49.3% mol Na_2O)	1100	–0.47
SiO_2–Na_2O glass (44.5% mol Na_2O)	1200	–0.46
SiO_2–Na_2O glass (39.5% mol Na_2O)	1400	–0.45
SiO_2–Na_2O glass (39.5% mol Na_2O)	1300	–0.39
SiO_2–Na_2O glass (44.5% mol Na_2O)	1100	–0.38
SiO_2–Na_2O glass (34.7% mol Na_2O)	1400	–0.33
SiO_2–Na_2O glass (39.5% mol Na_2O)	1200	–0.32
SiO_2–Na_2O glass (34.7% mol Na_2O)	1300	–0.27
SiO_2–Na_2O glass (39.5% mol Na_2O)	1100	–0.24
SiO_2–Na_2O glass (34.7% mol Na_2O)	1200	–0.20
SiO_2–Na_2O glass (29.7% mol Na_2O)	1400	–0.16
SiO_2–Na_2O glass (39.5% mol Na_2O)	1000	–0.13
SiO_2–Na_2O glass (34.7% mol Na_2O)	1100	–0.11
SiO_2–Na_2O glass (29.7% mol Na_2O)	1300	–0.10
SiO_2–Na_2O glass (29.7% mol Na_2O)	1200	–0.02
SiO_2–Na_2O glass (34.7% mol Na_2O)	1000	0.00
SiO_2–Na_2O glass (39.5% mol Na_2O)	900	0.00
SiO_2–Na_2O glass (29.7% mol Na_2O)	1100	0.08

Selecting Volume Resistivity of Glass (Continued)

Glass	Temperature (°C)	Resistivity (log Ω cm)
SiO$_2$–Na$_2$O glass (34.7% mol Na$_2$O)	900	0.12
SiO$_2$–Na$_2$O glass (39.5% mol Na$_2$O)	800	0.13
SiO$_2$–Na$_2$O glass (24.8% mol Na$_2$O)	1200	0.17
SiO$_2$–Na$_2$O glass (29.7% mol Na$_2$O)	1000	0.20
SiO$_2$–Na$_2$O glass (24.8% mol Na$_2$O)	1100	0.26
SiO$_2$–PbO glass (66.7% mol PbO)	1000	0.26
SiO$_2$–Na$_2$O glass (19.9% mol Na$_2$O)	1300	0.30
SiO$_2$–Na$_2$O glass (39.5% mol Na$_2$O)	700	0.33
SiO$_2$–Na$_2$O glass (29.7% mol Na$_2$O)	900	0.34
SiO$_2$–CaO glass (55.2% mol CaO)	1600	0.34
SiO$_2$–Na$_2$O glass (19.9% mol Na$_2$O)	1200	0.38
SiO$_2$–Na$_2$O glass (24.8% mol Na$_2$O)	1000	0.38
SiO$_2$–CaO glass (51.4% mol CaO)	1618	0.38
SiO$_2$–PbO glass (60% mol PbO)	1000	0.40
B$_2$O$_3$–Na$_2$O glass (32.8% mol Na$_2$O)	900	0.40
SiO$_2$–CaO glass (55.2% mol CaO)	1550	0.42–0.43
SiO$_2$–CaO glass (51.4% mol CaO)	1560	0.47
SiO$_2$–Na$_2$O glass (19.9% mol Na$_2$O)	1100	0.48
SiO$_2$–CaO glass (51.4% mol CaO)	1500	0.48–0.49
SiO$_2$–PbO glass (66.7% mol PbO)	900	0.50
SiO$_2$–CaO glass (55.2% mol CaO)	1499	0.51–0.53
SiO$_2$–Na$_2$O glass (24.8% mol Na$_2$O)	900	0.52
SiO$_2$–Na$_2$O glass (29.7% mol Na$_2$O)	800	0.52
SiO$_2$–CaO glass (45.4% mol CaO)	1622	0.52

Selecting Volume Resistivity of Glass (Continued)

Glass	Temperature (°C)	Resistivity (log Ω cm)
SiO_2–PbO glass (51.6% mol PbO)	1200	0.54
SiO_2–Na_2O glass (15% mol Na$_2$O)	1500	0.56
SiO_2–CaO glass (45.4% mol CaO)	1585	0.58–0.59
SiO_2–PbO glass (50.0% mol PbO)	1200	0.60
B_2O_3–Na_2O glass (32.8% mol Na$_2$O)	800	0.60
SiO_2–Na_2O glass (15% mol Na$_2$O)	1400	0.61
SiO_2–Na_2O glass (19.9% mol Na$_2$O)	1000	0.61
SiO_2–CaO glass (45.4% mol CaO)	1550	0.65
B_2O_3–Na_2O glass (21.9% mol Na$_2$O)	1000	0.65
SiO_2–Na_2O glass (39.5% mol Na$_2$O)	600	0.67
SiO_2–CaO glass (41.3% mol CaO)	1600	0.67–0.68
SiO_2–PbO glass (51.6% mol PbO)	1100	0.70
B_2O_3–Na_2O glass (27.5% mol Na$_2$O)	900	0.70
B_2O_3–CaO glass (40.0% mol CaO)	1250	0.75
SiO_2–Na_2O glass (19.9% mol Na$_2$O)	900	0.76
SiO_2–PbO glass (60% mol PbO)	900	0.76
SiO_2–CaO glass (41.3% mol CaO)	1550	0.76
SiO_2–Na_2O glass (29.7% mol Na$_2$O)	700	0.78
SiO_2–CaO glass (33.6% mol CaO)	1600	0.79–0.80
SiO_2–PbO glass (50.0% mol PbO)	1100	0.80
SiO_2–PbO glass (44.7% mol PbO)	1300	0.82
SiO_2–PbO glass (66.7% mol PbO)	800	0.82
SiO_2–CaO glass (41.3% mol CaO)	1519	0.82
B_2O_3–CaO glass (33.3% mol CaO)	1250	0.85

Selecting Volume Resistivity of Glass (Continued)

Glass	Temperature (°C)	Resistivity (log Ω cm)
B_2O_3–Na_2O glass (17.3% mol Na_2O)	1000	0.89
SiO_2–Na_2O glass (39.5% mol Na_2O)	550	0.91
SiO_2–Na_2O glass (10% mol Na_2O)	1600	0.92
SiO_2–PbO glass (51.6% mol PbO)	1000	0.92
SiO_2–CaO glass (33.6% mol CaO)	1560	0.93–0.94
B_2O_3–Na_2O glass (21.9% mol Na_2O)	900	0.94
SiO_2–Na_2O glass (19.9% mol Na_2O)	800	0.96
SiO_2–CaO glass (33.6% mol CaO)	1500	0.97
SiO_2–PbO glass (44.7% mol PbO)	1200	0.98
B_2O_3–CaO glass (40.0% mol CaO)	1150	0.98
B_2O_3–Na_2O glass (27.5% mol Na_2O)	800	1.00
SiO_2–PbO glass (50.0% mol PbO)	1000	1.02
B_2O_3–Na_2O glass (32.8% mol Na_2O)	700	1.02
SiO_2–Na_2O glass (10% mol Na_2O)	1500	1.03
SiO_2–PbO glass (38.5% mol PbO)	1300	1.04
SiO_2–PbO glass (60% mol PbO)	800	1.07
B_2O_3–CaO glass (33.3% mol CaO)	1150	1.10
SiO_2–PbO glass (44.7% mol PbO)	1100	1.15
SiO_2–Na_2O glass (29.7% mol Na_2O)	600	1.16
B_2O_3–Na_2O glass (17.3% mol Na_2O)	900	1.18
SiO_2–PbO glass (51.6% mol PbO)	900	1.20
B_2O_3–CaO glass (55.4% mol CaO)	1150	1.22
SiO_2–PbO glass (38.5% mol PbO)	1200	1.26
B_2O_3–Na_2O glass (21.9% mol Na_2O)	800	1.29

Selecting Volume Resistivity of Glass (Continued)

Glass	Temperature (°C)	Resistivity (log Ω cm)
SiO_2–Na_2O glass (29.7% mol Na_2O)	550	1.31
SiO_2–PbO glass (66.7% mol PbO)	700	1.32
SiO_2–Na_2O glass (19.9% mol Na_2O)	700	1.34
SiO_2–PbO glass (50.0% mol PbO)	900	1.36
B_2O_3–Na_2O glass (17.3% mol Na_2O)	850	1.39
SiO_2–PbO glass (44.7% mol PbO)	1000	1.40
B_2O_3–CaO glass (40.0% mol CaO)	1050	1.40
B_2O_3–Na_2O glass (12.1% mol Na_2O)	900	1.48
B_2O_3–CaO glass (33.3% mol CaO)	1050	1.52
SiO_2–PbO glass (38.5% mol PbO)	1100	1.56
SiO_2–PbO glass (51.6% mol PbO)	800	1.62
SiO_2–Na_2O glass (19.9% mol Na_2O)	600	1.68
B_2O_3–CaO glass (55.4% mol CaO)	1050	1.70
SiO_2–PbO glass (60% mol PbO)	650	1.72
SiO_2–PbO glass (60% mol PbO)	700	1.74
SiO_2–PbO glass (44.7% mol PbO)	900	1.82
B_2O_3–Na_2O glass (12.1% mol Na_2O)	800	1.89
SiO_2–PbO glass (50.0% mol PbO)	800	1.90
SiO_2–PbO glass (38.5% mol PbO)	1000	1.94
B_2O_3–Na_2O glass (3.63% mol Na_2O)	1000	2.00
B_2O_3–CaO glass (40.0% mol CaO)	950	2.06
B_2O_3–CaO glass (33.3% mol CaO)	950	2.25
SiO_2 glass (0.5 atm Ar pressure)	2100	2.30
B_2O_3–Na_2O glass (3.63% mol Na_2O)	900	2.30

Selecting Volume Resistivity of Glass (Continued)

Glass	Temperature (˚C)	Resistivity (log Ω cm)
SiO_2–Na_2O glass (45% mol Na_2O)	350	2.35
SiO_2–PbO glass (44.7% mol PbO)	800	2.38
B_2O_3–Na_2O glass (12.1% mol Na_2O)	700	2.43
B_2O_3–CaO glass (55.4% mol CaO)	950	2.46
SiO_2–PbO glass (38.5% mol PbO)	900	2.47
SiO_2–Na_2O glass (48% mol Na_2O)	300	2.58
SiO_2–Na_2O glass (40% mol Na_2O)	350	2.66
SiO_2–Na_2O glass (45% mol Na_2O)	300	2.69
SiO_2 glass (0.5 atm Ar pressure)	2000	2.70
B_2O_3–Na_2O glass (3.63% mol Na_2O)	800	2.70
SiO_2–Na_2O glass (35% mol Na_2O)	350	2.92
SiO_2–Na_2O glass (40% mol Na_2O)	300	2.97
B_2O_3–CaO glass (40.0% mol CaO)	850	2.97
SiO_2 glass (0.5 atm Ar pressure)	1900	3.00
B_2O_3–CaO glass (33.3% mol CaO)	850	3.10
SiO_2–PbO glass (38.5% mol PbO)	800	3.20
SiO_2–Al_2O_3 glass (5.51% wt Al_2O_3)	1900	3.20
SiO_2–Al_2O_3 glass (10.86% wt Al_2O_3)	1900	3.20
SiO_2–Na_2O glass (36% mol Na_2O)	300	3.22
SiO_2–Al_2O_3 glass (2.83% wt Al_2O_3)	1900	3.28
SiO_2–Na_2O glass (45% mol Na_2O)	250	3.30
SiO_2–Na_2O glass (33.3% mol Na_2O)	300	3.34
SiO_2–Al_2O_3 glass (10.86% wt Al_2O_3)	1700	3.34
SiO_2–Al_2O_3 glass (5.51% wt Al_2O_3)	1700	3.36

Selecting Volume Resistivity of Glass (Continued)

Glass	Temperature (°C)	Resistivity (log Ω cm)
SiO_2–Na_2O glass (30% mol Na_2O)	350	3.46
SiO_2–Al_2O_3 glass (2.83% wt Al_2O_3)	1700	3.46
SiO_2 glass (0.5 atm Ar pressure)	1800	3.48
SiO_2–Na_2O glass (25% mol Na_2O)	350	3.52
SiO_2–Al_2O_3 glass (10.86% wt Al_2O_3)	1500	3.52
SiO_2–B_2O_3 glass (2.74% wt B_2O_3)	1900	3.56
SiO_2–Al_2O_3 glass (5.51% wt Al_2O_3)	1500	3.56
SiO_2–Na_2O glass (40% mol Na_2O)	250	3.59
SiO_2–Na_2O glass (30% mol Na_2O)	300	3.64–3.78
SiO_2–Al_2O_3 glass (2.83% wt Al_2O_3)	1500	3.67
SiO_2–Al_2O_3 glass (10.86% wt Al_2O_3)	1300	3.74
SiO_2–B_2O_3 glass (2.74% wt B_2O_3)	1700	3.76
SiO_2–Al_2O_3 glass (5.51% wt Al_2O_3)	1300	3.76
SiO_2–Na_2O glass (20% mol Na_2O)	350	3.80
SiO_2–B_2O_3 glass (19.37% wt B_2O_3)	1900	3.84
SiO_2–Na_2O glass (35% mol Na_2O)	250	3.85
B_2O_3–CaO glass (55.4% mol CaO)	850	3.86
SiO_2–Na_2O glass (27% mol Na_2O)	300	3.94
SiO_2–B_2O_3 glass (5.48% wt B_2O_3)	1900	3.94
SiO_2–Al_2O_3 glass (2.83% wt Al_2O_3)	1300	3.94
SiO_2–B_2O_3 glass (10.75% wt B_2O_3)	1900	3.98
SiO_2 glass (0.5 atm Ar pressure)	1700	4.00
SiO_2–B_2O_3 glass (19.37% wt B_2O_3)	1700	4.00
SiO_2–B_2O_3 glass (2.74% wt B_2O_3)	1500	4.02

Selecting Volume Resistivity of Glass (Continued)

Glass	Temperature (°C)	Resistivity (log Ω cm)
SiO_2–Al_2O_3 glass (10.86% wt Al_2O_3)	1100	4.02
SiO_2–Na_2O glass (25% mol Na_2O)	300	4.03
SiO_2–Na_2O glass (48% mol Na_2O)	150	4.09
SiO_2–B_2O_3 glass (5.48% wt B_2O_3)	1700	4.10
SiO_2–Al_2O_3 glass (5.51% wt Al_2O_3)	1100	4.15
SiO_2–B_2O_3 glass (10.75% wt B_2O_3)	1700	4.16
SiO_2–B_2O_3 glass (19.37% wt B_2O_3)	1500	4.22
SiO_2–Al_2O_3 glass (2.83% wt Al_2O_3)	1100	4.29
SiO_2–B_2O_3 glass (5.48% wt B_2O_3)	1500	4.30
SiO_2–Na_2O glass (15% mol Na_2O)	350	4.32
SiO_2–Na_2O glass (45% mol Na_2O)	150	4.33
SiO_2–Na_2O glass (20% mol Na_2O)	300	4.36–4.64
SiO_2 glass (0.5 atm Ar pressure)	1600	4.40
SiO_2–PbO glass (38.5% mol PbO)	700	4.40
SiO_2–B_2O_3 glass (2.74% wt B_2O_3)	1300	4.40
SiO_2–B_2O_3 glass (10.75% wt B_2O_3)	1500	4.40
SiO_2–Na_2O glass (30% mol Na_2O)	250	4.42
SiO_2–B_2O_3 glass (19.37% wt B_2O_3)	1300	4.48
SiO_2–Na_2O glass (25% mol Na_2O)	250	4.50
SiO_2–Al_2O_3 glass (10.86% wt Al_2O_3)	900	4.54
SiO_2–B_2O_3 glass (5.48% wt B_2O_3)	1300	4.56
SiO_2–Na_2O glass (40% mol Na_2O)	150	4.58
SiO_2–Al_2O_3 glass (5.51% wt Al_2O_3)	900	4.65
SiO_2 glass (0.5 atm Ar pressure)	1500	4.66

Selecting Volume Resistivity of Glass (Continued)

Glass	Temperature (°C)	Resistivity (log Ω cm)
SiO_2–B_2O_3 glass (10.75% wt B_2O_3)	1300	4.69
SiO_2–B_2O_3 glass (2.74% wt B_2O_3)	1100	4.72
SiO_2–Na_2O glass (13% mol Na_2O)	300	4.77–4.79
SiO_2–B_2O_3 glass (19.37% wt B_2O_3)	1100	4.82
SiO_2–Al_2O_3 glass (2.83% wt Al_2O_3)	900	4.82
SiO_2–Na_2O glass (20% mol Na_2O)	250	4.85
SiO_2–Na_2O glass (36% mol Na_2O)	150	4.89
SiO_2 glass	1500	4.90
SiO_2–Na_2O glass (10% mol Na_2O)	350	4.96
SiO_2 glass	1400	5.00
SiO_2–Na_2O glass (33.3% mol Na_2O)	150	5.06
SiO_2–B_2O_3 glass (10.75% wt B_2O_3)	1100	5.08
SiO_2 glass	1300	5.15
SiO_2–Na_2O glass (44.2% mol Na_2O)	100	5.15
SiO_2–B_2O_3 glass (5.48% wt B_2O_3)	1100	5.16
SiO_2–Na_2O glass (10% mol Na_2O)	300	5.18
SiO_2 glass	1200	5.30
SiO_2–Na_2O glass (7.5% mol Na_2O)	300	5.30
SiO_2–B_2O_3 glass (2.74% wt B_2O_3)	900	5.30
SiO_2–Al_2O_3 glass (5.51% wt Al_2O_3)	700	5.34
SiO_2–Al_2O_3 glass (10.86% wt Al_2O_3)	700	5.38
SiO_2–Na_2O glass (15% mol Na_2O)	250	5.44
SiO_2 glass	1100	5.46
SiO_2–Na_2O glass (30% mol Na_2O)	150	5.48–5.75

Selecting Volume Resistivity of Glass (Continued)

Glass	Temperature (°C)	Resistivity (log Ω cm)
B_2O_3 glass	840	5.5
SiO_2–B_2O_3 glass (5.48% wt B_2O_3)	900	5.64
SiO_2–B_2O_3 glass (19.37% wt B_2O_3)	900	5.65
SiO_2 glass	1000	5.66
SiO_2–B_2O_3 glass (10.75% wt B_2O_3)	900	5.74
SiO_2–Al_2O_3 glass (2.83% wt Al_2O_3)	700	5.74
B_2O_3 glass	780	5.8
SiO_2–Na_2O glass (27% mol Na_2O)	150	5.87
SiO_2 glass	900	5.90
SiO_2–Na_2O glass (25% mol Na_2O)	150	6.05
B_2O_3–CaO glass (55.4% mol CaO)	750	6.13
SiO_2–Na_2O glass (10% mol Na_2O)	250	6.14
B_2O_3 glass	730	6.2
SiO_2 glass	800	6.20
SiO_2–Na_2O glass (5% mol Na_2O)	350	6.37
SiO_2–Na_2O glass (20% mol Na_2O)	150	6.45–6.80
SiO_2 glass	700	6.56
SiO_2–Na_2O glass (30.2% mol Na_2O)	100	6.58
B_2O_3 glass	680	6.6
B_2O_3 glass	640	6.9
SiO_2–Na_2O glass (13% mol Na_2O)	150	6.90–6.96
SiO_2 glass	600	7.00
B_2O_3 glass	600	7.3
SiO_2–Na_2O glass (5% mol Na_2O)	300	7.33–8.25

Selecting Volume Resistivity of Glass (Continued)

Glass	Temperature (˚C)	Resistivity (log Ω cm)
SiO_2–Na_2O glass (10% mol Na_2O)	150	7.35
SiO_2–Na_2O glass (7.5% mol Na_2O)	150	7.59
B_2O_3 glass	560	7.6
SiO_2–Na_2O glass (5% mol Na_2O)	250	7.63
SiO_2 glass	500	7.80
SiO_2–PbO glass (65% mol PbO)	300	7.81
SiO_2–PbO glass (60% mol PbO)	300	8.11
SiO_2–Na_2O glass (15.1% mol Na_2O)	100	8.15
B_2O_3–Na_2O glass (40% mol Na_2O)	100	8.46
SiO_2 glass	400	8.5–10.80
SiO_2–CaO glass (50% mol CaO)	400	8.70
SiO_2–PbO glass (50% mol PbO)	300	8.80–9.2
B_2O_3–Na_2O glass (30% mol Na_2O)	100	8.82
B_2O_3–Na_2O glass (40% mol Na_2O)	80	9.08
B_2O_3–Na_2O glass (30% mol Na_2O)	80	9.43
SiO_2–PbO glass (40% mol PbO)	300	9.48
B_2O_3–Na_2O glass (40% mol Na_2O)	60	9.73
SiO_2–PbO glass (65% mol PbO)	200	9.76
SiO_2–Na_2O glass (7.8% mol Na_2O)	100	9.89
SiO_2–PbO glass (35% mol PbO)	300	9.89
SiO_2–PbO glass (60% mol PbO)	200	10.04
SiO_2–PbO glass (57.1% mol PbO)	172	10.14
B_2O_3–Na_2O glass (30% mol Na_2O)	60	10.14
SiO_2–PbO glass (63.2% mol PbO)	159	10.34

Selecting Volume Resistivity of Glass (Continued)

Glass	Temperature (°C)	Resistivity (log Ω cm)
SiO_2–PbO glass (30% mol PbO)	300	10.44
SiO_2–Na_2O glass (5% mol Na_2O)	150	10.45–11.71
B_2O_3–Na_2O glass (40% mol Na_2O)	40	10.48
SiO_2–PbO glass (50% mol PbO)	200	10.69
SiO_2 glass	250	11.0–13.6
B_2O_3–Na_2O glass (20% mol Na_2O)	100	11.28
SiO_2–PbO glass (40% mol PbO)	200	11.54
SiO_2–PbO glass (51.4% mol PbO)	139	11.59
B_2O_3–Na_2O glass (10% mol Na_2O)	100	11.61
SiO_2–PbO glass (40.2% mol PbO)	175	11.70
SiO_2–PbO glass (47.3% mol PbO)	149	11.74
B_2O_3–Na_2O glass (30% mol Na_2O)	40	11.90
B_2O_3–Na_2O glass (20% mol Na_2O)	80	12.05
SiO_2–PbO glass (35% mol PbO)	200	12.10
SiO_2–CaO glass (50% mol CaO)	300	12.2
B_2O_3–Na_2O glass (10% mol Na_2O)	80	12.40
B_2O_3–Na_2O glass (20% mol Na_2O)	60	12.91
SiO_2–PbO glass (30% mol PbO)	200	12.94
B_2O_3–CaO glass (33.3% mol CaO)	300	13.16
B_2O_3–Na_2O glass (10% mol Na_2O)	60	13.21
B_2O_3–CaO glass (33.3% mol CaO)	250	13.50
B_2O_3–Na_2O glass (16% mol Na_2O)	100	13.58
SiO_2–PbO glass (33.8% mol PbO)	135	13.68
SiO_2–PbO glass (57.1% mol PbO)	77	13.70

Selecting Volume Resistivity of Glass (Continued)

Glass	Temperature (°C)	Resistivity (log Ω cm)
B_2O_3–Na_2O glass (20% mol Na_2O)	40	13.86
B_2O_3–CaO glass (33.3% mol CaO)	200	13.92
B_2O_3–Na_2O glass (10% mol Na_2O)	40	14.20
SiO_2–PbO glass (63.2% mol PbO)	57	14.29
B_2O_3–Na_2O glass (16% mol Na_2O)	80	14.32
B_2O_3–CaO glass (33.3% mol CaO)	150	14.40
SiO_2–PbO glass (47.3% mol PbO)	79	14.48
SiO_2–PbO glass (51.4% mol PbO)	65	14.52
SiO_2–PbO glass (40.2% mol PbO)	78	14.85
B_2O_3–Na_2O glass (16% mol Na_2O)	60	15.08
B_2O_3–Na_2O glass (16% mol Na_2O)	40	15.89
SiO_2–PbO glass (33.8% mol PbO)	66	16.14

Source: data compiled by J. S. Park *from* O. V. Mazurin, M. V. Streltsina and T. P. Shvaiko–Shvaikovskaya, Handbook of Glass Data, Part A and Part B, Elsevier, New York, 1983

SELECTING TANGENT LOSS IN GLASS

Glass	Frequency (Hz)	Temperature	Tangent Loss (tan δ)
SiO_2 glass	100 Hz	25°C	0.00002
SiO_2 glass	1 kHz	25°C	0.00002
SiO_2 glass	10 kHz	25°C	0.00002
SiO_2 glass	10 kHz	200°C	0.00004
B_2O_3 glass	32 kHz	50K	0.00005
B_2O_3 glass	32 kHz	100K	0.00011
SiO_2 glass	1 kHz	200°C	0.00012
B_2O_3 glass	32 kHz	300K	0.0003
B_2O_3-Na_2O glass (10% mol Na_2O)	1 kHz	134.5°C	0.0003
B_2O_3 glass	1 MHz	100°C	0.0004
B_2O_3 glass	1 MHz	200°C	0.0005
B_2O_3-Na_2O glass (12.5% mol Na_2O)	1 kHz	134.5°C	0.0005
SiO_2 glass	100 Hz	200°C	0.00052
B_2O_3 glass	32 kHz	150K	0.0007
SiO_2 glass	10 kHz	300°C	0.00072
B_2O_3 glass	32 kHz	250K	0.0008
B_2O_3 glass	1 MHz	300°C	0.0009
B_2O_3-Na_2O glass (10% mol Na_2O)	1 kHz	214°C	0.0009
B_2O_3-Na_2O glass (20% mol Na_2O)	1 kHz	16°C	0.0009
B_2O_3-CaO glass (33.3% mol CaO)	2 MHz	25°C	0.001
B_2O_3 glass	32 kHz	200K	0.0010
SiO_2-B_2O_3 glass (46.3% mol B_2O_3)	10 GHz		0.0014
SiO_2 glass	9.4 GHz	20°C	0.0015
B_2O_3-Na_2O glass (15% mol Na_2O)	1 kHz	134.5°C	0.0015

Selecting Tangent Loss in Glass (Continued)

Glass	Frequency (Hz)	Temperature	Tangent Loss (tan δ)
SiO_2 glass	9.4 GHz	200°C	0.0018
SiO_2 glass	9.4 GHz	400°C	0.002
B_2O_3-CaO glass (33.3% mol CaO)	2 MHz	100°C	0.002
SiO_2-Al_2O_3 glass (0.5% mol Al_2O_3)	100 K	100 K	0.0021
B_2O_3-Na_2O glass (10% mol Na_2O)	1MHz	room temp.	0.0022
B_2O_3-Na_2O glass (12.5% mol Na_2O)	1 kHz	214°C	0.0022
B_2O_3-Na_2O glass (25% mol Na_2O)	1 kHz	16°C	0.0022
SiO_2-Al_2O_3 glass (0.5% mol Al_2O_3)	50 K	50 K	0.0025
B_2O_3-Na_2O glass (8% mol Na_2O)	1MHz	room temp.	0.0025
B_2O_3-CaO glass (33.3% mol CaO)	2 MHz	200°C	0.0025
SiO_2-Al_2O_3 glass (0.5% mol Al_2O_3)	150 K	150 K	0.0026
B_2O_3-Na_2O glass (20% mol Na_2O)	1 kHz	90.5°C	0.0026
SiO_2 glass	9.4 GHz	600°C	0.0029
B_2O_3-Na_2O glass (16% mol Na_2O)	1MHz	room temp.	0.0031
B_2O_3-CaO glass (33.3% mol CaO)	2 MHz	300°C	0.0035
B_2O_3-Na_2O glass (10% mol Na_2O)	1 kHz	277°C	0.0038
B_2O_3-CaO glass (33.3% mol CaO)	2 MHz	400°C	0.0045
SiO_2 glass	9.4 GHz	800°C	0.0048
SiO_2-PbO glass (40% mol PbO)	100 GHz	room temp.	0.005
B_2O_3-CaO glass (33.3% mol CaO)	2 MHz	500°C	0.0055
SiO_2-Na_2O glass (16% mol Na_2O)	4.5×10^8 Hz	20°C	0.0058
B_2O_3-Na_2O glass (25% mol Na_2O)	1MHz	room temp.	0.0063
B_2O_3-Na_2O glass (15% mol Na_2O)	1 kHz	214°C	0.0064
B_2O_3-Na_2O glass (10% mol Na_2O)	1 kHz	298°C	0.0066

Selecting Tangent Loss in Glass (Continued)

Glass	Frequency (Hz)	Temperature	Tangent Loss (tan δ)
B_2O_3-CaO glass (33.3% mol CaO)	2 MHz	550°C	0.007
SiO_2 glass	1 kHz	300°C	0.0072
SiO_2-Na_2O glass (20% mol Na_2O)	4.5×10^8 Hz	20°C	0.0073
SiO_2-Na_2O glass (22.2% mol Na_2O)	4.5×10^8 Hz	20°C	0.0081
B_2O_3-Na_2O glass (28% mol Na_2O)	1MHz	room temp.	0.0081
B_2O_3-Na_2O glass (12.5% mol Na_2O)	1 kHz	277°C	0.0100
SiO_2-Na_2O glass (28.6% mol Na_2O)	4.5×10^8 Hz	20°C	0.0102
SiO_2 glass	9.4 GHz	1000°C	0.011
B_2O_3-Na_2O glass (20% mol Na_2O)	1 kHz	157°C	0.0149
SiO_2-PbO glass (40% mol PbO)	32 GHz	-150°C	0.015
B_2O_3-Na_2O glass (25% mol Na_2O)	1 kHz	90.5°C	0.0150
SiO_2-Na_2O glass (36% mol Na_2O)	4.5×10^8 Hz	20°C	0.0162
B_2O_3-Na_2O glass (12.5% mol Na_2O)	1 kHz	298°C	0.0170
SiO_2-PbO glass (40% mol PbO)	32 GHz	-100°C	0.018
SiO_2-PbO glass (40% mol PbO)	32 GHz	-50°C	0.020
SiO_2 glass	10 kHz	400°C	0.022
SiO_2-PbO glass (40% mol PbO)	32 GHz	0°C	0.022
SiO_2-PbO glass (40% mol PbO)	32 GHz	50°C	0.024
SiO_2 glass	9.4 GHz	1200°C	0.025
SiO_2-Na_2O glass (19.5% mol Na_2O)	300 kHz	room temp.	0.0295
B_2O_3-Na_2O glass (15% mol Na_2O)	1 kHz	277°C	0.0296
SiO_2-Na_2O glass (19.5% mol Na_2O)	100 kHz	room temp.	0.0364
SiO_2-Na_2O glass (24.4% mol Na_2O)	300 kHz	room temp.	0.0369
SiO_2-Na_2O glass (19.5% mol Na_2O)	50 kHz	room temp.	0.0428

Selecting Tangent Loss in Glass (Continued)

Glass	Frequency (Hz)	Temperature	Tangent Loss (tan δ)
SiO_2-Na_2O glass (24.4% mol Na_2O)	100 kHz	room temp.	0.0456
SiO_2 glass	9.4 GHz	1400°C	0.046
B_2O_3-Na_2O glass (15% mol Na_2O)	1 kHz	298°C	0.0477
SiO_2-Na_2O glass (19.5% mol Na_2O)	30 kHz	room temp.	0.0492
SiO_2-PbO glass (40% mol PbO)	1000 GHz	room temp.	0.050
SiO_2-Na_2O glass (24.4% mol Na_2O)	50 kHz	room temp.	0.0563
SiO_2-Na_2O glass (29.4% mol Na_2O)	300 kHz	room temp.	0.0568
SiO_2-Na_2O glass (24.4% mol Na_2O)	30 kHz	room temp.	0.0652
SiO_2-Na_2O glass (19.5% mol Na_2O)	10 kHz	room temp.	0.0656
SiO_2-Na_2O glass (29.4% mol Na_2O)	100 kHz	room temp.	0.0758
SiO_2 glass	100 Hz	300°C	0.080
SiO_2-Na_2O glass (19.5% mol Na_2O)	5 kHz	room temp.	0.0832
B_2O_3-Na_2O glass (20% mol Na_2O)	1 kHz	219°C	0.0890
SiO_2-Na_2O glass (24.4% mol Na_2O)	10 kHz	room temp.	0.0916
SiO_2-Na_2O glass (34.3% mol Na_2O)	300 kHz	room temp.	0.0936
SiO_2-Na_2O glass (29.4% mol Na_2O)	50 kHz	room temp.	0.0972
SiO_2-Na_2O glass (19.5% mol Na_2O)	3 kHz	room temp.	0.0984
SiO_2-Na_2O glass (34.3% mol Na_2O)	1kHz	room temp.	0.10324
B_2O_3-Na_2O glass (25% mol Na_2O)	1 kHz	157°C	0.1080
SiO_2-Na_2O glass (29.4% mol Na_2O)	30 kHz	room temp.	0.1172
SiO_2-Na_2O glass (24.4% mol Na_2O)	5 kHz	room temp.	0.1194
SiO_2-Na_2O glass (34.3% mol Na_2O)	100 kHz	room temp.	0.1388
SiO_2-Na_2O glass (39.3% mol Na_2O)	300 kHz	room temp.	0.1402
SiO_2-Na_2O glass (19.5% mol Na_2O)	1kHz	room temp.	0.144

Selecting Tangent Loss in Glass (Continued)

Glass	Frequency (Hz)	Temperature	Tangent Loss (tan δ)
SiO_2-Na_2O glass (24.4% mol Na_2O)	3 kHz	room temp.	0.1455
SiO_2-Na_2O glass (29.4% mol Na_2O)	10 kHz	room temp.	0.1764
SiO_2-Na_2O glass (34.3% mol Na_2O)	50 kHz	room temp.	0.1864
SiO_2 glass	1 kHz	400°C	0.2
SiO_2-Na_2O glass (39.3% mol Na_2O)	100 kHz	room temp.	0.2144
SiO_2-Na_2O glass (24.4% mol Na_2O)	1kHz	room temp.	0.2207
SiO_2-Na_2O glass (34.3% mol Na_2O)	30 kHz	room temp.	0.2314
SiO_2-Na_2O glass (29.4% mol Na_2O)	5 kHz	room temp.	0.2426
B_2O_3-Na_2O glass (20% mol Na_2O)	1 kHz	274°C	0.2480
SiO_2-Na_2O glass (29.4% mol Na_2O)	3 kHz	room temp.	0.3027
SiO_2-Na_2O glass (39.3% mol Na_2O)	50 kHz	room temp.	0.3032
SiO_2-Na_2O glass (34.3% mol Na_2O)	10 kHz	room temp.	0.3752
SiO_2-Na_2O glass (39.3% mol Na_2O)	30 kHz	room temp.	0.3835
SiO_2-Na_2O glass (29.4% mol Na_2O)	1kHz	room temp.	0.4923
SiO_2-Na_2O glass (34.3% mol Na_2O)	5 kHz	room temp.	0.5280
SiO_2-Na_2O glass (39.3% mol Na_2O)	10 kHz	room temp.	0.6338
SiO_2-Na_2O glass (34.3% mol Na_2O)	3 kHz	room temp.	0.6520
SiO_2 glass	100 Hz	400°C	1.0

Source: data compiled by J.S. Park *from* O. V. Mazurin, M. V. Streltsina and T. P. Shvaiko-Shvaikovskaya, *Handbook of Glass Data, Part A and Part B*, Elsevier, New York, 1983.

SELECTING TANGENT LOSS IN GLASS BY TEMPERATURE

Temperature	Glass	Frequency (Hz)	Tangent Loss (tan δ)
-100°C	SiO_2-PbO glass (40% mol PbO)	32 GHz	0.018
-150°C	SiO_2-PbO glass (40% mol PbO)	32 GHz	0.015
-50°C	SiO_2-PbO glass (40% mol PbO)	32 GHz	0.020
0°C	SiO_2-PbO glass (40% mol PbO)	32 GHz	0.022
16°C	B_2O_3-Na_2O glass (20% mol Na_2O)	1 kHz	0.0009
16°C	B_2O_3-Na_2O glass (25% mol Na_2O)	1 kHz	0.0022
20°C	SiO_2 glass	9.4 GHz	0.0015
20°C	B_2O_3-Na_2O glass (10% mol Na_2O)	1MHz	0.0022
20°C	B_2O_3-Na_2O glass (8% mol Na_2O)	1MHz	0.0025
20°C	B_2O_3-Na_2O glass (16% mol Na_2O)	1MHz	0.0031
20°C	SiO_2-PbO glass (40% mol PbO)	100 GHz	0.005
20°C	SiO_2-Na_2O glass (16% mol Na_2O)	4.5×10^8 Hz	0.0058
20°C	B_2O_3-Na_2O glass (25% mol Na_2O)	1MHz	0.0063
20°C	SiO_2-Na_2O glass (20% mol Na_2O)	4.5×10^8 Hz	0.0073
20°C	B_2O_3-Na_2O glass (28% mol Na_2O)	1MHz	0.0081
20°C	SiO_2-Na_2O glass (22.2% mol Na_2O)	4.5×10^8 Hz	0.0081
20°C	SiO_2-Na_2O glass (28.6% mol Na_2O)	4.5×10^8 Hz	0.0102
20°C	SiO_2-Na_2O glass (36% mol Na_2O)	4.5×10^8 Hz	0.0162
20°C	SiO_2-Na_2O glass (19.5% mol Na_2O)	300 kHz	0.0295
20°C	SiO_2-Na_2O glass (19.5% mol Na_2O)	100 kHz	0.0364
20°C	SiO_2-Na_2O glass (24.4% mol Na_2O)	300 kHz	0.0369
20°C	SiO_2-Na_2O glass (19.5% mol Na_2O)	50 kHz	0.0428

Selecting Tangent Loss in Glass by Temperature (Continued)

Temperature	Glass	Frequency (Hz)	Tangent Loss (tan δ)
20°C	SiO_2-Na_2O glass (24.4% mol Na_2O)	100 kHz	0.0456
20°C	SiO_2-Na_2O glass (19.5% mol Na_2O)	30 kHz	0.0492
20°C	SiO_2-PbO glass (40% mol PbO)	1000 GHz	0.050
20°C	SiO_2-Na_2O glass (24.4% mol Na_2O)	50 kHz	0.0563
20°C	SiO_2-Na_2O glass (29.4% mol Na_2O)	300 kHz	0.0568
20°C	SiO_2-Na_2O glass (24.4% mol Na_2O)	30 kHz	0.0652
20°C	SiO_2-Na_2O glass (19.5% mol Na_2O)	10 kHz	0.0656
20°C	SiO_2-Na_2O glass (29.4% mol Na_2O)	100 kHz	0.0758
20°C	SiO_2-Na_2O glass (19.5% mol Na_2O)	5 kHz	0.0832
20°C	SiO_2-Na_2O glass (24.4% mol Na_2O)	10 kHz	0.0916
20°C	SiO_2-Na_2O glass (34.3% mol Na_2O)	300 kHz	0.0936
20°C	SiO_2-Na_2O glass (29.4% mol Na_2O)	50 kHz	0.0972
20°C	SiO_2-Na_2O glass (19.5% mol Na_2O)	3 kHz	0.0984
20°C	SiO_2-Na_2O glass (34.3% mol Na_2O)	1kHz	0.10324
20°C	SiO_2-Na_2O glass (29.4% mol Na_2O)	30 kHz	0.1172
20°C	SiO_2-Na_2O glass (24.4% mol Na_2O)	5 kHz	0.1194
20°C	SiO_2-Na_2O glass (34.3% mol Na_2O)	100 kHz	0.1388
20°C	SiO_2-Na_2O glass (39.3% mol Na_2O)	300 kHz	0.1402
20°C	SiO_2-Na_2O glass (19.5% mol Na_2O)	1kHz	0.144
20°C	SiO_2-Na_2O glass (24.4% mol Na_2O)	3 kHz	0.1455
20°C	SiO_2-Na_2O glass (29.4% mol Na_2O)	10 kHz	0.1764
20°C	SiO_2-Na_2O glass (34.3% mol Na_2O)	50 kHz	0.1864
20°C	SiO_2-Na_2O glass (39.3% mol Na_2O)	100 kHz	0.2144
20°C	SiO_2-Na_2O glass (24.4% mol Na_2O)	1kHz	0.2207

Selecting Tangent Loss in Glass by Temperature (Continued)

Temperature	Glass	Frequency (Hz)	Tangent Loss (tan δ)
20°C	SiO_2-Na_2O glass (34.3% mol Na_2O)	30 kHz	0.2314
20°C	SiO_2-Na_2O glass (29.4% mol Na_2O)	5 kHz	0.2426
20°C	SiO_2-Na_2O glass (29.4% mol Na_2O)	3 kHz	0.3027
20°C	SiO_2-Na_2O glass (39.3% mol Na_2O)	50 kHz	0.3032
20°C	SiO_2-Na_2O glass (34.3% mol Na_2O)	10 kHz	0.3752
20°C	SiO_2-Na_2O glass (39.3% mol Na_2O)	30 kHz	0.3835
20°C	SiO_2-Na_2O glass (29.4% mol Na_2O)	1kHz	0.4923
20°C	SiO_2-Na_2O glass (34.3% mol Na_2O)	5 kHz	0.5280
20°C	SiO_2-Na_2O glass (39.3% mol Na_2O)	10 kHz	0.6338
20°C	SiO_2-Na_2O glass (34.3% mol Na_2O)	3 kHz	0.6520
25°C	SiO_2 glass	100 Hz	0.00002
25°C	SiO_2 glass	1 kHz	0.00002
25°C	SiO_2 glass	10 kHz	0.00002
25°C	B_2O_3-CaO glass (33.3% mol CaO)	2 MHz	0.001
50°C	SiO_2-PbO glass (40% mol PbO)	32 GHz	0.024
90.5°C	B_2O_3-Na_2O glass (20% mol Na_2O)	1 kHz	0.0026
90.5°C	B_2O_3-Na_2O glass (25% mol Na_2O)	1 kHz	0.0150
100°C	B_2O_3 glass	1 MHz	0.0004
100°C	B_2O_3-CaO glass (33.3% mol CaO)	2 MHz	0.002
134.5°C	B_2O_3-Na_2O glass (10% mol Na_2O)	1 kHz	0.0003
134.5°C	B_2O_3-Na_2O glass (12.5% mol Na_2O)	1 kHz	0.0005
134.5°C	B_2O_3-Na_2O glass (15% mol Na_2O)	1 kHz	0.0015

Selecting Tangent Loss in Glass by Temperature (Continued)

Temperature	Glass	Frequency (Hz)	Tangent Loss (tan δ)
157°C	B_2O_3-Na_2O glass (20% mol Na_2O)	1 kHz	0.0149
157°C	B_2O_3-Na_2O glass (25% mol Na_2O)	1 kHz	0.1080
200°C	SiO_2 glass	10 kHz	0.00004
200°C	SiO_2 glass	1 kHz	0.00012
200°C	B_2O_3 glass	1 MHz	0.0005
200°C	SiO_2 glass	100 Hz	0.00052
200°C	SiO_2 glass	9.4 GHz	0.0018
200°C	B_2O_3-CaO glass (33.3% mol CaO)	2 MHz	0.0025
214°C	B_2O_3-Na_2O glass (10% mol Na_2O)	1 kHz	0.0009
214°C	B_2O_3-Na_2O glass (12.5% mol Na_2O)	1 kHz	0.0022
214°C	B_2O_3-Na_2O glass (15% mol Na_2O)	1 kHz	0.0064
219°C	B_2O_3-Na_2O glass (20% mol Na_2O)	1 kHz	0.0890
274°C	B_2O_3-Na_2O glass (20% mol Na_2O)	1 kHz	0.2480
277°C	B_2O_3-Na_2O glass (10% mol Na_2O)	1 kHz	0.0038
277°C	B_2O_3-Na_2O glass (12.5% mol Na_2O)	1 kHz	0.0100
277°C	B_2O_3-Na_2O glass (15% mol Na_2O)	1 kHz	0.0296
298°C	B_2O_3-Na_2O glass (10% mol Na_2O)	1 kHz	0.0066
298°C	B_2O_3-Na_2O glass (12.5% mol Na_2O)	1 kHz	0.0170
298°C	B_2O_3-Na_2O glass (15% mol Na_2O)	1 kHz	0.0477
300°C	SiO_2 glass	10 kHz	0.00072
300°C	B_2O_3 glass	1 MHz	0.0009
300°C	B_2O_3-CaO glass (33.3% mol CaO)	2 MHz	0.0035
300°C	SiO_2 glass	1 kHz	0.0072
300°C	SiO_2 glass	100 Hz	0.080

Selecting Tangent Loss in Glass by Temperature (Continued)

Temperature	Glass	Frequency (Hz)	Tangent Loss (tan δ)
323°C	B_2O_3 glass	32 kHz	0.00005
373°C	B_2O_3 glass	32 kHz	0.00011
373°C	SiO_2-Al_2O_3 glass (0.5% mol Al_2O_3)	100 K	0.0021
323°C	SiO_2-Al_2O_3 glass (0.5% mol Al_2O_3)	50 K	0.0025
400°C	SiO_2 glass	9.4 GHz	0.002
400°C	B_2O_3-CaO glass (33.3% mol CaO)	2 MHz	0.0045
400°C	SiO_2 glass	10 kHz	0.022
400°C	SiO_2 glass	1 kHz	0.2
400°C	SiO_2 glass	100 Hz	1.0
423°C	SiO_2-Al_2O_3 glass (0.5% mol Al_2O_3)	150 K	0.0026
423°C	B_2O_3 glass	32 kHz	0.0007
473°C	B_2O_3 glass	32 kHz	0.0010
500°C	B_2O_3-CaO glass (33.3% mol CaO)	2 MHz	0.0055
523°C	B_2O_3 glass	32 kHz	0.0008
550°C	B_2O_3-CaO glass (33.3% mol CaO)	2 MHz	0.007
573°C	B_2O_3 glass	32 kHz	0.0003
600°C	SiO_2 glass	9.4 GHz	0.0029
800°C	SiO_2 glass	9.4 GHz	0.0048
1000°C	SiO_2 glass	9.4 GHz	0.011
1200°C	SiO_2 glass	9.4 GHz	0.025
1400°C	SiO_2 glass	9.4 GHz	0.046

Source: data compiled by J.S. Park *from* O. V. Mazurin, M. V. Streltsina and T. P. Shvaiko-Shvaikovskaya, *Handbook of Glass Data, Part A and Part B*, Elsevier, New York, 1983.

SELECTING TANGENT LOSS IN GLASS BY FREQUENCY

Frequency (Hz)	Glass	Temperature	Tangent Loss (tan δ)
100 Hz	SiO_2 glass	25°C	0.00002
100 Hz	SiO_2 glass	200°C	0.00052
100 Hz	SiO_2 glass	300°C	0.080
100 Hz	SiO_2 glass	400°C	1.0
1 kHz	SiO_2 glass	25°C	0.00002
1 kHz	SiO_2 glass	200°C	0.00012
1 kHz	B_2O_3-Na_2O glass (10% mol Na_2O)	134.5°C	0.0003
1 kHz	B_2O_3-Na_2O glass (12.5% mol Na_2O)	134.5°C	0.0005
1 kHz	B_2O_3-Na_2O glass (10% mol Na_2O)	214°C	0.0009
1 kHz	B_2O_3-Na_2O glass (20% mol Na_2O)	16°C	0.0009
1 kHz	B_2O_3-Na_2O glass (15% mol Na_2O)	134.5°C	0.0015
1 kHz	B_2O_3-Na_2O glass (12.5% mol Na_2O)	214°C	0.0022
1 kHz	B_2O_3-Na_2O glass (25% mol Na_2O)	16°C	0.0022
1 kHz	B_2O_3-Na_2O glass (20% mol Na_2O)	90.5°C	0.0026
1 kHz	B_2O_3-Na_2O glass (10% mol Na_2O)	277°C	0.0038
1 kHz	B_2O_3-Na_2O glass (15% mol Na_2O)	214°C	0.0064
1 kHz	B_2O_3-Na_2O glass (10% mol Na_2O)	298°C	0.0066
1 kHz	SiO_2 glass	300°C	0.0072
1 kHz	B_2O_3-Na_2O glass (12.5% mol Na_2O)	277°C	0.0100
1 kHz	B_2O_3-Na_2O glass (20% mol Na_2O)	157°C	0.0149
1 kHz	B_2O_3-Na_2O glass (25% mol Na_2O)	90.5°C	0.0150
1 kHz	B_2O_3-Na_2O glass (12.5% mol Na_2O)	298°C	0.0170
1 kHz	B_2O_3-Na_2O glass (15% mol Na_2O)	277°C	0.0296
1 kHz	B_2O_3-Na_2O glass (15% mol Na_2O)	298°C	0.0477

Selecting Tangent Loss in Glass by Frequency (Continued)

Frequency (Hz)	Glass	Temperature	Tangent Loss (tan δ)
1 kHz	B_2O_3-Na_2O glass (20% mol Na_2O)	219°C	0.0890
1 kHz	SiO_2-Na_2O glass (34.3% mol Na_2O)	room temp.	0.10324
1 kHz	B_2O_3-Na_2O glass (25% mol Na_2O)	157°C	0.1080
1 kHz	SiO_2-Na_2O glass (19.5% mol Na_2O)	room temp.	0.144
1 kHz	SiO_2 glass	400°C	0.2
1 kHz	SiO_2-Na_2O glass (24.4% mol Na_2O)	room temp.	0.2207
1 kHz	B_2O_3-Na_2O glass (20% mol Na_2O)	274°C	0.2480
1 kHz	SiO_2-Na_2O glass (29.4% mol Na_2O)	room temp.	0.4923
3 kHz	SiO_2-Na_2O glass (19.5% mol Na_2O)	room temp.	0.0984
3 kHz	SiO_2-Na_2O glass (24.4% mol Na_2O)	room temp.	0.1455
3 kHz	SiO_2-Na_2O glass (29.4% mol Na_2O)	room temp.	0.3027
3 kHz	SiO_2-Na_2O glass (34.3% mol Na_2O)	room temp.	0.6520
5 kHz	SiO_2-Na_2O glass (19.5% mol Na_2O)	room temp.	0.0832
5 kHz	SiO_2-Na_2O glass (24.4% mol Na_2O)	room temp.	0.1194
5 kHz	SiO_2-Na_2O glass (29.4% mol Na_2O)	room temp.	0.2426
5 kHz	SiO_2-Na_2O glass (34.3% mol Na_2O)	room temp.	0.5280
10 kHz	SiO_2 glass	25°C	0.00002
10 kHz	SiO_2 glass	200°C	0.00004
10 kHz	SiO_2 glass	300°C	0.00072
10 kHz	SiO_2 glass	400°C	0.022
10 kHz	SiO_2-Na_2O glass (19.5% mol Na_2O)	room temp.	0.0656
10 kHz	SiO_2-Na_2O glass (24.4% mol Na_2O)	room temp.	0.0916

Selecting Tangent Loss in Glass by Frequency (Continued)

Frequency (Hz)	Glass	Temperature	Tangent Loss (tan δ)
10 kHz	SiO_2-Na_2O glass (29.4% mol Na_2O)	room temp.	0.1764
10 kHz	SiO_2-Na_2O glass (34.3% mol Na_2O)	room temp.	0.3752
10 kHz	SiO_2-Na_2O glass (39.3% mol Na_2O)	room temp.	0.6338
30 kHz	SiO_2-Na_2O glass (19.5% mol Na_2O)	room temp.	0.0492
30 kHz	SiO_2-Na_2O glass (24.4% mol Na_2O)	room temp.	0.0652
30 kHz	SiO_2-Na_2O glass (29.4% mol Na_2O)	room temp.	0.1172
30 kHz	SiO_2-Na_2O glass (34.3% mol Na_2O)	room temp.	0.2314
30 kHz	SiO_2-Na_2O glass (39.3% mol Na_2O)	room temp.	0.3835
32 kHz	B_2O_3 glass	50K	0.00005
32 kHz	B_2O_3 glass	100K	0.00011
32 kHz	B_2O_3 glass	300K	0.0003
32 kHz	B_2O_3 glass	150K	0.0007
32 kHz	B_2O_3 glass	250K	0.0008
32 kHz	B_2O_3 glass	200K	0.0010
50 kHz	SiO_2-Na_2O glass (19.5% mol Na_2O)	room temp.	0.0428
50 kHz	SiO_2-Na_2O glass (24.4% mol Na_2O)	room temp.	0.0563
50 kHz	SiO_2-Na_2O glass (29.4% mol Na_2O)	room temp.	0.0972
50 kHz	SiO_2-Na_2O glass (34.3% mol Na_2O)	room temp.	0.1864
50 kHz	SiO_2-Na_2O glass (39.3% mol Na_2O)	room temp.	0.3032
100 kHz	SiO_2-Na_2O glass (19.5% mol Na_2O)	room temp.	0.0364
100 kHz	SiO_2-Na_2O glass (24.4% mol Na_2O)	room temp.	0.0456
100 kHz	SiO_2-Na_2O glass (29.4% mol Na_2O)	room temp.	0.0758

Selecting Tangent Loss in Glass by Frequency (Continued)

Frequency (Hz)	Glass	Temperature	Tangent Loss (tan δ)
100 kHz	SiO_2-Na_2O glass (34.3% mol Na_2O)	room temp.	0.1388
100 kHz	SiO_2-Na_2O glass (39.3% mol Na_2O)	room temp.	0.2144
300 kHz	SiO_2-Na_2O glass (19.5% mol Na_2O)	room temp.	0.0295
300 kHz	SiO_2-Na_2O glass (24.4% mol Na_2O)	room temp.	0.0369
300 kHz	SiO_2-Na_2O glass (29.4% mol Na_2O)	room temp.	0.0568
300 kHz	SiO_2-Na_2O glass (34.3% mol Na_2O)	room temp.	0.0936
300 kHz	SiO_2-Na_2O glass (39.3% mol Na_2O)	room temp.	0.1402
1 MHz	B_2O_3 glass	100°C	0.0004
1 MHz	B_2O_3 glass	200°C	0.0005
1 MHz	B_2O_3 glass	300°C	0.0009
1 MHz	B_2O_3-Na_2O glass (10% mol Na_2O)	room temp.	0.0022
1 MHz	B_2O_3-Na_2O glass (8% mol Na_2O)	room temp.	0.0025
1 MHz	B_2O_3-Na_2O glass (16% mol Na_2O)	room temp.	0.0031
1 MHz	B_2O_3-Na_2O glass (25% mol Na_2O)	room temp.	0.0063
1 MHz	B_2O_3-Na_2O glass (28% mol Na_2O)	room temp.	0.0081
2 MHz	B_2O_3-CaO glass (33.3% mol CaO)	25°C	0.001
2 MHz	B_2O_3-CaO glass (33.3% mol CaO)	100°C	0.002
2 MHz	B_2O_3-CaO glass (33.3% mol CaO)	200°C	0.0025
2 MHz	B_2O_3-CaO glass (33.3% mol CaO)	300°C	0.0035
2 MHz	B_2O_3-CaO glass (33.3% mol CaO)	400°C	0.0045
2 MHz	B_2O_3-CaO glass (33.3% mol CaO)	500°C	0.0055
2 MHz	B_2O_3-CaO glass (33.3% mol CaO)	550°C	0.007

Selecting Tangent Loss in Glass by Frequency (Continued)

Frequency (Hz)	Glass	Temperature	Tangent Loss (tan δ)
4.5×10^8 Hz	SiO_2-Na_2O glass (16% mol Na_2O)	20°C	0.0058
4.5×10^8 Hz	SiO_2-Na_2O glass (20% mol Na_2O)	20°C	0.0073
4.5×10^8 Hz	SiO_2-Na_2O glass (22.2% mol Na_2O)	20°C	0.0081
4.5×10^8 Hz	SiO_2-Na_2O glass (28.6% mol Na_2O)	20°C	0.0102
4.5×10^8 Hz	SiO_2-Na_2O glass (36% mol Na_2O)	20°C	0.0162
9.4 GHz	SiO_2 glass	20°C	0.0015
9.4 GHz	SiO_2 glass	200°C	0.0018
9.4 GHz	SiO_2 glass	400°C	0.002
9.4 GHz	SiO_2 glass	600°C	0.0029
9.4 GHz	SiO_2 glass	800°C	0.0048
9.4 GHz	SiO_2 glass	1000°C	0.011
9.4 GHz	SiO_2 glass	1200°C	0.025
9.4 GHz	SiO_2 glass	1400°C	0.046
10 GHz	SiO_2-B_2O_3 glass (46.3% mol B_2O_3)		0.0014
32 GHz	SiO_2-PbO glass (40% mol PbO)	-150°C	0.015
32 GHz	SiO_2-PbO glass (40% mol PbO)	-100°C	0.018
32 GHz	SiO_2-PbO glass (40% mol PbO)	-50°C	0.020
32 GHz	SiO_2-PbO glass (40% mol PbO)	0°C	0.022
32 GHz	SiO_2-PbO glass (40% mol PbO)	50°C	0.024
100 GHz	SiO_2-PbO glass (40% mol PbO)	room temp.	0.005
1000 GHz	SiO_2-PbO glass (40% mol PbO)	room temp.	0.050

Source: data compiled by J.S. Park *from* O. V. Mazurin, M. V. Streltsina and T. P. Shvaiko-Shvaikovskaya, *Handbook of Glass Data, Part A and Part B*, Elsevier, New York, 1983.

SELECTING ELECTRICAL PERMITTIVITY OF GLASS

Glass	Frequency (Hz)	Temperature (°C)	Electrical Permittivity
B_2O_3 glass	50 kHz	800	3.04
B_2O_3 glass	50 kHz	620	3.05
B_2O_3 glass	50 kHz	750	3.06
B_2O_3 glass	50 kHz	700	3.09
B_2O_3 glass	50 kHz	500	3.10
B_2O_3 glass	50 kHz	650	3.10
B_2O_3 glass	50 kHz	580	3.115
B_2O_3 glass	50 kHz	550	3.12
B_2O_3 glass	10 kHz	500	3.13
B_2O_3 glass	10 kHz	550	3.14
B_2O_3 glass	10 kHz	580	3.145
B_2O_3 glass	3 kHz	500	3.15
B_2O_3 glass	10 kHz	620	3.15
B_2O_3 glass	10 kHz	650	3.15
B_2O_3 glass	10 kHz	700	3.16
B_2O_3 glass	1 kHz	500	3.17
B_2O_3 glass	3 kHz	550	3.17
B_2O_3 glass	3 kHz	580	3.18
B_2O_3 glass	1 kHz	550	3.21
B_2O_3 glass	3 kHz	620	3.21
B_2O_3 glass	3 kHz	650	3.25
B_2O_3 glass	1 kHz	580	3.27
SiO_2–Al_2O_3 glass (46.3% mol B_2O_3)	10 GHz		3.55
B_2O_3–Na_2O glass (4.08% mol Na_2O)	56.8 MHz	room temp.	3.72

Selecting Electrical Permittivity of Glass (Continued)

Glass	Frequency (Hz)	Temperature (°C)	Electrical Permittivity
SiO_2 glass	9.4 GHz	20	3.81
SiO_2 glass	10 GHz	20	3.82
SiO_2 glass	10 GHz	220	3.82
SiO_2 glass	9.4 GHz	200	3.83
SiO_2 glass	9.4 GHz	400	3.84
SiO_2 glass	9.4 GHz	600	3.86
SiO_2 glass	9.4 GHz	800	3.88
SiO_2 glass	9.4 GHz	1000	3.91
SiO_2 glass	10 GHz	888	3.91
SiO_2 glass	9.4 GHz	1200	3.93
SiO_2 glass	9.4 GHz	1400	3.96
SiO_2 glass	10 GHz	1170	3.98
SiO_2 glass	100 Hz	25	4.0
SiO_2 glass	100 Hz	200	4.0
SiO_2 glass	100 Hz	300	4.0
SiO_2 glass	1 kHz	25	4.0
SiO_2 glass	1 kHz	200	4.0
SiO_2 glass	1 kHz	300	4.0
SiO_2 glass	10 kHz	25	4.0
SiO_2 glass	10 kHz	200	4.0
SiO_2 glass	10 kHz	300	4.0
SiO_2 glass	10 kHz	400	4.0
SiO_2 glass	10 GHz	1764	4.04
SiO_2 glass	10 GHz	1335	4.05

Selecting Electrical Permittivity of Glass (Continued)

Glass	Frequency (Hz)	Temperature (°C)	Electrical Permittivity
SiO_2 glass	10 GHz	1764	4.05
SiO_2 glass	10 GHz	1420	4.07
SiO_2 glass	10 GHz	1480	4.09
SiO_2 glass	1 kHz	400	4.1
SiO_2 glass	10 GHz	1526	4.11
SiO_2 glass	10 GHz	1584	4.12
SiO_2 glass	10 GHz	1647	4.12
SiO_2 glass	10 GHz	1602	4.15
B_2O_3–Na_2O glass (7.35% mol Na_2O)	56.8 MHz	room temp.	4.20
SiO_2–PbO glass (40% mol PbO)	32 GHz	−150	4.25
SiO_2–PbO glass (40% mol PbO)	32 GHz	−100	4.30
SiO_2–PbO glass (40% mol PbO)	32 GHz	−50	4.40
SiO_2–PbO glass (40% mol PbO)	32 GHz	0	4.45
B_2O_3–Na_2O glass (14.15% mol Na_2O)	56.8 MHz	room temp.	4.94
SiO_2–PbO glass (40% mol PbO)	32 GHz	50	5.00
B_2O_3–Na_2O glass (10% mol Na_2O)	1 kHz	73	5.00
B_2O_3–Na_2O glass (10% mol Na_2O)	1 kHz	134.5	5.05
B_2O_3–Na_2O glass (10% mol Na_2O)	1 kHz	214	5.15
B_2O_3–Na_2O glass (17.31% mol Na_2O)	56.8 MHz	room temp.	5.27
B_2O_3–Na_2O glass (10% mol Na_2O)	1 kHz	277	5.45
B_2O_3–Na_2O glass (12.5% mol Na_2O)	1 kHz	73	5.45
SiO_2 glass	100 Hz	400	5.5
B_2O_3–Na_2O glass (10% mol Na_2O)	1 kHz	298	5.60
B_2O_3–Na_2O glass (12.5% mol Na_2O)	1 kHz	134.5	5.60

Selecting Electrical Permittivity of Glass (Continued)

Glass	Frequency (Hz)	Temperature (°C)	Electrical Permittivity
B_2O_3–Na_2O glass (12.5% mol Na_2O)	1 kHz	214	5.75
B_2O_3–Na_2O glass (15% mol Na_2O)	1 kHz	73	5.80
B_2O_3–Na_2O glass (15% mol Na_2O)	1 kHz	134.5	6.00
SiO_2–Na_2O glass (16% mol Na_2O)	4.5×10^8 Hz	20	6.01
B_2O_3–Na_2O glass (20% mol Na_2O)	1 kHz	16	6.15
B_2O_3–Na_2O glass (24.77% mol Na_2O)	56.8 MHz	room temp.	6.24
B_2O_3–Na_2O glass (12.5% mol Na_2O)	1 kHz	277	6.30
B_2O_3–Na_2O glass (20% mol Na_2O)	1 kHz	90.5	6.43
SiO_2–Na_2O glass (20% mol Na_2O)	4.5×10^8 Hz	20	6.48
B_2O_3–Na_2O glass (15% mol Na_2O)	1 kHz	214	6.50
B_2O_3–Na_2O glass (12.5% mol Na_2O)	1 kHz	298	6.65
SiO_2–Na_2O glass (22.2% mol Na_2O)	4.5×10^8 Hz	20	6.85
B_2O_3–Na_2O glass (31.98% mol Na_2O)	56.8 MHz	room temp.	7.03
B_2O_3–Na_2O glass (20% mol Na_2O)	1 kHz	157	7.45
B_2O_3–Na_2O glass (25% mol Na_2O)	1 kHz	16	7.50
SiO_2–Na_2O glass (19.5% mol Na_2O)	300 kHz	room temp.	7.62
SiO_2–Na_2O glass (28.6% mol Na_2O)	4.5×10^8 Hz	20	7.62
SiO_2–Na_2O glass (19.5% mol Na_2O)	100 kHz	room temp.	7.74
B_2O_3–Na_2O glass (15% mol Na_2O)	1 kHz	277	7.80
SiO_2–Na_2O glass (19.5% mol Na_2O)	50 kHz	room temp.	7.88
SiO_2–Na_2O glass (19.5% mol Na_2O)	30 kHz	room temp.	8.00
SiO_2–Na_2O glass (19.5% mol Na_2O)	10 kHz	room temp.	8.26
SiO_2–Na_2O glass (19.5% mol Na_2O)	5 kHz	room temp.	8.56
B_2O_3–Na_2O glass (15% mol Na_2O)	1 kHz	298	8.60

Selecting Electrical Permittivity of Glass (Continued)

Glass	Frequency (Hz)	Temperature (°C)	Electrical Permittivity
SiO_2–Na_2O glass (24.4% mol Na_2O)	300 kHz	room temp.	8.75
B_2O_3–Na_2O glass (25% mol Na_2O)	1 kHz	90.5	8.90
SiO_2–Na_2O glass (24.4% mol Na_2O)	100 kHz	room temp.	8.91
SiO_2–Na_2O glass (19.5% mol Na_2O)	3 kHz	room temp.	8.97
SiO_2–Na_2O glass (24.4% mol Na_2O)	50 kHz	room temp.	9.14
SiO_2–Na_2O glass (24.4% mol Na_2O)	30 kHz	room temp.	9.30
SiO_2–Na_2O glass (19.5% mol Na_2O)	1kHz	room temp.	9.40
SiO_2–Na_2O glass (36% mol Na_2O)	4.5×10^8 Hz	20	9.40
SiO_2–Na_2O glass (24.4% mol Na_2O)	10 kHz	room temp.	9.74
SiO_2–Na_2O glass (29.4% mol Na_2O)	300 kHz	room temp.	10.15
SiO_2–Na_2O glass (24.4% mol Na_2O)	5 kHz	room temp.	10.21
SiO_2–Na_2O glass (29.4% mol Na_2O)	100 kHz	room temp.	10.47
SiO_2–Na_2O glass (24.4% mol Na_2O)	3 kHz	room temp.	10.61
SiO_2–Na_2O glass (29.4% mol Na_2O)	50 kHz	room temp.	10.86
SiO_2–Na_2O glass (34.3% mol Na_2O)	300 kHz	room temp.	11.14
SiO_2–Na_2O glass (29.4% mol Na_2O)	30 kHz	room temp.	11.21
SiO_2–Na_2O glass (24.4% mol Na_2O)	1kHz	room temp.	11.62
SiO_2–Na_2O glass (34.3% mol Na_2O)	100 kHz	room temp.	11.78
B_2O_3–Na_2O glass (20% mol Na_2O)	1 kHz	219	11.85
SiO_2–Na_2O glass (29.4% mol Na_2O)	10 kHz	room temp.	12.08
SiO_2–Na_2O glass (39.3% mol Na_2O)	300 kHz	room temp.	12.43
SiO_2–Na_2O glass (34.3% mol Na_2O)	50 kHz	room temp.	12.57
SiO_2–Na_2O glass (29.4% mol Na_2O)	5 kHz	room temp.	13.19
SiO_2–Na_2O glass (34.3% mol Na_2O)	30 kHz	room temp.	13.28

Selecting Electrical Permittivity of Glass (Continued)

Glass	Frequency (Hz)	Temperature (°C)	Electrical Permittivity
SiO_2–Na_2O glass (39.3% mol Na_2O)	100 kHz	room temp.	13.55
SiO_2–Na_2O glass (29.4% mol Na_2O)	3 kHz	room temp.	14.23
SiO_2–Na_2O glass (39.3% mol Na_2O)	50 kHz	room temp.	15.06
SiO_2–Na_2O glass (34.3% mol Na_2O)	10 kHz	room temp.	15.22
SiO_2–Na_2O glass (39.3% mol Na_2O)	30 kHz	room temp.	16.56
B_2O_3–Na_2O glass (25% mol Na_2O)	1 kHz	157	17.30
SiO_2–Na_2O glass (29.4% mol Na_2O)	1kHz	room temp.	17.52
SiO_2–Na_2O glass (34.3% mol Na_2O)	5 kHz	room temp.	18.13
SiO_2–Na_2O glass (34.3% mol Na_2O)	3 kHz	room temp.	21.30
SiO_2–Na_2O glass (39.3% mol Na_2O)	10 kHz	room temp.	22.08
B_2O_3–Na_2O glass (20% mol Na_2O)	1 kHz	274	31.00
SiO_2–Na_2O glass (34.3% mol Na_2O)	1kHz	room temp.	38.61

Source: data compiled by J.S. Park *from* O. V. Mazurin, M. V. Streltsina and T. P. Shvaiko-Shvaikovskaya, *Handbook of Glass Data, Part A and Part B*, Elsevier, New York, 1983.

SELECTING ELECTRICAL PERMITTIVITY OF GLASS BY FREQUENCY

Frequency (Hz)	Glass	Temperature (°C)	Electrical Permittivity
100 Hz	SiO_2 glass	25	4.0
100 Hz	SiO_2 glass	200	4.0
100 Hz	SiO_2 glass	300	4.0
100 Hz	SiO_2 glass	400	5.5
1 kHz	B_2O_3 glass	500	3.17
1 kHz	B_2O_3 glass	550	3.21
1 kHz	B_2O_3 glass	580	3.27
1 kHz	SiO_2 glass	25	4.0
1 kHz	SiO_2 glass	200	4.0
1 kHz	SiO_2 glass	300	4.0
1 kHz	SiO_2 glass	400	4.1
1 kHz	B_2O_3–Na_2O glass (10% mol Na_2O)	73	5.00
1 kHz	B_2O_3–Na_2O glass (10% mol Na_2O)	134.5	5.05
1 kHz	B_2O_3–Na_2O glass (10% mol Na_2O)	214	5.15
1 kHz	B_2O_3–Na_2O glass (10% mol Na_2O)	277	5.45
1 kHz	B_2O_3–Na_2O glass (12.5% mol Na_2O)	73	5.45
1 kHz	B_2O_3–Na_2O glass (10% mol Na_2O)	298	5.60
1 kHz	B_2O_3–Na_2O glass (12.5% mol Na_2O)	134.5	5.60
1 kHz	B_2O_3–Na_2O glass (12.5% mol Na_2O)	214	5.75
1 kHz	B_2O_3–Na_2O glass (15% mol Na_2O)	73	5.80
1 kHz	B_2O_3–Na_2O glass (15% mol Na_2O)	134.5	6.00
1 kHz	B_2O_3–Na_2O glass (20% mol Na_2O)	16	6.15
1 kHz	B_2O_3–Na_2O glass (12.5% mol Na_2O)	277	6.30
1 kHz	B_2O_3–Na_2O glass (20% mol Na_2O)	90.5	6.43

Selecting Electrical Permittivity of Glass by Frequency (Continued)

Frequency (Hz)	Glass	Temperature (°C)	Electrical Permittivity
1 kHz	B_2O_3–Na_2O glass (15% mol Na_2O)	214	6.50
1 kHz	B_2O_3–Na_2O glass (12.5% mol Na_2O)	298	6.65
1 kHz	B_2O_3–Na_2O glass (20% mol Na_2O)	157	7.45
1 kHz	B_2O_3–Na_2O glass (25% mol Na_2O)	16	7.50
1 kHz	B_2O_3–Na_2O glass (15% mol Na_2O)	277	7.80
1 kHz	B_2O_3–Na_2O glass (15% mol Na_2O)	298	8.60
1 kHz	B_2O_3–Na_2O glass (25% mol Na_2O)	90.5	8.90
1 kHz	SiO_2–Na_2O glass (19.5% mol Na_2O)	room temp.	9.40
1 kHz	SiO_2–Na_2O glass (24.4% mol Na_2O)	room temp.	11.62
1 kHz	B_2O_3–Na_2O glass (20% mol Na_2O)	219	11.85
1 kHz	B_2O_3–Na_2O glass (25% mol Na_2O)	157	17.30
1 kHz	SiO_2–Na_2O glass (29.4% mol Na_2O)	room temp.	17.52
1 kHz	B_2O_3–Na_2O glass (20% mol Na_2O)	274	31.00
1 kHz	SiO_2–Na_2O glass (34.3% mol Na_2O)	room temp.	38.61
3 kHz	B_2O_3 glass	500	3.15
3 kHz	B_2O_3 glass	550	3.17
3 kHz	B_2O_3 glass	580	3.18
3 kHz	B_2O_3 glass	620	3.21
3 kHz	B_2O_3 glass	650	3.25
3 kHz	SiO_2–Na_2O glass (19.5% mol Na_2O)	room temp.	8.97
3 kHz	SiO_2–Na_2O glass (24.4% mol Na_2O)	room temp.	10.61
3 kHz	SiO_2–Na_2O glass (29.4% mol Na_2O)	room temp.	14.23
3 kHz	SiO_2–Na_2O glass (34.3% mol Na_2O)	room temp.	21.30

Selecting Electrical Permittivity of Glass by Frequency (Continued)

Frequency (Hz)	Glass	Temperature (°C)	Electrical Permittivity
5 kHz	SiO_2–Na_2O glass (19.5% mol Na_2O)	room temp.	8.56
5 kHz	SiO_2–Na_2O glass (24.4% mol Na_2O)	room temp.	10.21
5 kHz	SiO_2–Na_2O glass (29.4% mol Na_2O)	room temp.	13.19
5 kHz	SiO_2–Na_2O glass (34.3% mol Na_2O)	room temp.	18.13
10 kHz	B_2O_3 glass	500	3.13
10 kHz	B_2O_3 glass	550	3.14
10 kHz	B_2O_3 glass	580	3.145
10 kHz	B_2O_3 glass	620	3.15
10 kHz	B_2O_3 glass	650	3.15
10 kHz	B_2O_3 glass	700	3.16
10 kHz	SiO_2 glass	25	4.0
10 kHz	SiO_2 glass	200	4.0
10 kHz	SiO_2 glass	300	4.0
10 kHz	SiO_2 glass	400	4.0
10 kHz	SiO_2–Na_2O glass (19.5% mol Na_2O)	room temp.	8.26
10 kHz	SiO_2–Na_2O glass (24.4% mol Na_2O)	room temp.	9.74
10 kHz	SiO_2–Na_2O glass (29.4% mol Na_2O)	room temp.	12.08
10 kHz	SiO_2–Na_2O glass (34.3% mol Na_2O)	room temp.	15.22
10 kHz	SiO_2–Na_2O glass (39.3% mol Na_2O)	room temp.	22.08
30 kHz	SiO_2–Na_2O glass (19.5% mol Na_2O)	room temp.	8.00
30 kHz	SiO_2–Na_2O glass (24.4% mol Na_2O)	room temp.	9.30
30 kHz	SiO_2–Na_2O glass (29.4% mol Na_2O)	room temp.	11.21
30 kHz	SiO_2–Na_2O glass (34.3% mol Na_2O)	room temp.	13.28
30 kHz	SiO_2–Na_2O glass (39.3% mol Na_2O)	room temp.	16.56

Selecting Electrical Permittivity of Glass by Frequency (Continued)

Frequency (Hz)	Glass	Temperature (°C)	Electrical Permittivity
50 kHz	B_2O_3 glass	800	3.04
50 kHz	B_2O_3 glass	620	3.05
50 kHz	B_2O_3 glass	750	3.06
50 kHz	B_2O_3 glass	700	3.09
50 kHz	B_2O_3 glass	500	3.10
50 kHz	B_2O_3 glass	650	3.10
50 kHz	B_2O_3 glass	580	3.115
50 kHz	B_2O_3 glass	550	3.12
50 kHz	SiO_2–Na_2O glass (19.5% mol Na_2O)	room temp.	7.88
50 kHz	SiO_2–Na_2O glass (24.4% mol Na_2O)	room temp.	9.14
50 kHz	SiO_2–Na_2O glass (29.4% mol Na_2O)	room temp.	10.86
50 kHz	SiO_2–Na_2O glass (34.3% mol Na_2O)	room temp.	12.57
50 kHz	SiO_2–Na_2O glass (39.3% mol Na_2O)	room temp.	15.06
100 kHz	SiO_2–Na_2O glass (19.5% mol Na_2O)	room temp.	7.74
100 kHz	SiO_2–Na_2O glass (24.4% mol Na_2O)	room temp.	8.91
100 kHz	SiO_2–Na_2O glass (29.4% mol Na_2O)	room temp.	10.47
100 kHz	SiO_2–Na_2O glass (34.3% mol Na_2O)	room temp.	11.78
100 kHz	SiO_2–Na_2O glass (39.3% mol Na_2O)	room temp.	13.55
300 kHz	SiO_2–Na_2O glass (19.5% mol Na_2O)	room temp.	7.62
300 kHz	SiO_2–Na_2O glass (24.4% mol Na_2O)	room temp.	8.75
300 kHz	SiO_2–Na_2O glass (29.4% mol Na_2O)	room temp.	10.15
300 kHz	SiO_2–Na_2O glass (34.3% mol Na_2O)	room temp.	11.14
300 kHz	SiO_2–Na_2O glass (39.3% mol Na_2O)	room temp.	12.43

Selecting Electrical Permittivity of Glass by Frequency (Continued)

Frequency (Hz)	Glass	Temperature (°C)	Electrical Permittivity
56.8 MHz	B_2O_3–Na_2O glass (4.08% mol Na_2O)	room temp.	3.72
56.8 MHz	B_2O_3–Na_2O glass (7.35% mol Na_2O)	room temp.	4.20
56.8 MHz	B_2O_3–Na_2O glass (14.15% mol Na_2O)	room temp.	4.94
56.8 MHz	B_2O_3–Na_2O glass (17.31% mol Na_2O)	room temp.	5.27
56.8 MHz	B_2O_3–Na_2O glass (24.77% mol Na_2O)	room temp.	6.24
56.8 MHz	B_2O_3–Na_2O glass (31.98% mol Na_2O)	room temp.	7.03
4.5×10^8 Hz	SiO_2–Na_2O glass (16% mol Na_2O)	20	6.01
4.5×10^8 Hz	SiO_2–Na_2O glass (20% mol Na_2O)	20	6.48
4.5×10^8 Hz	SiO_2–Na_2O glass (22.2% mol Na_2O)	20	6.85
4.5×10^8 Hz	SiO_2–Na_2O glass (28.6% mol Na_2O)	20	7.62
4.5×10^8 Hz	SiO_2–Na_2O glass (36% mol Na_2O)	20	9.40
9.4 GHz	SiO_2 glass	20	3.81
9.4 GHz	SiO_2 glass	200	3.83
9.4 GHz	SiO_2 glass	400	3.84
9.4 GHz	SiO_2 glass	600	3.86
9.4 GHz	SiO_2 glass	800	3.88
9.4 GHz	SiO_2 glass	1000	3.91
9.4 GHz	SiO_2 glass	1200	3.93
9.4 GHz	SiO_2 glass	1400	3.96
10 GHz	SiO_2–Al_2O_3 glass (46.3% mol B_2O_3)		3.55
10 GHz	SiO_2 glass	20	3.82
10 GHz	SiO_2 glass	220	3.82
10 GHz	SiO_2 glass	888	3.91

Selecting Electrical Permittivity of Glass by Frequency (Continued)

Frequency (Hz)	Glass	Temperature (°C)	Electrical Permittivity
10 GHz	SiO_2 glass	1170	3.98
10 GHz	SiO_2 glass	1764	4.04
10 GHz	SiO_2 glass	1335	4.05
10 GHz	SiO_2 glass	1764	4.05
10 GHz	SiO_2 glass	1420	4.07
10 GHz	SiO_2 glass	1480	4.09
10 GHz	SiO_2 glass	1526	4.11
10 GHz	SiO_2 glass	1584	4.12
10 GHz	SiO_2 glass	1647	4.12
10 GHz	SiO_2 glass	1602	4.15
32 GHz	SiO_2–PbO glass (40% mol PbO)	−150	4.25
32 GHz	SiO_2–PbO glass (40% mol PbO)	−100	4.30
32 GHz	SiO_2–PbO glass (40% mol PbO)	−50	4.40
32 GHz	SiO_2–PbO glass (40% mol PbO)	0	4.45
32 GHz	SiO_2–PbO glass (40% mol PbO)	50	5.00

Source: data compiled by J.S. Park *from* O. V. Mazurin, M. V. Streltsina and T. P. Shvaiko-Shvaikovskaya, *Handbook of Glass Data, Part A and Part B*, Elsevier, New York, 1983.

Selecting Ceramics and Glasses: Optical Properties

SELECTING TRANSMISSION RANGE OF OPTICAL MATERIALS

Material & Crystal Structure	Transmission Region (mm, at 298 K)
Magnesium Fluoride (Single Crystal)	0.1 – 9.7
Silica (High Purity Crystalline)	0.12 – 4.5
Silica (High Purity Fused)	0.12 – 4.5
Lithium Fluoride (Single Crystal)	0.12 – 9.0
Ammonium Dihydrogen Phosphate (ADP, Single Crystal)	0.13 – 1.7
Calcium Fluoride (Single Crystal)	0.13 – 12
Alumina (Sapphire, Single Crystal)	0.15 – 6.5
Sodium Fluoride (Single Crystal)	0.19 – 15
Magnesium Fluoride (Film)	0.2 – 5.0
Calcium Carbonate (Calcite, Single Crystal)	0.2 – 5.5
Thallium Chloribromide (KRS–6, Mixed Crystal)	0.21 – 35
Magnesium Oxide (Single Crystal)	0.25 – 8.5
Barium Fluoride (Single Crystal)	0.25 – 15
Potassium Bromide (Single Crystal)	0.25 – 35
Potassium Iodide (Single Crystal)	0.25 – 45
Cesium Iodide (Single Crystal)	0.25 – 80
Cesium Bromide (Single Crystal)	0.3 – 55
Lithium Niobate (Single Crystal)	0.33 – 5.2
Strontium Titanate (Single Crystal)	0.39 – 6.8
Silver Chloride (Single Crystal)	0.4 – 2.8
Cuprous Chloride (Single Crystal)	0.4 – 19
Titanium Dioxide (Rutile, Single Crystal)	0.43 – 6.2
Silver Bromide (Single Crystal)	0.45 – 35
Cadmium Sulfide (Bulk and Hexagonal Single Crystal)	0.5 – 16

Selecting Transmission Range of Optical Materials (Continued)

Material & Crystal Structure	Transmission Region (mm, at 298 K)
Zinc Selenide (Single Crystal, Cubic)	~0.5 – 22
Arsenic Trisulfade (Glass)	0.6 – 13
Zinc Sulfide (Single Crystal, Cubic)	~0.6 – 15.6
Thallium Bromoiodide (KRS–5, Mixed Crystal)	0.6 – 40
Cadmium Telluride (Hot Pressed Polycrystalline)	0.9 – 16
Gallium Arsenide (Intrinsic Single Crystal)	1.0 – 15
Selenium (Amorphous)	1.0 – 20
Silicon (Single Crystal)	1.2 – 15
Germanium (Intrinsic Single Crystal)	1.8 – 23
Lead Sulfide (Single Crystal)	3.0 – 7.0
Tellurium (Polycrystalline Film)	3.5 – 8.0
Tellurium (Single Crystal)	3.5 – 8.0
Indium Arsenide (Single Crystal)	3.8 – 7.0

External transmittance ≥ 10% with 2.0 mm thickness.

Source: Data compiled by J.S. Park *from various sources.*

SELECTING REFRACTIVE INDICES OF GLASSES

Glass	Wavelength (λ)	Temperature (°C)	Refractive Index (n_D)
B_2O_3 glass	5461 Å	700	1.4130
SiO_2 glass	3.245 µm	26	1.41353
B_2O_3 glass	5461 Å	650	1.4155
B_2O_3 glass	5461 Å	600	1.4180
B_2O_3 glass	5461 Å	550	1.4210
SiO_2 glass	3.245 µm	828	1.42243
B_2O_3 glass	5461 Å	500	1.4240
B_2O_3 glass	5461 Å	450	1.4270
SiO_2 glass	2.553 µm	26	1.42949
B_2O_3 glass	5461 Å	400	1.4315
SiO_2 glass	2.553 µm	471	1.43450
B_2O_3 glass	5461 Å	350	1.4365
SiO_2 glass	2.553 µm	828	1.43854
SiO_2 glass	1.981 µm	26	1.43863
B_2O_3 glass	5461 Å	300	1.4420
SiO_2 glass	1.660 µm	26	1.44307
SiO_2 glass	1.981 µm	471	1.44361
SiO_2 glass	1.470 µm	26	1.44524
SiO_2 glass	1.981 µm	828	1.44734
SiO_2 glass	1.254 µm	26	1.44772
SiO_2 glass	1.660 µm	471	1.44799
SiO_2–B_2O_3 glass (quenched, 13.5% mol B_2O_3)	1.002439 µm	23	1.4485
SiO_2–B_2O_3 glass (annealed, 13.5% mol B_2O_3)	1.002439 µm	23	1.4493
SiO_2 glass	1.470 µm	471	1.45031

Selecting Refractive Indices of Glasses (Continued)

Glass	Wavelength (λ)	Temperature (°C)	Refractive Index (n_D)
SiO_2 glass	1.01398 μm	26	1.45039
B_2O_3 glass	5461 Å	250	1.4505
SiO_2–B_2O_3 glass (quenched, 13.5% mol B_2O_3)	0.852111 μm	23	1.4507
SiO_2–B_2O_3 glass (annealed, 13.5% mol B_2O_3)	0.852111 μm	23	1.4515
SiO_2 glass	1.660 μm	828	1.45174
SiO_2–B_2O_3 glass (quenched, 13.5% mol B_2O_3)	0.734620 μm	23	1.4528
SiO_2 glass	1.254 μm	471	1.45283
SiO_2–B_2O_3 glass (annealed, 13.5% mol B_2O_3)	0.734620 μm	23	1.4537
SiO_2 glass	1.470 μm	828	1.45440
SiO_2 glass	1.01398 μm	471	1.45562
SiO_2–B_2O_3 glass (quenched, 13.5% mol B_2O_3)	0.589263 μm	23	1.4570
SiO_2 glass	1.254 μm	828	1.45700
SiO_2–B_2O_3 glass (annealed, 13.5% mol B_2O_3)	0.589263 μm	23	1.4579
SiO_2–B_2O_3 glass (20% mol B_2O_3)	5145 Å		1.4582
SiO_2–B_2O_3 glass (15% mol B_2O_3)	5145 Å		1.4584
SiO_2–B_2O_3 glass (30% mol B_2O_3)	5145 Å		1.4588
SiO_2–B_2O_3 glass (10% mol B_2O_3)	5145 Å		1.4592
SiO_2–Al_2O_3 glass (1.4% mol Al_2O_3)	589.262 nm		1.4595
SiO_2 glass	1.01398 μm	828	1.45960
SiO_2 glass	0.54607 μm	26	1.46028
SiO_2–B_2O_3 glass (50% mol B_2O_3)	5145 Å		1.4604
B_2O_3 glass	5461 Å	200	1.4605
SiO_2–B_2O_3 glass (quenched, 13.5% mol B_2O_3)	0.508582 μm	23	1.4606
SiO_2–B_2O_3 glass (75% mol B_2O_3)	5145 Å		1.4612

Selecting Refractive Indices of Glasses (Continued)

Glass	Wavelength (λ)	Temperature (°C)	Refractive Index (n_D)
SiO_2–B_2O_3 glass (annealed, 13.5% mol B_2O_3)	0.508582 μm	23	1.4615
SiO_2–B_2O_3 glass (90% mol B_2O_3)	5145 Å		1.4617
B_2O_3 glass	5461 Å	150	1.4625
SiO_2–Al_2O_3 glass (3.1% mol Al_2O_3)	589.262 nm		1.4630
B_2O_3 glass	5461 Å	100	1.4635
B_2O_3 glass	5461 Å	20	1.4650
SiO_2–Al_2O_3 glass (3.7% mol Al_2O_3)	589.262 nm		1.4652–1.4667
B_2O_3–Na_2O glass (0.01% mol Na_2O)		25	1.46536
SiO_2–B_2O_3 glass (quenched, 13.5% mol B_2O_3)	0.435833 μm	23	1.4657
SiO_2 glass	0.54607 μm	471	1.46575
SiO_2–B_2O_3 glass (annealed, 13.5% mol B_2O_3)	0.435833 μm	23	1.4665
B_2O_3 glass	5461 Å	0	1.467
B_2O_3 glass	5461 Å	−100	1.469
SiO_2 glass	0.40466 μm	26	1.46978
SiO_2 glass	0.54607 μm	828	1.47004
SiO_2 glass	0.40466 μm	471	1.47575
SiO_2 glass	0.33415 μm	26	1.48000
SiO_2 glass	0.40466 μm	828	1.48033
SiO_2–Na_2O glass (15% mol Na_2O)			1.4822
B_2O_3–Na_2O glass (4.4% mol Na_2O)		25	1.48387
SiO_2 glass	0.33415 μm	471	1.48633
SiO_2 glass	0.30215 μm	26	1.48738
SiO_2 glass	3.245 μm	471	1.4893
SiO_2–Na_2O glass (20% mol Na_2O)			1.4906

Selecting Refractive Indices of Glasses (Continued)

Glass	Wavelength (λ)	Temperature (°C)	Refractive Index (n_D)
SiO_2 glass	0.28936 µm	26	1.49121
SiO_2 glass	0.33415 µm	828	1.49135
SiO_2 glass	0.30215 µm	471	1.49407
B_2O_3–Na_2O glass (8.7% mol Na_2O)		25	1.49442
SiO_2 glass	0.27528 µm	26	1.49615
B_2O_3–Na_2O glass (11.5% mol Na_2O)		25	1.49662
SiO_2 glass	0.28936 µm	471	1.49818
SiO_2–Na_2O glass (25% mol Na_2O)			1.4983
B_2O_3–Na_2O glass (13.7% mol Na_2O)		25	1.49841
SiO_2 glass	0.30215 µm	828	1.49942
B_2O_3–Na_2O glass (16.2% mol Na_2O)		25	1.49984
B_2O_3–Na_2O glass (15.8% mol Na_2O)		25	1.50024
B_2O_3–Na_2O glass (17.4% mol Na_2O)		25	1.50155
B_2O_3–Na_2O glass (18.4% mol Na_2O)		25	1.50210
SiO_2 glass	0.27528 µm	471	1.50327
SiO_2 glass	0.28936 µm	828	1.50358
SiO_2–Na_2O glass (30% mol Na_2O)			1.5041
B_2O_3–Na_2O glass (19.6% mol Na_2O)		25	1.50468
B_2O_3–Na_2O glass (20.0% mol Na_2O)		25	1.50500
SiO_2–Na_2O glass (33.3% mol Na_2O)			1.5061
B_2O_3–Na_2O glass (22.5% mol Na_2O)		25	1.50806
SiO_2 glass	0.24827 µm	26	1.50865
SiO_2 glass	0.27528 µm	828	1.50889
B_2O_3–Na_2O glass (23.6% mol Na_2O)		25	1.50979

Selecting Refractive Indices of Glasses (Continued)

Glass	Wavelength (λ)	Temperature (°C)	Refractive Index (n_D)
SiO_2–Na_2O glass (39.3% mol Na_2O)			1.5099
SiO_2 glass	0.2407 μm	26	1.51361
SiO_2–Na_2O glass (45.1% mol Na_2O)			1.5137
B_2O_3–Na_2O glass (28.9% mol Na_2O)		25	1.51611
SiO_2 glass	0.24827 μm	471	1.51665
SiO_2–Na_2O glass (50% mol Na_2O)			1.517
SiO_2 glass	0.23021 μm	26	1.52034
SiO_2 glass	0.2407 μm	471	1.52201
SiO_2 glass	0.24827 μm	828	1.52289
SiO_2 glass	0.2407 μm	828	1.52832
SiO_2 glass	0.23021 μm	471	1.52908
SiO_2 glass	0.23021 μm	828	1.53584
SiO_2–CaO glass (39.0% mol CaO)			1.5905
B_2O_3–CaO glass (35% mol CaO)			1.6021
SiO_2–CaO glass (44.6% mol CaO)			1.6120
SiO_2–PbO glass (20.78% mol PbO)			1.6174
SiO_2–Al_2O_3 glass (70.2% mol Al_2O_3)			1.629
SiO_2–CaO glass (50.0% mol CaO)			1.6295
SiO_2–Al_2O_3 glass (77.0% mol Al_2O_3)			1.634
SiO_2–CaO glass (52.9% mol CaO)			1.6350
SiO_2–CaO glass (57.5% mol CaO)			1.6455
SiO_2–PbO glass (24.90% mol PbO)			1.6509
B_2O_3–CaO glass (64.1% mol CaO)			1.6525
SiO_2–PbO glass (29.71% mol PbO)			1.6948

Selecting Refractive Indices of Glasses (Continued)

Glass	Wavelength (λ)	Temperature (°C)	Refractive Index (n_D)
SiO_2–Al_2O_3 glass (84.1% mol Al_2O_3)			1.720
SiO_2–PbO glass (33.01% mol PbO)			1.7270
SiO_2–Al_2O_3 glass (91.8% mol Al_2O_3)			1.728
SiO_2–PbO glass (36.64% mol PbO)			1.7632
SiO_2–PbO glass (40.80% mol PbO)			1.8092
SiO_2–PbO glass (44.07% mol PbO)			1.8457
SiO_2–PbO glass (47.83% mol PbO)			1.8865
SiO_2–PbO glass (50.50% mol PbO)			1.9189
SiO_2–PbO glass (53.46% mol PbO)			1.9545
SiO_2–PbO glass (56.43% mol PbO)			1.9894
SiO_2–PbO glass (61.38% mol PbO)			2.0460–2.0512
SiO_2–PbO glass (65.97% mol PbO)			2.1030

Source: data compiled by J.S. Park *from* O. V. Mazurin, M. V. Streltsina and T. P. Shvaiko–Shvaikovskaya, *Handbook of Glass Data, Part A and Part B*, Elsevier, New York, 1983.

Selecting Polymeric Materials

Continued

Selecting Polymeric Materials: Table of Contents (Continued)

Selecting Polymeric Materials: Table of Contents (Continued)

PERIODIC TABLE OF ELEMENTS IN POLYMERIC MATERIALS

The Elements in Polymeric Materials

1 IA	2 IIA	3	4	5	6	7	8	9	10	11	12	13	14	15	16	17	18 VIIA
IA	IIA	IIIB	IVB	VB	VIB	VIIB	----- VIII -----		IB	IIB	IIIA	IVA	VA	VIA	VIIA	
1 H													6 C	7 N	8 O	9 F	
													14 Si				

Selecting Polymers: Thermal Properties

SELECTING SPECIFIC HEAT OF POLYMERS

Polymer	Specific Heat (Btu/lb/°F)
Polymide: Glass reinforced	0.15—0.27
Reinforced polyester moldings: Sheet molding compounds, general purpose	0.20—0.25
Standard Epoxies: High strength laminate	0.21
Polytrifluoro chloroethylene (PTFCE)	0.22
Standard Epoxies: Filament wound composite	0.24
Phenylene oxides (Noryl): Standard	0.24
Silicone: Woven glass fabric/ silicone laminate	0.246
Polytetrafluoroethylene (PTFE)	0.25
Polymide: Unreinforced	0.25—0.35
Reinforced polyester moldings: High strength (glass fibers)	0.25—0.35
Polystyrenes; Molded: Glass fiber -30% reinforced	0.256
Polyphenylene sulfide: Standard	0.26
Phenolics; Molded; General: Arc resistant—mineral filled	0.27—0.37
Fluorinated ethylene propylene(FEP)	0.28
Nylon, Type 6: Type 12	0.28
Phenolics; Molded; General: Very high shock: glass fiber filled	0.28—0.32
PVC–acrylic sheet	0.293
Thermoset Carbonate: Allyl diglycol carbonate	0.3
Chlorinated polyvinyl chloride	0.3
Polycarbonate	0.3
Phenolics; Molded; General: High shock: chopped fabric or cord filled	0.30—0.35
Polystyrenes; Molded: General purpose	0.30—0.35
Polystyrenes; Molded: Medium impact	0.30—0.35
Polystyrenes; Molded: High impact	0.30—0.35

Selecting Specific Heat of Polymers (Continued)

Polymer	Specific Heat (Btu/lb/°F)
Polyesters: Thermoset Cast; Rigid	0.30—0.55
Cellulose Acetate Butyrate; Molded, Extruded; ASTM Grade: H4	0.3—0.4
Cellulose Acetate Butyrate; Molded, Extruded; ASTM Grade: MH	0.3—0.4
Cellulose Acetate Butyrate; Molded, Extruded; ASTM Grade: S2	0.3—0.4
Cellusose Acetate Propionate; Molded, Extruded; ASTM Grade: 1	0.3—0.4
Cellusose Acetate Propionate; Molded, Extruded; ASTM Grade: 3	0.3—0.4
Cellusose Acetate Propionate; Molded, Extruded; ASTM Grade: 6	0.3—0.4
Cellulose Acetate; Molded, Extruded; ASTM Grade: H6—1	0.3—0.42
Cellulose Acetate; Molded, Extruded; ASTM Grade: H4—1	0.3—0.42
Cellulose Acetate; Molded, Extruded; ASTM Grade: H2—1	0.3—0.42
Cellulose Acetate; Molded, Extruded; ASTM Grade: MH—1, MH—2	0.3—0.42
Cellulose Acetate; Molded, Extruded; ASTM Grade: MS—1, MS—2	0.3—0.42
Cellulose Acetate; Molded, Extruded; ASTM Grade: S2—1	0.3—0.42
6/6 Nylon: General purpose molding	0.3—0.5
6/6 Nylon: General purpose extrusion	0.3—0.5
6/10 Nylon: General purpose	0.3—0.5
Vinylidene chloride	0.32
Polyvinylidene— fluoride (PVDF)	0.33
Rubber phenolic—woodflour or flock	0.33
Styrene acrylonitrile (SAN)	0.33
Acrylic Moldings: High impact grade	0.34
Acrylics; Cast Resin Sheets, Rods: General purpose, type I	0.35
Acrylics; Cast Resin Sheets, Rods: General purpose, type II	0.35
Acrylic Moldings: Grades 5, 6, 8	0.35

Selecting Specific Heat of Polymers (Continued)

Polymer	Specific Heat (Btu/lb/°F)
Polyacetal: Standard	0.35
Polyacetal Copolymer: Standard	0.35
Polyacetal Copolymer: High flow	0.35
ABS Resins; Molded, Extruded; Low temperature impact	0.35—0.38
Phenolics; Molded; General: woodflour and flock filled	0.35—0.40
ABS Resins; Molded, Extruded; Medium impact	0.36—0.38
ABS Resins; Molded, Extruded; High impact	0.36—0.38
ABS Resins; Molded, Extruded; Very high impact	0.36—0.38
ABS Resins; Molded, Extruded; Heat resistant	0.37—0.39
Nylon, Type 6: General purpose	0.4
Nylon, Type 6: Cast	0.4
Nylon, Type 6: Type 8	0.4
Standard Epoxies: Cast rigid	0.4-0.5
Polypropylene: General purpose	0.45
Polypropylene: High impact	0.45—0.48
Polyethylenes; Molded, Extruded; Type III: Melt index 0.2—0.9	0.46—0.55
Polyethylenes; Molded, Extruded; Type III: Melt index 0.l—12.0	0.46—0.55
Polyethylenes; Molded, Extruded; Type III: Melt index 1.5—15	0.46—0.55
Polyethylenes; Molded, Extruded; Type I: Melt index 0.3—3.6	0.53—0.55
Polyethylenes; Molded, Extruded; Type I: Melt index 6—26	0.53—0.55
Polyethylenes; Molded, Extruded; Type I: Melt index 200	0.53—0.55

Selecting Specific Heat of Polymers (Continued)

Polymer	Specific Heat (Btu/lb/°F)
Polyethylenes; Molded, Extruded; Type II: Melt index 20	0.53—0.55
Polyethylenes; Molded, Extruded; Type II: Melt index 1.0—1.9	0.53—0.55
Nylon, Type 6: Type 11	0.58

Source: data compiled by J.S. Park *from* Charles T. Lynch, *CRC Handbook of Materials Science*, Vol. 3, CRC Press, Boca Raton, Florida and *Engineered Materials Handbook*, Vol.2, Engineering Plastics, ASM International, Metals Park, Ohio, 1988.

SELECTING THERMAL CONDUCTIVITY OF POLYMERS

Polymer	Thermal Conductivity (ASTM C177) Btu / (hr • ft • °F)
ABS Resins; Molded, Extruded: Very high impact	0.01—0.14
Polystyrene: Medium impact	0.024—0.090
Polystyrene: High impact	0.024—0.090
Rubber Phenolic: Asbestos Filled	0.04
Rubber Phenolic: Chopped Fabric Filled	0.05
Vinylidene Chloride	0.053
Polystyrene: General purpose	0.058—0.090
Polyvinyl Chloride & Copolymers: Nonrigid—General	0.07—0.10
Polyvinyl Chloride & Copolymers: Nonrigid—Electrical	0.07—0.10
Polyvinyl Chloride & Copolymers: Rigid—Normal Impact	0.07—0.10
Silicone: Woven Glass Fabric/ Silicone Laminate	0.075—0.125
ABS Resins; Molded, Extruded: Low temperature impact	0.08—0.14
ABS Resins; Molded, Extruded: Medium impact	0.08—0.18
Phenolics; High Shock: Chopped Fabric or Cord Filled	0.097—0.170
Phenolics; Molded: General: Woodflour and Flock Filled	0.097—0.3
Polyester, Thermoset: Cast Rigid	0.10—0.12
Cellulose Acetate, ASTM Grade: H6—1	0.10—0.19
Cellulose Acetate, ASTM Grade: H4—1	0.10—0.19
Cellulose Acetate, ASTM Grade: H2—1	0.10—0.19
Cellulose Acetate, ASTM Grade: MH—1, MH—2	0.10—0.19
Cellulose Acetate, ASTM Grade: MS—1, MS—2	0.10—0.19
Cellulose Acetate, ASTM Grade: S2—1	0.10—0.19
Cellulose Acetate Butyrate; ASTM Grade: H4	0.10—0.19
Cellulose Acetate Butyrate; ASTM Grade: MH	0.10—0.19

Selecting Thermal Conductivity of Polymers (Continued)

Polymer	Thermal Conductivity (ASTM C177) Btu / (hr • ft • °F)
Cellulose Acetate Butyrate; ASTM Grade: S2	0.10—0.19
Cellulose Acetate Propionate, ASTM Grade: 1	0.10—0.19
Cellulose Acetate Propionate, ASTM Grade: 3	0.10—0.19
Cellulose Acetate Propionate, ASTM Grade: 6	0.10—0.19
Phenolics; Molded: Shock: Paper, Flock, or Pulp Filled	0.1—0.16
Epoxy, Standard: Cast rigid	0.1—0.3
Epoxy, Standard: Molded	0.1—0.5
Polycarbonate	0.11
Polystyrene: Glass fiber -30% reinforced	0.117
Acrylic Cast Resin Sheets, Rods: General purpose, type I	0.12
Acrylic Cast Resin Sheets, Rods: General purpose, type II	0.12
Acrylic Moldings: Grades 5, 6, 8	0.12
Acrylic Moldings: High impact grade	0.12
Fluorinated ethylene propylene(FEP)	0.12
Rubber Phenolic: Woodflour or Flock Filled	0.12
ABS Resins; Molded, Extruded: High impact	0.12—0.16
ABS Resins; Molded, Extruded: Heat resistant	0.12—0.20
Polycarbonate (40% glass fiber reinforced)	0.13
Polyacetal Homopolymer: Standard	0.13
Polytetrafluoroethylene (PTFE)	0.14
Polyvinylidene— fluoride (PVDF)	0.14
Polytrifluoro chloroethylene (PTFCE)	0.145
Polyacetal Copolymer: Standard	0.16
Melamine; Molded: Cellulose Electrical	0.17—0.20

Selecting Thermal Conductivity of Polymers (Continued)

Polymer	Thermal Conductivity (ASTM C177) Btu / (hr • ft • °F)
Urea; Molded: Alpha—Cellulose Filled (ASTM Type l)	0.17—0.244
Silicone: Fibrous (Glass) Reinforced	0.18
Polyethylene; Molded, Extruded; Type I: Melt Index 0.3—3.6	0.19
Polyethylene; Molded, Extruded; Type I: Melt Index 6—26	0.19
Polyethylene; Molded, Extruded; Type I: Melt Index 200	0.19
Polyethylene; Molded, Extruded; Type II: Melt Index 20	0.19
Polyethylene; Molded, Extruded; Type II: Melt Index L.0—1.9	0.19
Polyethylene; Molded, Extruded; Type III: Melt Index 0.2—0.9	0.19
Polyethylene; Type III: Melt Melt Index 0.1—12.0	0.19
Polyethylene; Molded, Extruded; Type III: Melt Index 1.5—15	0.19
Polyethylene; Molded, Extruded; Type III: High Molecular Weight	0.19
Phenolics; Molded: Very High Shock: Glass Fiber Filled	0.2
Alkyds; Molded: Glass reinforced (heavy duty parts)	0.20—0.30
Phenolics; Molded: Arc Resistant—Mineral	0.24—0.34
Silicone: Granular (Silica) Reinforced	0.25—0.5
Melamine; Molded: Glass Fiber Filled	0.28
Alkyds; Molded: Putty (encapsulating)	0.35—0.60
Alkyds; Molded: Rope (general purpose)	0.35—0.60
Alkyds; Molded: Granular (high speed molding)	0.35—0.60
Polyester Injection Moldings: General purpose grade	0.36—0.55
Chlorinated polyether	0.91
Chlorinated polyvinyl chloride	0.95
PVC–Acrylic Injection Molded	0.98
PVC–Acrylic Sheet	1.01

Selecting Thermal Conductivity of Polymers (Continued)

Polymer	Thermal Conductivity (ASTM C177) Btu / (hr • ft • °F)
Phenylene Oxide: SE—100	1.1
Polyarylsulfone	1.1
Phenylene Oxide: Glass fiber reinforced	1.1–1.15
Nylon; Molded, Extruded Type 6: General purpose	1.2—1.69
Nylon; Type 6: Cast	1.2—1.7
Polypropylene: General Purpose	1.21—1.36
Polyester, Thermoset: High strength (glass fiber filled)	1.32—1.68
Thermoset Allyl diglycol Carbonate	1.45
Nylon: Type 11	1.5
6/10 Nylon: General purpose	1.5
Phenylene Oxide: SE—1	1.5
6/6 Nylon: Glass fiber reinforced	1.5— 3.3
Polyacetal Copolymer: High flow	1.6
6/6 Nylon: General purpose molding	1.69—1.7
Nylon; Molded, Extruded Type 6: Glass fiber (30%) reinforced	1.69—3.27
Nylon: Type 12	1.7
6/6 Nylon: General purpose extrusion	1.7
Polypropylene: High Impact	1.72
Phenylene oxides (Noryl): Standard	1.8
Polyphenylene Sulfide: Standard	2

Selecting Thermal Conductivity of Polymers (Continued)

Polymer	Thermal Conductivity (ASTM C177) Btu / (hr • ft • °F)
Polyphenylene Sulfide: 40% Glass Reinforced	2
Epoxy, Standard: High strength laminate	2.35
ABS–Polycarbonate Alloy	2.46
6/10 Nylon: Glass fiber (30%) reinforced	3.5
Polymide: Glass Reinforced	3.59
Polymide: Unreinforced	3.8–6.78

Source: data compiled by J.S. Park from Charles T. Lynch, *CRC Handbook of Materials Science, Vol. 3*, CRC Press, Boca Raton, Florida, 1975 and *Engineered Materials Handbook, Vol.2*, Engineering Plastics, ASM International, Metals Park, Ohio, 1988.

SELECTING THERMAL EXPANSION OF POLYMERS

Polymer	Thermal Expansion Coefficient ASTM D696 ($°F^{-1}$)
Polymides: Glass Reinforced	0.8×10^{-6}
Polycarbonate (40% Glass Fiber Reinforced)	$1.0—1.1 \times 10^{-6}$
Epoxy Novolacs: Cast, Rigid	$1.6—3.0 \times 10^{-6}$
Epoxies: High Performance Resins: Molded	$1.7—2.2 \times 10^{-6}$
Polymides: Unreinforced	$2.5—4.5 \times 10^{-6}$
ABS Resin; Molded, Extruded: Heat Resistant	$3.0—4.0 \times 10^{-6}$
Acrylic Moldings: Grades 5, 6, 8	$3—4 \times 10^{-6}$
ABS Resin; Molded, Extruded: Medium Impact	$3.2—4.8 \times 10^{-6}$
Standard Epoxies: General Purpose Glass Cloth Laminate	$3.3—4.8 \times 10^{-6}$
Standard Epoxies: High Strength Laminate	$3.3—4.8 \times 10^{-6}$
Polycarbonate	3.75×10^{-6}
Acrylic Moldings: High Impact Grade	$4—6 \times 10^{-6}$
Chlorinated Polyvinyl Chloride	4.4×10^{-6}
Acrylics; Cast Resin Sheets, Rods: General Purpose, Type I	4.5×10^{-6}
Acrylics; Cast Resin Sheets, Rods: General Purpose, Type II	4.5×10^{-6}
ABS Resin; Molded, Extruded: Very High Impact	$5.0—6.0 \times 10^{-6}$
ABS Resin; Molded, Extruded: Low Temperature Impact	$5.0—6.0 \times 10^{-6}$
ABS Resin; Molded, Extruded: High Impact	$5.5—6.0 \times 10^{-6}$
Chlorinated Polyether	6.6×10^{-6}
Melamines; Molded: Glass Fiber Filled	0.82×10^{-5}
Rubber Phenolic—Woodflour or Flock	$0.83—2.20 \times 10^{-5}$
Phenolics, Molded; General: Very High Shock: Glass Fiber Filled	0.88×10^{-5}
Standard Epoxies: Molded	$1—2 \times 10^{-5}$
Melamines; Molded: Cellulose Filled Electrical	$1.11—2.78 \times 10^{-5}$

Selecting Thermal Expansion of Polymers (Continued)

Polymer	Thermal Expansion Coefficient ASTM D696 ($°F^{-1}$)
Nylon; Molded, Extruded; Type 6: Glass Fiber (30%) Reinforced	1.2×10^{-5}
Phenylene Oxides (Noryl): Glass Fiber Reinforced	$1.2–1.6 \times 10^{-5}$
Ureas; Molded: Alpha—Cellulose Filled (ASTM Type l)	$1.22—1.50 \times 10^{-5}$
Alkyds; Molded: Putty (encapsulating)	1.3×10^{-5}
Alkyds; Molded: Rope (general purpose)	1.3×10^{-5}
Alkyds; Molded: Granular (high speed molding)	1.3×10^{-5}
Alkyds; Molded: Glass reinforced (heavy duty parts)	1.3×10^{-5}
Reinforced Polyester Moldings: High Strength (Glass Fibers)	$13—19 \times 10^{-6}$
Phenylene Oxides: Glass Fiber Reinforced	$1.4–2.0 \times 10^{-5}$
6/10 Nylon: General purpose	1.5×10^{-5}
6/6 Nylon; General Purpose Molding: Glass Fiber Reinforced	$1.5—3.3 \times 10^{-5}$
Glass Fiber (30%) Reinforced SAN	1.6×10^{-5}
Phenolics, General: High Shock: Chopped Fabric or Cord Filled	$1.60—2.22 \times 10^{-5}$
Phenolics, Molded; General: Shock: Paper, Flock, or Pulp	$1.6—2.3 \times 10^{-5}$
Polypropylene: Glass Reinforced	$1.6—2.4 \times 10^{-5}$
Phenolics, Molded; General: Woodflour And Flock Filled	$1.66—2.50 \times 10^{-5}$
6/6 Nylon; General Purpose Molding	$1.69—1.7 \times 10^{-5}$
6/6 Nylon; General Purpose Extrusion	1.7×10^{-5}
Rubber Phenolic—Chopped Fabric	1.7×10^{-5}
Polytetrafluoroethylene (PTFE), Ceramic Reinforced	$1.7—2.0 \times 10^{-5}$
Polystyrenes; Molded: Glass Fiber -30% Reinforced	1.8×10^{-5}
Polymide Homopolymer: 20% Glass Reinforced	$2.0—4.5 \times 10^{-5}$
Polypropylene: Asbestos Filled	$2—3 \times 10^{-5}$
Standard Epoxies: Filament Wound Composite	$2—6 \times 10^{-5}$

Selecting Thermal Expansion of Polymers (Continued)

Polymer	Thermal Expansion Coefficient ASTM D696 ($°F^{-1}$)
Diallyl Phthalates; Molded: Glass Fiber Filled	$2.2.—2.6 \times 10^{-5}$
Rubber Phenolic—Asbestos	2.2×10^{-5}
Polymide Copolymer: 25% Glass Reinforced	$2.2—4.7 \times 10^{-5}$
Polystyrenes; Molded: High Impact	$2.2—5.6 \times 10^{-5}$
Silicones; Molded, Laminated: Granular (Silica) Reinforced	$2.5—5.0 \times 10^{-5}$
Polyarylsulfone	2.6×10^{-5}
Polyester; Injection Moldings: Glass Reinforced Grades	$2.7—3.3 \times 10^{-5}$
Polyvinyl Chloride; Molded, Extruded: Rigid—normal impact	$2.8—3 .3 \times 10^{-5}$
Polyphenylene Sulfide: Standard	$3.0—4.9 \times 10^{-5}$
Standard Epoxies: Cast Flexible	$3—5 \times 10^{-5}$
Phenylene Oxides (Noryl): Standard	3.1×10^{-5}
Silicones; Molded, Laminated: Fibrous (Glass) Reinforced	$3.17—3.23 \times 10^{-5}$
Standard Epoxies: Cast rigid	3.3×10^{-5}
Phenylene Oxides: SE—1	3.3×10^{-5}
Polystyrenes; Molded: Medium Impact	$3.3—4.7 \times 10^{-5}$
Polystyrenes; Molded: General Purpose	$3.3—4.8 \times 10^{-5}$
6/10 Nylon: Glass fiber (30%) reinforced	3.5×10^{-5}
PVC–Acrylic Alloy Sheet	3.5×10^{-5}
Polyester; Injection Moldings: Glass Reinforced Self Extinguishing	3.5×10^{-5}
Styrene Acrylonitrile (SAN)	$3.6—3.7 \times 10^{-5}$
Phenylene Oxides: SE—100	3.8×10^{-5}
Polypropylene: General Purpose	$3.8—5.8 \times 10^{-5}$
Polytrifluoro chloroethylene (PTFCE)	3.88×10^{-5}
Thermoset Cast Polyyester: Rigid	$3.9—5.6 \times 10^{-5}$

Selecting Thermal Expansion of Polymers (Continued)

Polymer	Thermal Expansion Coefficient ASTM D696 ($°F^{-1}$)
Polyphenylene Sulfide: 40% Glass Reinforced	4×10^{-5}
Diallyl Phthalates; Molded: Asbestos Filled	4.0×10^{-5}
Polypropylene: High Impact	$4.0—5.9 \times 10^{-5}$
Nylon; Type 6: Cast	4.4×10^{-5}
Cellulose Acetate; Molded, Extruded; ASTM Grade: H6—1	$4.4—9.0 \times 10^{-5}$
Cellulose Acetate; Molded, Extruded; ASTM Grade: H4—1	$4.4—9.0 \times 10^{-5}$
Cellulose Acetate; Molded, Extruded; ASTM Grade: H2—1	$4.4—9.0 \times 10^{-5}$
Cellulose Acetate; ASTM Grade: MH—1, MH—2	$4.4—9.0 \times 10^{-5}$
Cellulose Acetate; ASTM Grade: MS—1, MS—2	$4.4—9.0 \times 10^{-5}$
Cellulose Acetate; Molded, Extruded; ASTM Grade: S2—1	$4.4—9.0 \times 10^{-5}$
Polymide Homopolymer: Standard	4.5×10^{-5}
Polymide Homopolymer: 22% TFE Reinforced	4.5×10^{-5}
Polymide Copolymer: Standard	4.7×10^{-5}
Polymide Copolymer: High Flow	4.7×10^{-5}
Nylon; Molded, Extruded; Type 6: General Purpose	4.8×10^{-5}
Polyester; Injection Moldings: General Purpose Grade	$4.9—13.0 \times 10^{-5}$
Diallyl Phthalates; Molded: Orlon Filled	5.0×10^{-5}
Diallyl Phthalates; Molded: Dacron Filled	5.2×10^{-5}
Polyester; Thermoplastic Injection Moldings: General Purpose Grade	5.3×10^{-5}
Nylon; Type 11	5.5×10^{-5}
Thermoset Carbonate: Allyl diglycol carbonate	6×10^{-5}
Cellulose Acetate Butyrate; Molded, Extruded; ASTM Grade: H4	$6—9 \times 10^{-5}$
Cellulose Acetate Butyrate; Molded, Extruded; ASTM Grade: MH	$6—9 \times 10^{-5}$
Cellulose Acetate Butyrate; Molded, Extruded; ASTM Grade: S2	$6—9 \times 10^{-5}$

Selecting Thermal Expansion of Polymers (Continued)

Polymer	Thermal Expansion Coefficient ASTM D696 ($^\circ F^{-1}$)
Cellusose Acetate Propionate; Molded, Extruded; ASTM Grade: 1	$6—9 \times 10^{-5}$
Cellusose Acetate Propionate; Molded, Extruded; ASTM Grade: 3	$6—9 \times 10^{-5}$
Cellusose Acetate Propionate; Molded, Extruded; ASTM Grade: 6	$6—9 \times 10^{-5}$
ABS–Polycarbonate Alloy	6.12×10^{-5}
Nylon; Type 12	7.2×10^{-5}
Fluorinated Ethylene Propylene(FEP)	$8.3—10.5 \times 10^{-5}$
Polyethylene; Molded, Extruded; Type II: Melt Index 20	$8.3—16.7 \times 10^{-5}$
Polyethylene; Molded, Extruded; Type II: Melt index 1.0—1.9	$8.3—16.7 \times 10^{-5}$
Polyethylene; Molded, Extruded; Type III: Melt Index 0.2—0.9	$8.3—16.7 \times 10^{-5}$
Polyethylene; Type III: Melt Melt Index 0.1—12.0	$8.3—16.7 \times 10^{-5}$
Polyethylene; Molded, Extruded; Type III: Melt Index 1.5—15	$8.3—16.7 \times 10^{-5}$
Polyvinylidene— Fluoride (PVDF)	8.5×10^{-5}
Vinylidene chloride	8.78×10^{-5}
Polyethylene; Molded, Extruded; Type I: Melt Index 0.3—3.6	$8.9—11.0 \times 10^{-5}$
Polyethylene; Molded, Extruded; Type I: Melt Index 6—26	$8.9—11.0 \times 10^{-5}$
Polyethylene; Molded, Extruded; Type I: Melt Index 200	11×10^{-5}
Polytetrafluoroethylene (PTFE)	55×10^{-5}

Source: *data compiled by* J.S. Park from Charles T. Lynch, *CRC Handbook of Materials Science, Vol. 3*, CRC Press, Boca Raton, Florida and *Engineered Materials Handbook, Vol.2*, Engineering Plastics, ASM International, Metals Park, Ohio, 1988.

Selecting Polymers: Mechanical Properties

SELECTING TENSILE STRENGTHS OF POLYMERS

Polymer	Tensile Strength (ASTM D638) (10^3 psi)
Olefin Copolymer: EEA (ethylene ethyl acrylate)	0.2
Olefin Copolymer: Ethylene butene	0.35
Olefin Copolymer: EVA (ethylene vinyl acetate)	0.36
Propylene–ethylene	0.4
Ethylene Ionomer	0.4
Fluorocarbons: Ceramic reinforced (PTFE)	0.75—2.5
Polyethylene, Type I, low density: Melt index 200	0.9—1.1 (ASTM D412)
Polyvinyl Chloride & Copolymer: Nonrigid—general	1—3.5 (ASTM D412)
Polyesters, cast thermoset: Flexible	1—8
6/6 Nylon: General purpose extrusion	1.26, 8.6
Polyethylene, Type I, low density: Melt index 6—26	1.4—2.0 (ASTM D412)
Polyethylene, Type I, low density: Melt index 0.3—3.6	1.4—2.5 (ASTM D412)
Standard Epoxy: Cast flexible	1.4—7.6
Polyethylene, Type II, medium density: Melt index 20	2
Polyvinyl Chloride & Copolymer: Nonrigid—electrical	2—3.2 (ASTM D412)
Polyethylene, Type II, medium density: Melt index 1.0—1.9	2.3—2.4
Fluorocarbons: Fluorinated ethylene propylene(FEP)	2.5—4.0
Fluorocarbons: Polytetrafluoroethylene (PTFE)	2.5—6.5
Polyethylene, Type III, higher density: Melt Melt index 0.1—12.0	2.9—4.0
Cellulose Acetate Butyrate, ASTM Grade: S2	3.0—4.0 at Fracture
Cellulose Acetate; ASTM Grade: S2—1	3.0—4.4 at Fracture
Alkyd; Molded: Granular (high speed molding)	3—4
Ethylene Polyallomer	3—4.3
Phenolics: Rubber phenolic—chopped fabric	3—5 (ASTM D651)

Selecting Tensile Strengths of Polymers (Continued)

Polymer	Tensile Strength (ASTM D638) $(10^3$ psi$)$
Polystyrene: High impact	3.3—5.1
Cellulose Acetate; ASTM Grade: MS—1, MS—2	3.9—5.3 at Fracture
Cellulose Acetate Propionate, ASTM Grade: 6	4
Phenolics: Rubber phenolic—asbestos	4 (ASTM D651)
Polystyrene: Medium impact	4.0—6.0
Alkyd; Molded: Putty (encapsulating)	4—5
ABS Resin; Molded, Extruded: Low temperature impact	4—6
Reinforced polyester moldings: Heat & chemical resistant (asbestos)	4—6
Silicone: Granular (silica) reinforced	4—6 (ASTM D651)
Diallyl Phthalates, Molded: Asbestos filled	4—6.5
Polyvinyl Chloride & Copolymer: Vinylidene chloride	4—8,15—40 (ASTM D412)
Polyethylene, Type III, higher density: Melt index 0.2—0.9	4.4
Polyethylene, Type III, higher density: Melt index 1.5—15	4.4
Diallyl Phthalates, Molded: Orlon filled	4.5—6
ABS Resin; Molded, Extruded: Very high impact	4.5—6.0
Polypropylene: general purpose	4.5—6.0
Phenolics: Rubber phenolic—woodflour or flock	4.5—9 (ASTM D651)
Fluorocarbons: Polytrifluoro chloroethylene (PTFCE)	4.6—5.7
Diallyl Phthalates, Molded: Dacron filled	4.6—6.2
Cellulose Acetate; ASTM Grade: MH—1, MH—2	4.8—6.3 at Fracture
Polystyrene: General purpose	5.0—10
ABS Resin; Molded, Extruded: High impact	5.0—6.0
Cellulose Acetate Butyrate, ASTM Grade: MH	5.0—6.0 at Fracture
Phenolics, General: woodflour and flock filler	5.0—8.5 (ASTM D651)

Selecting Tensile Strengths of Polymers (Continued)

Polymer	Tensile Strength (ASTM D638) (10^3 psi)
Phenolics, Shock: paper, flock, or pulp filler	5.0—8.5 (ASTM D651)
Reinforced polyester moldings: High strength (glass fibers)	5—10
Urea: Alpha, cellulose filled (ASTM Type 1)	5—10
Phenolics, Very high shock: glass fiber filler	5—10 (ASTM D651)
Polyesters, cast thermoset: Rigid	5—15
Allyl diglycol carbonate (thermoset)	5—6
Melamine, molded: Alpha cellulose and mineral filler	5—8
Alkyd; Molded: Glass reinforced (heavy duty parts)	5—9
Melamine, molded: Cellulose electrical filler	5—9
Phenolics, High shock: chopped fabric or cord filler	5—9 (ASTM D651)
Cellulose Acetate Propionate, ASTM Grade: 3	5.1—5.9
Epoxiy, (cycloaliphatic diepoxides): Molded	5.2—5.3
Fluorocarbons: Polyvinylidene— fluoride (PVDF)	5.2—8.6
Polyethylene, Type III, higher density, high molecular weight	5.4
Diallyl Phthalates, Molded: Glass fiber filled	5.5—11
Polyvinyl Chloride & Copolymer: Rigid—normal impact	5.5—8 (ASTM D412)
Acrylic Moldings: High impact grade	5.5—8.0
Cellulose Acetate; ASTM Grade: H2—1	5.8—7.2 at Fracture
Cellulose Acetate Propionate, ASTM Grade: 1	5.9—6.5
Chlorinated polyether	6
Phenolics: Arc resistant—mineral	6 (ASTM D651)
Acrylic Cast Resin Sheets, Rods: General purpose, type I	6—9
Melamine, molded: Glass fiber filler	6—9
ABS Resin; Molded, Extruded: Medium impact	6.3—8.0

Selecting Tensile Strengths of Polymers (Continued)

Polymer	Tensile Strength (ASTM D638) (10^3 psi)
Silicone: Fibrous (glass) reinforced	6.5 (ASTM D651)
Polyacetal homopolymer: 22% TFE reinforced	6.9
Cellulose Acetate Butyrate, ASTM Grade: H4	6.9 at Fracture
ABS Resin; Molded, Extruded: Heat resistant	7.0—8.0
Alkyd; Molded: Rope (general purpose)	7—8
Cellulose Acetate; ASTM Grade: H4—1	7—8 at Fracture
Nylon, Type 12	7.1—8.5
6/10 Nylon: General purpose	7.1—8.5
Chlorinated polyvinyl chloride	7.3
Nylon, Type 6: Flexible copolymers	7.5—10.0
Acrylic Cast Resin Sheets, Rods: General purpose, type II	8—10
Standard Epoxy: Molded	8—11
Epoxiy, (cycloaliphatic diepoxides): Cast, rigid	8—12
ABS–Polycarbonate Alloy	8.2
Polystyrene: Styrene acrylonitrile (SAN)	8.3—12.0
Polyacetal homopolymer: 20% glass reinforced	8.5
Polyacetal copolymer: Standard	8.8
Polyacetal copolymer: High flow	8.8
Acrylic Moldings: Grades 5, 6, 8	8.8—10.5
Polycarbonate	9.5
Standard Epoxy: Cast rigid	9.5-11.5
Nylon, Type 6: General purpose	9.5—12.5
Epoxy novolacs: Cast, rigid	9.6—12.0
Polyacetal homopolymer: Standard	10

Selecting Tensile Strengths of Polymers (Continued)

Polymer	Tensile Strength (ASTM D638) (10^3 psi)
6/6 Nylon: General purpose molding	11.2—11.8
Nylon, Type 6: Cast	12.8
Polyarylsulfone	13
Polystyrene: Glass fiber -30% reinforced	14
Reinforced polyester: Sheet molding, general purpose	15—17
Polycarbonate (40% glass fiber reinforced)	18
Polystyrene: Glass fiber (30%) reinforced SAN	18
Polyacetal copolymer: 25% glass reinforced	18.5
6/10 Nylon: Glass fiber (30%) reinforced	19
6/6 Nylon: Glass fiber Molybdenum disulfide filled	19—22
Nylon, Type 6: Glass fiber (30%) reinforced	21—24
6/6 Nylon: Glass fiber reinforced	25—30
Silicone: Woven glass fabric / silicone laminate	30—35 (ASTM D651)
Epoxy: Glass cloth laminate	50-58
Epoxiy, (cycloaliphatic diepoxides): Glass cloth laminate	50—52
Epoxy novolacs: Glass cloth laminate	59.2
Epoxy: Glass cloth: High strength laminate	160
Epoxy: Glass cloth laminate: Filament wound composite	230-240 (hoop)

To convert **psi** to **MPa**, multiply by **145**.

Source: *data compiled by* J.S. Park *from* Charles T. Lynch, *CRC Handbook of Materials Science, Vol. 3*, CRC Press, Boca Raton, Florida, 1975 and *Engineered Materials Handbook, Vol.2*, Engineering Plastics, ASM International, Metals Park, Ohio, 1988.

SELECTING COMPRESSIVE STRENGTHS OF POLYMERS

Polymer	Compressive Strength (1000 psi)
ABS Resins; Molded, Extruded: Medium impact	0.5—11.0
Polyester, Cast Thermoset: Flexible	1—17
Styrene acrylonitrile (SAN), Glass fiber (30%) reinforced	2.3
Polystyrene, Molded: Medium impact	4—9
Polystyrene, Molded: High impact	4—9
PVC–acrylic injection molded	6.2
ABS Resins; Molded, Extruded: High impact	7.0—9.0
PVC–acrylic sheet	8.4
Chlorinated polyether	9
ABS Resins; Molded, Extruded: Heat resistant	9.3—11.0
Silicone, Molded: Fibrous (glass) reinforced silicones	10—12.5
Rubber phenolic, Molded: , chopped fabric filled	10—15
Rubber phenolic, Molded: , asbestos filled	10—20
Silicone, Molded: Granular (silica) reinforced silicones	10.6—17
Polyvinyl Chloride: Rigid—normal impact	11—12
ABS–Polycarbonate Alloy	11.1—11.8
Polystyrene, Molded: General purpose	11.5—16.0
Phenylene Oxide: SE—100	12
Rubber phenolic, Molded: woodflour or flock filled	12—20
Polyester, Cast Thermoset: Rigid	12—37
Polycarbonate	12.5
Polyester; Thermoplastic Moldings: General purpose grade	13
Phenylene oxide (Noryl): Standard	13.9—14
Silicone, Laminated with woven glass fabric	15—24

Selecting Compressive Strengths of Polymers (Continued)

Polymer	Compressive Strength (1000 psi)
Phenolic; Molded: High shock, chopped fabric or cord filled	15—30
Polyester; Thermoplastic Moldings: Glass reinforced grades	16—18
Alkyds; Molded: Granular (high speed molding)	16—20
Phenylene Oxide: SE—1	16.4
Epoxy, Standard : Cast rigid	16.5—24
Epoxy, High performance resins: Cast, rigid	17—19
Phenolic; Molded: Very high shock, glass fiber filled	17—30
Phenylene Oxide: Glass fiber reinforced	17.6—17.9
Polyarylsulfone	17.8
Polyester; Thermoplastic: Glass reinforced, self extinguishing	18
Diallyl Phthalate; Molded: Asbestos filled	18—25
Polymide: Unreinforced	18.4, 27.4
Polycarbonate (40% glass fiber reinforced)	18.5
Polystyrene, Molded: Glass fiber -30% reinforced	19
Alkyds; Molded: Putty (encapsulating)	20—25
Diallyl Phthalate; Molded: Orlon filled	20—25
Polyester: Heat and chemical resistsnt (asbestos reinforced)	20—25
Polyester: High strength, (glass fibers reinforced)	20—26
Diallyl Phthalate; Molded: Dacron filled	20—30
Phenolic, Molded: Arc resistant, mineral filled	20—30
Melamine; Molded: Glass fiber filled	20—42
Epoxy, High performance resins: Molded	22—26
Phenolic; Molded: General, woodflour and flock filled	22—36
Polyester: Sheet molding compounds, general purpose	22—36

Selecting Compressive Strengths of Polymers (Continued)

Polymer	Compressive Strength (1000 psi)
Thermoset Carbonate: Allyl diglycol carbonate	22.5
Alkyds; Molded: Glass reinforced (heavy duty parts)	24—30
Phenolic; Molded: Shock, paper, flock, or pulp filled	24—35
Diallyl Phthalate; Molded: Glass fiber filled	25
Melamine; Molded: Cellulose electrical filled	25—35
Urea, Molded: Woodflour filled	25—35
Urea, Molded: Alpha—cellulose filled (ASTM Type l)	25—38
Melamine; Molded: Mineral filled	26—30
Alkyds; Molded: Rope (general purpose)	28
Epoxy novolac: Cast, rigid	30—50
Epoxy, Standard : Molded	34-38
Melamine; Molded: Unfilled	40—45
Melamine; Molded: Alpha cellulose filled	40—45
Polymide: Glass reinforced	42
Epoxy novolac: Glass cloth laminate	48—57
Epoxy, Standard : General purpose glass cloth laminate	50-60
Epoxy, High performance resins: Glass cloth laminate	67—71
Epoxy, Standard : High strength laminate	80-90 (edgewise)

To convert **psi** to **MPa**, multiply by **145**.

Source: *data compiled by* J.S. Park *from* Charles T. Lynch, *CRC Handbook of Materials Science, Vol. 3*, CRC Press, Boca Raton, Florida, 1975 and *Engineered Materials Handbook, Vol.2*, Engineering Plastics, ASM International, Metals Park, Ohio, 1988.

SELECTING YIELD STRENGTHS OF POLYMERS

Polymer	Yield Strength, (ASTM D638) (10^3 psi)
Polypropylene: High impact	2.8—4.3
Polystyrene, Molded: High impact	2.8—5.3
Polypropylene: Asbestos filled	3.3—8.2
Polypropylene: Flame retardant	3.6—4.2
Polystyrene, Molded: Medium impact	3.7—6.0
Nylon; Molded or Extruded: Type 8	3.9
Polypropylene: General purpose	4.5—6.0
Polystyrene, Molded: General purpose	5.0—10
Polymide: Unreinforced	5—7.5
PVC–acrylic injection molded	5.5
Nylon; Molded or Extruded: Type 12	5.5—6.5
Chlorinated Polyether	5.9
PVC–acrylic sheet	6.5
Polypropylene: Glass reinforced	7—11
Nylon, Type 6/10; Molded or Extruded: General purpose	7.1—8.5
Nylon; Molded or Extruded: Flexible copolymers	7.5—10.0
Polyester Injection Moldings: General purpose grade	7.5—8
Phenylene Oxide: SE—100	7.8
Nylon, Type 6/6: General purpose molding	8.0—11.8
Polyarylsulfone	8—12
ABS–Polycarbonate Alloy	8.2
Polyester: General purpose grade	8.2
Polycarbonate	8.5
Nylon; Molded or Extruded: Type 11	8.5

Selecting Yield Strengths of Polymers (Continued)

Polymer	Yield Strength, (ASTM D638) (10^3 psi)
Nylon; Molded or Extruded: General purpose	8.5—12.5
Nylon, Type 6/6: General purpose extrusion	8.6—12.6
Polyacetal Copolymer: Standard	8.8
Polyacetal Copolymer: High flow	8.8
Polyphenylene sulfide: Standard	9.511
Phenylene Oxide: SE—1	9.6
Polyacetal Homopolymer: Standard	10
Phenylene oxide (Noryl): Standard	10.2
Polyester: Asbestos filled grade	12
Nylon; Molded or Extruded: Cast	12.8
Polyester: Glass reinforced grade	14
Polystyrene, Molded: Glass fiber 30% reinforced	14
Phenylene Oxide: Glass fiber reinforced	14.5—17.0
Polyester Moldings: Glass reinforced self extinguishing	17
Phenylene oxide (Noryl): Glass fiber reinforced	17—19
Polyester Injection Moldings: Glass reinforced grades	17—25
Styrene acrylonitrile (SAN): Glass fiber (30%) reinforced	18
Polyacetal Copolymer: 25% glass reinforced	18.5
Polyphenylene sulfide: 40% glass reinforced	20—21
Nylon, Type 6/6; Molded or Extruded: Glass fiber reinforced	25
Polymide: Glass reinforced	28

To convert **psi** to **MPa**, multiply by **145**.

Source: *data compiled by* J.S. Park *from* Charles T. Lynch, *CRC Handbook of Materials Science, Vol. 3*, CRC Press, Boca Raton, Florida, 1975 and *Engineered Materials Handbook, Vol.2*, Engineering Plastics, ASM International, Metals Park, Ohio, 1988.

SELECTING COMPRESSIVE YIELD STRENGTHS
OF POLYMERS

Polymer	Compressive Yield Strength (ASTM D690 or D695) (0.1% offset, 1000 psi)
Polytetrafluoroethylene (PTFE)	0.7—1.8
Ceramic reinforced (PTFE)	1.4—1.8
Fluorinated ethylene propylene(FEP)	1.6
Polytrifluoro chloroethylene (PTFCE)	2
Cellulose Acetate Butyrate, ASTM Grade: S2	2.6—4.3
6/10 Nylon: General purpose	3.0
Cellulose Acetate, ASTM Grade: S2—1	3.15—6.1
Cellulose Acetate, ASTM Grade: MS—1, MS—2	3.2—7.2
Cellulose Acetate, ASTM Grade: H2—1	4.3—9.6
Polypropylene: High impact	4.4
Cellulose Acetate, ASTM Grade: MH—1, MH—2	4.4—8.4
Polyacetal Homopolymer: 22% TFE reinforced	4.5
Polyacetal Copolymer: Standard	4.5
Polyacetal Copolymer: High flow	4.5
6/6 Nylon: General purpose molding	4.9
6/6 Nylon: General purpose extrusion	4.9
Cellusose Acetate Propionate, ASTM Grade: 3	4.9—5.8
Polyacetal Homopolymer: Standard	5.2
Polyacetal Homopolymer: 20% glass reinforced	5.2
Cellulose Acetate Butyrate, ASTM Grade: MH	5.3—7.1
Polypropylene: General purpose	5.5—6.5
Cellusose Acetate Propionate, ASTM Grade: 1	6.2—7.3
Cellulose Acetate, ASTM Grade: H4—1	6.5—10.6
Polypropylene: Glass reinforced	6.5—7

Selecting Compressive Yield Strengths of Polymers (Continued)

Polymer	Compressive Yield Strength (ASTM D690 or D695) (0.1% offset, 1000 psi)
Polypropylene: Asbestos filled	7
Acrylic Moldings: High impact grade	7.3—12.0
Cellulose Acetate Butyrate, ASTM Grade: H4	8.8
Nylon, Type 6: General purpose	9.7
Polyvinyl Chloride: Rigid—normal impact	10—11
Acrylic Cast Resin Sheets, Rods: General purpose, type I	12—14
Polyvinylidene— fluoride (PVDF)	12.8—14.2
Nylon, Type 6: Cast	14
Acrylic Cast Resin Sheets, Rods: General purpose, type II	14—18
Acrylic Moldings: Grades 5, 6, 8	14.5—17
6/10 Nylon: Glass fiber (30%) reinforced	18
Nylon, Type 6: Glass fiber (30%) reinforced	19—20
6/6 Nylon: Glass fiber reinforced	20—24
Vinylidene chloride	75—85

To convert from **psi** to **MPa**, multiply by **145**.

Source: *data compiled by* J.S. Park *from* Charles T. Lynch, CRC Handbook of Materials Science, Vol. 3, CRC Press, Boca Raton, Florida, 1975 and *Engineered Materials Handbook, Vol.2*, Engineering Plastics, ASM International, Metals Park, Ohio, 1988.

SELECTING FLEXURAL STRENGTHS OF POLYMERS

Polymer	Flexural Strength (ASTM D790) (10^3 psi)
Epoxy, Standard: Cast flexible	1.2—12.7
Cellulose Acetate Butyrate, ASTM Grade: S2	2.5—3.95 (yield)
Fluorinated ethylene propylene(FEP)	3 (0.1% offset)
Nylon, Type 6: Flexible copolymers	3.4—16.4
Polytrifluoro chloroethylene (PTFCE)	3.5 (0.1% offset)
Cellulose Acetate, ASTM Grade: S2—1	3.5—5.7 (yield)
Cellulose Acetate, ASTM Grade: MS—1, MS—2	3.8—7.1 (yield)
Polyesters, Cast Thermoset: Flexible	4—16
Polypropylene: High impact	4.1 (yield)
Cellulose Acetate, ASTM Grade: MH—1, MH—2	4.4—8.65 (yield)
Chlorinated polyether	5 (0.1% offset)
ABS Resins; Molded or Extruded: Low temperature impact	5—8
Cellusose Acetate Propionate, ASTM Grade: 3	5.6—6.2 (yield)
Cellulose Acetate Butyrate, ASTM Grade: MH	5.6—6.7 (yield)
Cellulose Acetate, ASTM Grade: H2—1	6.0—10.0 (yield)
ABS Resins; Molded or Extruded: Very high impact	6.0—9.8
Silicone: Granular (silica) reinforced	6—10
Melamines, Molded: Cellulose filled, electrical	6—15
Reinforced polyester: High strength (glass fibers)	6—26
Polypropylene: General purpose	6—7 (yield)
Polymide: Unreinforced	6.6—11
Cellusose Acetate Propionate, ASTM Grade: 1	6.8—7.9 (yield)
Rubber phenolic—chopped fabric filled	7
Rubber phenolic—asbestos filled	7

Selecting Flexural Strengths of Polymers (Continued)

Polymer	Flexural Strength (ASTM D790) (10^3 psi)
Alkyd, Molded: Granular (high speed molding)	7—10
Rubber phenolic—woodflour or flock filled	7—12
Diallyl Phthalate, Molded: Orlon filled	7.5—10.5
Urea, Molded: Woodflour filled	7.5—12.0
Urea, Molded: Cellulose filled (ASTM Type 2)	7.5—13
Polypropylene: Asbestos filled	7.5—9 (yield)
ABS Resins; Molded or Extruded: High impact	7.5—9.5
6/10 Nylon: General purpose	8
Phenolic: Shock: paper, flock, or pulp filled	8.0—11.5
Diallyl Phthalate, Molded: Asbestos filled	8—10
Alkyd, Molded: Putty (encapsulating)	8—11
Polypropylene: Glass reinforced	8—11 (yield)
Phenolic: High shock, chopped fabric or cord filled	8—15
Urea, Molded: Alpha—cellulose filled (ASTM Type l)	8—18
Polyesters, Cast Thermoset: Rigid	8—24
Cellulose Acetate, ASTM Grade: H4—1	8.1—11.15 (yield)
Phenolic: General, woodflour and flock filled	8.5—12
Polyvinylidene— fluoride (PVDF)	8.6—10.8 (0.1% offset)
PVC–acrylic injection molded	8.7
Acrylic Moldings: High impact grade	8.7—12.0
Cellulose Acetate Butyrate, ASTM Grade: H4	9 (yield)
Diallyl Phthalate, Molded: Dacron filled	9—11.5
Melamines, Molded: Unfilled	9.5—14
ABS Resins; Molded or Extruded: Medium impact	9.9—11.8

Selecting Flexural Strengths of Polymers (Continued)

Polymer	Flexural Strength (ASTM D790) (10^3 psi)
Epoxy, High performance resins: Molded	10—12
Phenolic: Arc resistant—mineral filled	10—13
Reinforced polyester: Heat and chemical resistant (asbestos)	10—13
Polystyrene: General purpose	10—15
Diallyl Phthalate, Molded: Glass fiber filled	10—18
Phenolic: Very high shock, glass fiber filled	10—45
PVC–acrylic sheet	10.7
ABS Resins; Molded or Extruded: Heat resistant	11.0—12.0
Epoxy, High performance resins: Cast, rigid	11—16
Melamines, Molded: Alpha cellulose filled	11—16
Polyvinyl Chloride And Copolymers: Rigid—normal impact	11—16
Polyester Injection Moldings: General purpose grade	12
Epoxy novolacs: Cast, rigid	12—13
Acrylic, Cast Resin Sheets, Rods: General purpose, type I	12—14
Alkyd, Molded: Glass reinforced (heavy duty parts)	12—17
Polyester Injection Moldings: General purpose grade	12.8
Phenylene Oxide: SE—100	12.8
Polyacetal Copolymer: Standard	13
Polyacetal Copolymer: High flow	13
Polycarbonate	13.5
Phenylene Oxide: SE—1	13.5
Epoxy, Standard: Cast rigid	14—18
Melamines, Molded: Glass fiber filled	14—18
Polyacetal Homopolymer: Standard	14.1

Selecting Flexural Strengths of Polymers (Continued)

Polymer	Flexural Strength (ASTM D790) (10^3 psi)
ABS–Polycarbonate Alloy	14.3
Chlorinated polyvinyl chloride	14.5
Acrylic Moldings: Grades 5, 6, 8	15—16
Acrylic, Cast Resin Sheets, Rods: General purpose, type II	15—17
Vinylidene chloride	15—17
Phenylene oxides (Noryl): Standard	15.4
Silicone: Fibrous (glass) reinforced	16—19
Polyarylsulfone	16.1—17.2
Nylon, Type 6: Cast	16.5
Polystyrene: Glass fiber —30% reinforced	17
Melamines, Molded: Alpha mineral filled	18—10
Polyester Injection Moldings: Glass reinforced grade	19
Polyester Injection Moldings: Asbestos—filled grade	19
Alkyd, Molded: Rope (general purpose)	19—20
Epoxy, Standard: Molded	19—22
Polyphenylene sulfide: Standard	20
Phenylene Oxide: Glass fiber reinforced	20.5—22
Styrene acrylonitrile (SAN): Glass fiber (30%) reinforced	22
Polyester Injection Moldings: Glass reinforced grades	22—24
6/10 Nylon: Glass fiber (30%) reinforced	23
Polyester Injection Moldings: Glass reinforced self extinguishing	23
Phenylene oxides (Noryl): Glass fiber reinforced	25—28
6/6 Nylon: Glass fiber Molybdenum disulfide filled	26—28
Reinforced polyester sheet molding: general purpose	26—32

Selecting Flexural Strengths of Polymers (Continued)

Polymer	Flexural Strength (ASTM D790) (10^3 psi)
Nylon, Type 6: Glass fiber (30%) reinforced	26—34
6/6 Nylon: Glass fiber reinforced	26—35
Polycarbonate (40% glass fiber reinforced)	27
Polyacetal Copolymer: 25% glass reinforced	28
Silicone: Woven glass fabric/ silicone laminate	33—47
Polyphenylene sulfide: 40% glass reinforced	37
Polymide: Glass reinforced	56
Epoxy, High performance resins: Glass cloth laminate	70—72
Epoxy, Standard: General purpose glass cloth laminate	80—90
Epoxy novolacs: Glass cloth laminate	84—89
Epoxy, Standard: High strength laminate	165—177
Epoxy, Standard: Filament wound composite	180—170
Nylon, Type 6: General purpose	Unbreakable
6/6 Nylon: General purpose molding	Unbreakable

To convert from **psi** to **MPa**, multiply by **145**.

Source: *data compiled by* J.S. Park *from* Charles T. Lynch, *CRC Handbook of Materials Science, Vol. 3*, CRC Press, Boca Raton, Florida, 1975 and *Engineered Materials Handbook, Vol.2*, Engineering Plastics, ASM International, Metals Park, Ohio, 1988.

SELECTING HARDNESS OF POLYMERS

Polymer	Hardness, (ASTM D785) (Rockwell)
Polyester, Thermoset: Flexible	6—40 (Barcol)
Polyester, Thermoset: Rigid	35—50 (Barcol)
Polyester: Heat & chemical resistant (asbestos reinforced)	40—70 (Barcol)
Polyester: Sheet molding compounds, general purpose	45—60 (Barcol)
Cellusose Acetate Propionate, ASTM Grade: 6	57
Polyethylene, Type III: High molecular weight	60—65 (Shore)
Alkyd, Molded: Putty (encapsulating)	60—70 (Barcol)
Alkyd, Molded: Granular (high speed molding)	60—70 (Barcol)
Polyester moldings: High strength (glass fibers) Reinforced	60—80 (Barcol)
Epoxy, Standard: Cast High strength laminate	70—72 (Barcol)
Alkyd, Molded: Rope (general purpose)	70—75 (Barcol)
Alkyd, Molded: Glass reinforced (heavy duty parts)	70—80 (Barcol)
Silicone: Woven glass fabric/ silicone laminate	75 (Barcol)
Epoxy, Standard: Cast Molded	75-80 (Barcol)
Epoxy, High performance resins: Glass cloth laminate	75—80
Cellusose Acetate Propionate, ASTM Grade: 3	92—96
Cellusose Acetate Propionate, ASTM Grade: 1	100—109
Epoxy, High performance resins: Cast, rigid	107—112
Polyvinyl Chloride: Nonrigid—general	A50—100 (Shore, ASTM D676)
Polyvinyl Chloride: Nonrigid—electrical	A78—100 (Shore, ASTM D676)
Vinylidene chloride	>A95 (Shore, ASTM D676)
Polyethylene, Type I: Melt index 6—26	C73, D47—53 (Shore)
Polyethylene, Type I: Melt index 0.3—3.6	C73, D50—52 (Shore)

Selecting Hardness of Polymers (Continued)

Polymer	Hardness, (ASTM D785) (Rockwell)
Olefin Copolymer, Molded: EEA (ethylene ethyl acrylate)	D35 (Shore)
Olefin Copolymer, Molded: EVA (ethylene vinyl acetate)	D36 (Shore)
Polyethylene, Type I: Melt index 200	D45 (Shore)
Polytetrafluoroethylene (PTFE)	D52
Polyethylene, Type II: Melt index 20	D55 (Shore)
Polyethylene, Type II: Melt index 1.0—1.9	D55—D56 (Shore)
Fluorinated ethylene propylene(FEP)	D57—58
Olefin Copolymer, Molded: Propylene—ethylene ionomer	D60 (Shore)
Polyethylene, Type III: Melt Melt index 0.1—12.0	D60—70 (Shore)
Olefin Copolymer, Molded: Ethylene butene	D65 (Shore)
Polyethylene, Type III: Melt index 0.2—0.9	D68—70 (Shore)
Polyethylene, Type III: Melt index 1.5—15	D68—70 (Shore)
Polyvinyl Chloride: Rigid—normal impact	D70—85 (Shore, ASTM D676)
Epoxy, High performance resins: Molded	D94—96
6/10 Nylon: Glass fiber (30%) reinforced	E40—50
Phenolic, Molded: Very high shock: glass fiber filled	E50—70
6/6 Nylon: Glass fiber reinforced	E60—E80
Phenolic, Molded: High shock: chopped fabric or cord filled	E80—90
Phenolic, Molded: General: woodflour and flock filled	E85—100
Phenolic, Molded: Shock: paper, flock, or pulp filled	E85—95
Urea, Molded: Alpha—cellulose filled (ASTM Type l)	E94—97
Melamine, Molded: Unfilled	E110
Polymide: Glass reinforced	E114
Phenylene Oxide: Glass fiber reinforced	L106, L108

Selecting Hardness of Polymers (Continued)

Polymer	Hardness, (ASTM D785) (Rockwell)
Polystyrene, Molded: High impact	M3—43
Acrylic Moldings: High impact grade	M38—45
Rubber phenolic—woodflour or flock filled	M40—90
Polystyrene, Molded: Medium impact	M47—65
Rubber phenolic—asbestos filled	M50
Epoxy, Standard: Cast Cast flexible	M50-100
Polyvinyl Chloride & Copolymers: Vinylidene chloride	M50—65
Rubber phenolic—chopped fabric filled	M57
Polycarbonate	M70
Silicone: Granular (silica) reinforced	M71—95
Polystyrene, Molded: General purpose	M72
Styrene acrylonitrile (SAN)	M75—85
Polyacetal Homopolymer: 22% TFE reinforced	M78
Polyacetal Copolymer: 25% glass reinforced	M79
Polyacetal Copolymer: Standard	M80
Polyacetal Copolymer: High flow	M80
Acrylic Moldings: Grades 5, 6, 8	M80—103
Acrylic Cast Resin Sheets, Rods: General purpose, type I	M80—90
Phenylene oxides (Noryl): Glass fiber reinforced	M84
Polyester, Thermoplastic Moldings: Asbestos—filled grade	M85
Polyarylsulfone	M85—110
Polystyrene, Molded: Glass fiber -30% reinforced	M85—95
Silicone: Fibrous (glass) reinforced	M87
Polyacetal Homopolymer: 20% glass reinforced	M90

Selecting Hardness of Polymers (Continued)

Polymer	Hardness, (ASTM D785) (Rockwell)
Glass fiber (30%) reinforced Styrene acrylonitrile (SAN)	M90—123
Polyacetal Homopolymer: Standard	M94
6/6 Nylon: Glass fiber Molybdenum disulfide filled	M95—100
Thermoset Carbonate: Allyl diglycol carbonate	M95—M100 (Barcol)
Acrylic Cast Resin Sheets, Rods: General purpose, type II	M96—102
Polycarbonate (40% glass fiber reinforced)	M97
Epoxy, Standard: Cast Filament wound composite	M98-120
Phenolic, Molded: Arc resistant—mineral	M105—115
Epoxy, Standard: Cast rigid	M106
Cellusose Acetate Propionate, ASTM Grade: Asbestos filled	M107
Cellusose Acetate Propionate, ASTM Grade: Orlon filled	M108
Cellusose Acetate Propionate, ASTM Grade: Glass fiber filled	M108
Epoxy, Standard: Cast General purpose glass cloth laminate	M115—117
Melamine, Molded: Cellulose filled electrical	M115—125
Urea, Molded: Alpha—cellulose filled (ASTM Type 1)	M116—120
Urea, Molded: Woodflour filled	M116—120
Cellulose Acetate Butyrate, ASTM Grade: S2	R23—42
Polypropylene: High impact	R28—95
Polytetrafluoroethylene (PTFE): Ceramic reinforced	R35—55
Cellulose Acetate, ASTM Grade: S2—1	R49—88
Cellulose Acetate, ASTM Grade: MS—1, MS—2	R54—96
Polypropylene: Flame retardant	R60—R105
Nylon, Type 6: Flexible copolymers	R72—Rll9
Cellulose Acetate, ASTM Grade: MH—1, MH—2	R74—104

Selecting Hardness of Polymers (Continued)

Polymer	Hardness, (ASTM D785) (Rockwell)
ABS Resin; Molded, Extruded: Low temperature impact	R75—95
Cellulose Acetate Butyrate, ASTM Grade: MH	R80—100
Polypropylene: General purpose	R80—R100
ABS Resin; Molded, Extruded: Very high impact	R85—105
Cellulose Acetate, ASTM Grade: H2—1	R89—112
Polypropylene: Asbestos filled	R90—R110
Polypropylene: Glass reinforced	R90—R115
Nylon, Type 6: Glass fiber (30%) reinforced	R93—121
ABS Resin; Molded, Extruded: High impact	R95—113
Chlorinated polyether	R100
Cellulose Acetate, ASTM Grade: H4—1	R103—120
PVC–acrylic injection molded	R104
PVC–acrylic sheet	R105
Nylon, Type 12	R106
ABS Resin; Molded, Extruded: Heat resistant	R107—116
ABS Resin; Molded, Extruded: Medium impact	R108—115
Polyvinylidene— fluoride (PVDF)	R109—110
Polytrifluoro chloroethylene (PTFCE)	R110—115
Polyvinyl Chloride & Copolymers: Rigid—normal impact	R110—120
6/10 Nylon: General purpose	R111
Cellulose Acetate Butyrate, ASTM Grade: H4	R114
Phenylene Oxide: SE—100	R115
Nylon, Type 6: Cast	R116
Polyester, Thermoplastic Moldings: General purpose grade	R117

Selecting Hardness of Polymers (Continued)

Polymer	Hardness, (ASTM D785) (Rockwell)
Polyester, Thermoplastic Moldings: General purpose grade	R117
Polyester, Thermoplastic Moldings: Glass reinforced grade	R117—M85
Chlorinated polyvinyl chloride	R118
ABS–Polycarbonate Alloy	R118
6/6 Nylon: General purpose extrusion	R118—108
Nylon, Type 6: General purpose	R118—R120
6/6 Nylon: General purpose molding	R118—120, R108
Polyester, Thermoplastic Moldings: Glass reinforced grades	R118—M90
Polyester, Thermoplastic: Glass reinforced self extinguishing	R119
Phenylene Oxide: SE—1	R119
Phenylene oxides (Noryl): Standard	R120
Polyphenylene sulfide: Standard	R120—124
Polyphenylene sulfide: 40% glass reinforced	R123
Nylon, Type 11	R100—R108

Source: data compiled by J.S. Park *from* Charles T. Lynch, *CRC Handbook of Materials Science, Vol. 3*, CRC Press, Boca Raton, Florida, 1975 and *Engineered Materials Handbook, Vol.2*, Engineering Plastics, ASM International, Metals Park, Ohio, 1988.

SELECTING COEFFICIENTS OF STATIC FRICTION FOR POLYMERS

Polymer	Coefficient of Static Friction (Against Self) (Dimensionless)
6/6 Nylon: General purpose molding	0.04—0.13
Polyacetal Homopolymer: 22% TFE reinforced	0.05—0.15 (against steel)
Polyarylsulfone	0.1—0.3
Polyacetal Homopolymer: Standard	0.1—0.3 (against steel)
Polyacetal Homopolymer: 20% glass reinforced	0.1—0.3 (against steel)
Polyester; Thermoplastic Moldings: General purpose grade	0.13 (against steel)
Polyester; Thermoplastic Moldings: Glass reinforced grades	0.14 (against steel)
Polyester; Thermoplastic : Glass reinforced self extinguishing	0.14 (against steel)
Polyacetal Copolymer: Standard	0.15 (against steel)
Polyacetal Copolymer: 25% glass reinforced	0.15 (against steel)
Polyacetal Copolymer: High flow	0.15 (against steel)
Polyester; Thermoplastic Moldings: Glass reinforced grades	0.16 (ASTM D1894)
Polyester; Thermoplastic: Glass reinforced self extinguishing	0.16 (ASTM D1894)
Polyester; Thermoplastic Moldings: General purpose grade	0.17 (ASTM D1894)
ABS–Polycarbonate Alloy	0.2
Nylon, Type 6: Cast	0.32 (dynamic)
Polycarbonate	0.52
Phenylene oxides (Noryl): Standard	0.67

Source: data compiled by J.S. Park *from* Charles T. Lynch, *CRC Handbook of Materials Science, Vol. 3*, CRC Press, Boca Raton, Florida, 1975 and *Engineered Materials Handbook, Vol.2*, Engineering Plastics, ASM International, Metals Park, Ohio, 1988.

SELECTING ABRASION RESISTANCE OF POLYMERS

Polymer	Abrasion Resistance (Taber, CS—17 wheel, ASTM D1044) (mg / 1000 cycles)
Polymide: Unreinforced	0.004—0.08
PVC–acrylic injection molded	0.0058 (CS—10 wheel)
PVC–acrylic sheet	0.073 (CS—10 wheel)
Nylon, Type 6: Cast	2.7
6/6 Nylon: General purpose extrusion	3—5
6/6 Nylon: General purpose molding	3—8
Nylon, Type 6: General purpose	5
Polyester Injection Moldings:General purpose grade	6.5
Polyacetal Homopolymer: 22% TFE reinforced	9
Polyester Injection Moldings:Glass reinforced grades	9—50
Polycarbonate	10
Polyester Injection Moldings:Glass reinforced self extinguishing	11
Polyacetal Copolymer:Standard	14
Polyacetal Copolymer:High flow	14
Polyacetal Homopolymer: Standard	14—20
Polymide: Glass reinforced	20
Phenylene Oxide: SE—1	20
Phenylene oxides (Noryl): Standard	20
Polyacetal Homopolymer: 20% glass reinforced	33
Phenylene Oxide: Glass fiber reinforced	35

Selecting Abrasion Resistance of Polymers (Continued)

Polymer	Abrasion Resistance (Taber, CS—17 wheel, ASTM D1044) (mg / 1000 cycles)
Polycarbonate (40% glass fiber reinforced)	40
Polyacetal Copolymer:25% glass reinforced	40
Polyarylsulfone	40
Phenylene Oxide: SE—100	100
Polystyrene, Molded: Glass fiber -30% reinforced	164
Polyvinylidene— fluoride (PVDF)	600—1200
Polytrifluoro chloroethylene (PTFCE)	8000

Source: data compiled by J.S. Park *from* Charles T. Lynch, *CRC Handbook of Materials Science, Vol. 3*, CRC Press, Boca Raton, Florida, 1975 and *Engineered Materials Handbook, Vol.2*, Engineering Plastics, ASM International, Metals Park, Ohio, 1988.

SELECTING IMPACT STRENGTHS OF POLYMERS

Polymer	Impact Strength (Izod notched, ASTM D256) (ft—lb / in.)
Thermoset Cast Polyyester: Rigid	0.18—0.40
Melamine, Molded: mineral filled	0.2
Urea, Molded: Cellulose filled (ASTM Type 2)	0.20—0.275
Urea, Molded: Alpha—cellulose filled (ASTM Type l)	0.20—0.35
Acrylic Moldings: Grades 5, 6, 8	0.2—0.4
Thermoset Allyl diglycol carbonate	0.2—0.4
Polystyrene, Molded: General purpose	0.2—0.4 (ASTM D638)
Epoxy, Standard: Cast rigid	0.2—0.5
Phenolic, Molded: General, woodflour and flock filled	0.24—0.50
Alkyd, Molded: Putty (encapsulating)	0.25—0.35
Urea, Molded: Woodflour filled	0.25—0.35
Melamine, Molded: Cellulose filled electrical	0.27—0.36
Styrene acrylonitrile (SAN)	0.29—0.54
Polyphenylene sulfide: Standard	0.3
Alkyd, Molded: Granular (high speed molding)	0.30—0.35
Melamine, Molded: Alpha cellulose filled	0.30—0.35
Phenolic, Molded: Arc resistant—mineral filled	0.30—0.45
Diallyl Phthalate, Molded: Asbestos filled	0.30—0.50
Epoxy, Standard: Cast flexible	0.3—0.2
Rubber phenolic—asbestos filled	0.3—0.4
Epoxy, High performance: Molded	0.3—0.5
Silicone, Molded: Granular (silica) reinforced	0.34
Rubber phenolic—woodflour or flock filled	0.34—1.0
Acrylic Cast Resin Sheets, Rods: General purpose, type I	0.4

Selecting Impact Strengths of Polymers (Continued)

Polymer	Impact Strength (Izod notched, ASTM D256) (ft—lb / in.)
Acrylic Cast Resin Sheets, Rods: General purpose, type II	0.4
Olefin Copolymers, Molded: Ethylene butene	0.4
Chlorinated polyether	0.4 (D758)
Epoxy, Standard: Molded	0.4—0.5
Phenolic, Molded: Shock: paper, flock, or pulp filled	0.4—1.0
Polypropylene: General purpose	0.4—2.2
Polyethylene, Type III: Melt Melt index 0.1—12.0	0.4—6.0
Reinforced polyester: Heat and chemical resistsnt (asbestos)	0.45—1.0
Epoxy, High performance: Cast, rigid	0.5
Polymide: Unreinforced	0.5
Polyester; Thermoplastic Moldings: Asbestos—filled grade	0.5
Diallyl Phthalate, Molded: Orlon filled	0.5—1.2
Polystyrene, Molded: Medium impact	0.5—1.2 (ASTM D638)
Polypropylene: Asbestos filled	0.5—1.5
Polyvinyl Chloride And Copolymers: Rigid—normal impact	0.5—10
Melamine, Molded: Glass fiber filled	0.5—12.0
Diallyl Phthalate, Molded: Glass fiber filled	0.5—15.0
Polypropylene: Glass reinforced	0.5—2
6/6 Nylon: General purpose molding	0.55—2.0 (ASTM D638)
Nylon Type 6: General purpose	0.6—1.2
6/10 Nylon: General purpose	0.6—1.6
Phenolic, Molded: High shock: chopped fabric or cord filled	0.6—8.0
Polyacetal Homopolymer: 22% TFE reinforced	0.7 (ASTM D638)
Polyacetal Homopolymer: 20% glass reinforced	0.8 (ASTM D638)

Selecting Impact Strengths of Polymers (Continued)

Polymer	Impact Strength (Izod notched, ASTM D256) (ft—lb / in.)
Polystyrene, Molded: High impact	0.8—1.8 (ASTM D638)
Acrylic Moldings: High impact grade	0.8—2.3
Polyacetal Copolymer: High flow	1
Polyester; Thermoplastic Moldings: General purpose grade	1.0—1.2
Polyester; Thermoplastic Moldings: Glass reinforced grades	1.0—2.2
Reinforced polyester moldings: High strength (glass fibers)	1—10
Polyphenylene sulfide: 40% glass reinforced	1.09
Olefin Copolymers, Molded: Propylene—ethylene	1.1
Nylon Type 6: Cast	1.2
Phenylene oxides (Noryl): Standard	1.2—1.3
Polyethylene, Type III: Melt index 1.5—15	1.2—2.5
Nylon: Type 12	1.2—4.2
Polyacetal Copolymer: Standard	1.3
6/6 Nylon: General purpose extrusion	1.3 (ASTM D638)
Glass fiber (30%) reinforced Styrene acrylonitrile (SAN)	1.35—3.0
Polyacetal Homopolymer: Standard	1.4 (ASTM D638)
Olefin Copolymers, Molded: Polyallomer	1.5
Polypropylene: High impact	1.5—12
Nylon Type 6: Flexible copolymers	1.5—19
Polyarylsulfone	1.6—5.0
Cellusose Acetate Propionate, ASTM Grade: 1	1.7—2.7
Diallyl Phthalate, Molded: Dacron filled	1.7—5.0

Selecting Impact Strengths of Polymers (Continued)

Polymer	Impact Strength (Izod notched, ASTM D256) (ft—lb / in.)
Polyacetal Copolymer: 25% glass reinforced	1.8
Polyester; Moldings: Glass reinforced self extinguishing	1.8
Phenylene oxides (Noryl): Glass fiber reinforced	1.8—2.0
Rubber phenolic—chopped fabric filled	2.0—2.3
ABS Resin: Medium impact	2.0—4.0
ABS Resin: Heat resistant	2.0—4.0
Polytetrafluoroethylene (PTFE)	2.0—4.0
Vinylidene chloride	2—8
Alkyd, Molded: Rope (general purpose)	2.2
Polypropylene: Flame retardant	2.2
Nylon Type 6: Glass fiber (30%) reinforced	2.2—3.4
Phenylene Oxide: Glass fiber reinforced	2.3 (ASTM D638)
Polystyrene, Molded: Glass fiber —30% reinforced	2.5
6/6 Nylon: Glass fiber reinforced	2.5—3.4 (ASTM D638)
Cellulose Acetate Butyrate, ASTM Grade: H4	3
Polyvinylidene— fluoride (PVDF)	3.0—10.3
ABS Resin: High impact	3.0—5.0
Nylon: Type 11	3.3—3.6
6/10 Nylon: Glass fiber (30%) reinforced	3.4
Polytrifluoro chloroethylene (PTFCE)	3.50—3.62
Cellusose Acetate Propionate, ASTM Grade: 3	3.5—5.6
Thermoset Cast Polyyester: Flexible	4

Selecting Impact Strengths of Polymers (Continued)

Polymer	Impact Strength (Izod notched, ASTM D256) (ft—lb / in.)
Polyethylene, Type III: Melt index 0.2—0.9	4.0—14
Cellulose Acetate Butyrate, ASTM Grade: MH	4.4—6.9
Phenylene Oxide: SE—100	5 (ASTM D638)
Phenylene Oxide: SE—1	5 (ASTM D638)
ABS Resin: Very high impact	5.0—7.5
Reinforced polyester Sheet molding, general purpose	5—15
ABS Resin: Low temperature impact	6—10
Chlorinated polyvinyl chloride	6.3
Cellulose Acetate Butyrate, ASTM Grade: S2	7.5—10.0
Alkyd, Molded: Glass reinforced (heavy duty parts)	8—12
Olefin Copolymers, Molded: Ionomer	9—14
Cellusose Acetate Propionate, ASTM Grade: 6	9.4
Silicone, Molded: Fibrous (glass) reinforced	10
ABS–Polycarbonate Alloy	10 (ASTM D638)
Silicone: Woven glass fabric/ silicone laminate	10—25
Phenolic, Molded: Very high shock: glass fiber filled	10—33
Epoxy, Standard: General purpose glass cloth laminate	12—15
Polycarbonate	12—16
Epoxy novolacs: Cast, rigid	13—17
PVC–acrylic sheet	15

Selecting Impact Strengths of Polymers (Continued)

Polymer	Impact Strength (Izod notched, ASTM D256) (ft—lb / in.)
PVC–acrylic injection molded	15
Polymide: Glass reinforced	17
Epoxy, Standard: High strength laminate	60—61
Nylon: Type 8	>16
Polyethylene, Type III: High molecular weight	>20
Fluorinated ethylene propylene(FEP)	No break
Polyvinyl Chloride And Copolymers: Nonrigid—general	Variable
Polyvinyl Chloride And Copolymers: Nonrigid—electrical	Variable

To convert **ft—lb / in**. to **N•m/m**, multiply by **53.38**

Source: data compiled by J.S. Park *from* Charles T. Lynch, *CRC Handbook of Materials Science, Vol. 3*, CRC Press, Boca Raton, Florida, 1975 and *Engineered Materials Handbook, Vol.2*, Engineering Plastics, ASM International, Metals Park, Ohio, 1988.

SELECTING MODULI OF ELASTICITY IN TENSION FOR POLYMERS

Polymer	Modulus of Elasticity in Tension (ASTM D638) (10^5 psi)
Polyester, Cast Thermoset: Flexible	0.001—0.10
Polyvinyl Chloride & Copolymers: Nonrigid—general	0.004—0.03 (ASTM D412)
Polyvinyl Chloride & Copolymers: Nonrigid—electrical	0.01—0.03 (ASTM D412)
Polyethylene, Type I: Melt index 6—26	0.20—0.24
Polyethylene, Type I: Melt index 0.3—3.6	0.21—0.27
Polytetrafluoroethylene (PTFE)	0.38—0.65
Fluorinated ethylene propylene(FEP)	0.5—0.7
Epoxy, Standard: Cast flexible	0.5—2.5
Vinylidene chloride	0.7—2.0 (ASTM D412)
Chlorinated polyether	1.5
Polystyrene, Molded: High impact	1.50—3.80 (D638)
Ceramic reinforced (PTFE)	1.5—2.0
Polyester, Cast Thermoset: Rigid	1.5—6.5
Polyvinylidene— fluoride (PVDF)	1.7—2
Polytrifluoro chloroethylene (PTFCE)	1.9—3.0
ABS Resin: Very high impact	2.0—3.1
ABS Resin: Low temperature impact	2.0—3.1
Acrylic Cast Resin Moldings: High impact grade	2.3—3.3
ABS Resin: High impact	2.6—3.2
Polystyrene, Molded: Medium impact	2.6—4.7 (D638)
Polyvinyl Chloride & Copolymers: Rigid—normal impact	3 5—4.0 (ASTM D412)
ABS Resin: Medium impact	3.3—4.0
Polycarbonate	3.45
ABS Resin: Heat resistant	3.5—4.2

Selecting Moduli of Elasticity in Tension for Polymers (Continued)

Polymer	Modulus of Elasticity in Tension (ASTM D638) (10^5 psi)
Acrylic Cast Resin Sheets, Rods: General purpose, type I	3.5—4.5
Acrylic Cast Resin Moldings: Grades 5, 6, 8	3.5—5.0
Rubber phenolic—chopped fabric filled	3.5—6
Chlorinated polyvinyl chloride	3.7
Acrylic Cast Resin Sheets, Rods: General purpose, type II	4.0—5.0
Styrene acrylonitrile (SAN)	4.0—5.2
Epoxy, High performance: Cast, rigid	4—5
Rubber phenolic—woodflour or flock filled	4—6
Epoxy, Standard: Cast rigid	4.5
Polystyrene, Molded: General purpose	4.6—5.0 (D638)
Epoxy novolacs: Cast, rigid	4.8—5.0
Rubber phenolic—asbestos filled	5—9
Diallyl Phthalate, Molded: Orlon filled	6
Phenolic, Shock: paper, flock, or pulp filled	8—12
Phenolic, General: woodflour and flock filled	8—13
Phenolic, High shock: chopped fabric or cord filled	9—14
Melamine; Molded: Cellulose filled electrical	10—11
Phenolic, Molded: Arc resistant—mineral filled	10—30
Urea, Molded: Woodflour filled	11—14
Diallyl Phthalate, Molded: Asbestos filled	12
Reinforced polyester moldings: Heat & chemical resistsnt (asbestos)	12—15
Polystyrene, Molded: Glass fiber -30% reinforced	12.1 (D638)
Urea, Molded: Alpha—cellulose filled (ASTM Type l)	13—16
Reinforced polyester Sheet molding: general purpose	15—20

Selecting Moduli of Elasticity in Tension for Polymers (Continued)

Polymer	Modulus of Elasticity in Tension (ASTM D638) (10^5 psi)
Reinforced polyester moldings: High strength (glass fibers)	16—20
Polycarbonate (40% glass fiber reinforced)	17
Glass fiber (30%) reinforced Styrene acrylonitrile (SAN)	17.5
Epoxy novolacs: Glass cloth laminate	27.5
Silicone: Woven glass fabric/ silicone laminate	28 (ASTM D651)
Phenolic, Very high shock: glass fiber filled	30—33
Epoxy, High performance Molded: Glass cloth laminate	32—33
Epoxy, Standard, Molded: General purpose glass cloth laminate	33—36
Epoxy, Standard, Molded: High strength laminate	57—58
Epoxy, Standard, Molded: Filament wound composite	64—72

To convert **psi** to **MPa**, multiply by **145**.

Source: *data compiled by* J.S. Park *from* Charles T. Lynch, *CRC Handbook of Materials Science, Vol. 3*, CRC Press, Boca Raton, Florida, 1975 and *Engineered Materials Handbook, Vol.2*, Engineering Plastics, ASM International, Metals Park, Ohio, 1988.

SELECTING MODULI OF ELASTICITY IN COMPRESSION FOR POLYMERS

Polymer	Modulus of Elasticity in Compression (ASTM D638) (10^5 psi)
Polytetrafluoroethylene (PTFE)	0 70—0.90
Fluorinated ethylene propylene(FEP)	0.6—0.8
Ceramic reinforced (PTFE)	1.5—2.0
Polyvinylidene— fluoride (PVDF)	1.7—2
Polytrifluoro chloroethylene (PTFCE)	1.8

To convert from **psi** to **MPa**, multiply by **145**.

Source: *data compiled by* J.S. Park *from* Charles T. Lynch, *CRC Handbook of Materials Science, Vol. 3*, CRC Press, Boca Raton, Florida, 1975 and *Engineered Materials Handbook, Vol.2*, Engineering Plastics, ASM International, Metals Park, Ohio, 1988.

SELECTING MODULI OF ELASTICITY IN FLEXURE OF POLYMERS

Polymer	Modulus of Elasticity in Flexure (ASTM D790) (10^5 psi)
Polyester, Thermoset Cast: Flexible	0.001—0.39
Olefin Copolymer, Molded: Propylene—ethylene	0.00140
Olefin Copolymer, Molded: Ethylene butene	0.00165
Polyethylene, Type I: Melt index 200	0.1 (ASTM D747)
Polyethylene, Type I: Melt index 6—26	0.12—0.3 (ASTM D747)
Polyethylene, Type I: Melt index 0.3—3.6	0.13—0.27 (ASTM D747)
Polyethylene, Type II: Melt index 20	0.35—0.5 (ASTM D747)
Polyethylene, Type II: Melt index 1.0—1.9	0.35—0.5 (ASTM D747)
Epoxy, Standard: Cast flexible	0.36—3.9
Nylon, Type 8	0.4
Polytetrafluoroethylene (PTFE)	0.6—1.1
Cellulose Acetate Butyrate, ASTM Grade: S2	0.70—0.90 (ASTM D747)
Olefin Copolymer, Molded: Polyallomer	0.7—1.3
Polyethylene, Type III: High molecular weight	0.75 (ASTM D747)
Fluorinated ethylene propylene(FEP)	0.8
Polyethylene, Type III: Melt Melt index 0.1—12.0	0.9—0.25 (ASTM D747)
Nylon, Type 6: Flexible copolymers	0.92—3.2
Nylon, Type 6: Glass fiber (30%) reinforced	1.0—1.4
Polypropylene: High impact	1.0—2.0
Polyester, Thermoset Cast: Rigid	1—9

Selecting Moduli of Elasticity in Flexure of Polymers (Continued)

Polymer	Modulus of Elasticity in Flexure (ASTM D790) (10^5 psi)
Cellulose Acetate, ASTM Grade: S2—1	1.05—1.65 (ASTM D747)
Cellusose Acetate Propionate, ASTM Grade: 6	1.1
Cellulose Acetate Butyrate, ASTM Grade: MH	1.20—1.40 (ASTM D747)
Cellulose Acetate, ASTM Grade: MS—1, MS—2	1.25—1.90 (ASTM D747)
Chlorinated polyether	1.3 (0.1% offset)
Polyethylene, Type III: Melt index 0.2—0.9	1.3—1.5 (ASTM D747)
Nylon, Type 6: General purpose	1.4—3.9
Cellusose Acetate Propionate, ASTM Grade: 3	1.45—1.55
Polyethylene, Type III: Melt index 1.5—15	1.5 (ASTM D747)
Cellulose Acetate, ASTM Grade: MH—1, MH—2	1.50—2.15 (ASTM D747)
Cellulose Acetate, ASTM Grade: H2—1	1.50—2.35 (ASTM D747)
Nylon, Type 11	1.51
6/10 Nylon: General purpose	1.6—2.8
Cellusose Acetate Propionate, ASTM Grade: 1	1.7—1.8
Polypropylene: General purpose	1.7—2.5
Polyvinylidene— fluoride (PVDF)	1.75—2.0
6/6 Nylon: General purpose extrusion	1.75—4.1
Cellulose Acetate Butyrate, ASTM Grade: H4	1.8 (ASTM D747)
Polypropylene: Flame retardant	1.9—6.1
Polytrifluoro chloroethylene (PTFCE)	2.0—2.5
Cellulose Acetate, ASTM Grade: H4—1	2.0—2.55 (ASTM D747)
ABS Resins; Molded, Extruded: Very high impact	2.0—3.2
ABS Resins; Molded, Extruded: Low temperature impact	2.0—3.2
Polystyrene; Molded: High impact	2.3—4.0

Selecting Moduli of Elasticity in Flexure of Polymers (Continued)

Polymer	Modulus of Elasticity in Flexure (ASTM D790) (10^5 psi)
ABS Resins; Molded, Extruded: High impact	2.5—3.2
Thermoset Allyl diglycol carbonate	2.5—3.3
Acrylic Moldings: High impact grade	2.7—3.6
PVC–acrylic injection molded	3
Polycarbonate	3.4
Polyester, Injection Moldings: General purpose grade	3.4
Polypropylene: Asbestos filled	3.4—6.5
Rubber phenolic—chopped fabric filled	3.5
ABS Resins; Molded, Extruded: Medium impact	3.5—4.0
ABS Resins; Molded, Extruded: Heat resistant	3.5—4.2
Acrylic Cast Resin Sheets, Rods: General purpose, type I	3.5—4.5
Acrylic Moldings: Grades 5, 6, 8	3.5—5.0
Polystyrene; Molded: Medium impact	3.5—5.0
Phenylene Oxide: SE—100	3.6
Phenylene Oxide: SE—1	3.6
Polyacetal Copolymer: Standard	3.75
Polyacetal Copolymer: High flow	3.75
Polyvinyl Chloride And Copolymers: Rigid—normal impact	3.8—5.4
Chlorinated polyvinyl chloride	3.85
Phenylene oxides (Noryl): Standard	3.9
ABS–Polycarbonate Alloy	4
PVC–acrylic sheet	4
Polyacetal Homopolymer: 22% TFE reinforced	4
Polyarylsulfone	4

Selecting Moduli of Elasticity in Flexure of Polymers (Continued)

Polymer	Modulus of Elasticity in Flexure (ASTM D790) (10^5 psi)
Acrylic Cast Resin Sheets, Rods: General purpose, type II	4.0—5.0
Epoxy, High performance: Cast, rigid	4—5
Polystyrene; Molded: General purpose	4—5
Rubber phenolic—woodflour or flock filled	4—6
Polypropylene: Glass reinforced	4—8.2
Polyacetal Homopolymer: Standard	4.1
6/6 Nylon: General purpose molding	4.1—4.5, 1.75
Epoxy novolacs: Cast, rigid	4.4—4.8
Epoxy, Standard: Cast rigid	4.5—5.4
Ceramic reinforced (PTFE)	4.64
Rubber phenolic—asbestos filled	5
Polymide: Unreinforced	5—7
Nylon, Type 6: Cast	5.05
Polyphenylene sulfide: Standard	5.5—6.0
Phenylene Oxide: Glass fiber reinforced	7.4—10.4
Phenolic, Molded: General: woodflour and flock filled	8—12
Phenolic, Molded: Shock: paper, flock, or pulp filled	8—12
6/10 Nylon: Glass fiber (30%) reinforced	8.5
Polyacetal Homopolymer: 20% glass reinforced	8.8
Phenolic, Molded: High shock: chopped fabric or cord filled	9—13
Melamine, Molded: Unfilled	10—13
Melamine, Molded: Cellulose filled electrical	10—13
6/6 Nylon: Glass fiber reinforced	10—18
Phenolics: Molded: Arc resistant—mineral	10—30

Selecting Moduli of Elasticity in Flexure of Polymers (Continued)

Polymer	Modulus of Elasticity in Flexure (ASTM D790) (10^5 psi)
Polyacetal Copolymer: 25% glass reinforced	11
6/6 Nylon: Glass fiber Molybdenum disulfide filled	11—13
Polycarbonate (40% glass fiber reinforced)	12
Polyester, Moldings: Glass reinforced self extinguishing	12
Polystyrene; Molded: Glass fiber -30% reinforced	12
Phenylene oxides (Noryl): Glass fiber reinforced	12, 15.5
Polyester, Thermoplastic Moldings: Glass reinforced grades	12—15
Silicone, Molded: Granular (silica) reinforced	14—17
Glass fiber (30%) reinforced Styrene acrylonitrile (SAN)	14.5
Reinforced polyester sheet molding: general purpose	15—18
Epoxy, Standard: Molded	15—25
Reinforced polyester moldings: High strength (glass fibers)	15—25
Polyphenylene sulfide: 40% glass reinforced	17—22
Alkyds, Molded Rope (general purpose)	22—27
Alkyds, Molded: Granular (high speed molding)	22—27
Alkyds, Molded: Glass reinforced (heavy duty parts)	22—28
Melamine, Molded: Glass fiber filled	24
Silicone, Molded: Fibrous (glass) reinforced	25
Silicone, Molded: Woven glass fabric/ silicone laminate	26—32
Epoxy, High performance: Glass cloth laminate	28—31
Phenolic, Molded: Very high shock: glass fiber filled	30—33
Epoxy novolacs: Glass cloth laminate	32—35
Polyester, Thermoplastic Moldings: General purpose grade	33
Epoxy, Standard: General purpose glass cloth laminate	36—39

Selecting Moduli of Elasticity in Flexure of Polymers (Continued)

Polymer	Modulus of Elasticity in Flexure (ASTM D790) (10^5 psi)
Polyimide: Glass reinforced	38.4
Epoxy, Standard: High strength laminate	53—55
Epoxy, Standard: Filament wound composite	69—75
Polyester, Thermoplastic Moldings: Glass reinforced grade	87
Polyester, Thermoplastic Moldings: Asbestos—filled grade	90

To convert from **psi** to **MPa**, multiply by **145**.

Source: data compiled by J.S. Park *from* Charles T. Lynch, *CRC Handbook of Materials Science, Vol. 3*, CRC Press, Boca Raton, Florida, 1975 and *Engineered Materials Handbook, Vol.2*, Engineering Plastics, ASM International, Metals Park, Ohio, 1988.

SELECTING TOTAL ELONGATION OF POLYMERS

Polymer	Elongation (in 2 in.), (ASTM D638) (%)
Polycarbonate (40% glass fiber reinforced)	0—5
Phenolic, Molded, Very high shock: glass fiber filled	0.2
Reinforced polyester moldings: High strength (glass fibers)	0.3—0.5
Phenolic, Molded, High shock: chopped fabric or cord filled	0.37—0.57
Phenolic, Molded, General: woodflour and flock filled	0.4—0.8
Styrene acrylonitrile (SAN)	0.5—4.5
Melamine, Molded: Cellulose electrical	0.6
Rubber phenolic—woodflour or flock filled	0.75—2.25
Polymide: Glass reinforced	<1
Polymide: Unreinforced	<1—1.2
Ureas; Molded: Alpha—cellulose filled (ASTM Type l)	1
Polystyrenes, Molded: General purpose	1.0—2.3
Polyvinyl Chloride & Copolymers: Rigid—normal impact	1—10
Polyester, Thermoplastic Moldings: Glass reinforced grades	1—5
Polystyrenes, Molded: Glass fiber -30% reinforced	1.1
Glass fiber (30%) reinforced Styrene acrylonitrile (SAN)	1.4—1.6
Epoxy, Standard: Cast flexible	1.5-60
Polyester, Thermoset Cast: Rigid	1.7—2.6
6/6 Nylon, Molded, Extruded: Glass fiber reinforced	1.8—2.2
6/10 Nylon: Glass fiber (30%) reinforced	1.9
Polypropylene: Glass reinforced	2—4
Epoxy, High performance: Cast, rigid	2—5

Selecting Total Elongation of Polymers (Continued)

Polymer	Elongation (in 2 in.), (ASTM D638) (%)
Acrylic Cast Resin Sheets, Rods: General purpose, type I	2—7
Acrylic Cast Resin Sheets, Rods: General purpose, type II	2—7
Nylon, Type 6: Glass fiber (30%) reinforced	2.2—3.6
Epoxy novolacs: Glass cloth laminate	2.2—4.8
Silicones: Fibrous (glass) reinforced silicones	<3 (ASTM D651)
Silicone: Granular (silica) reinforced	<3 (ASTM D651)
6/6 Nylon: Glass fiber Molybdenum disulfide filled	3
Polyacetal Copolymer: 25% glass reinforced	3
Polyphenylene sulfide: Standard	3
Polystyrenes, Molded: Medium impact	3.0—40
Polypropylene: Flame retardant	3—15
Polypropylene: Asbestos filled	3—20
Acrylic Moldings: Grades 5, 6, 8	3—5
Polyphenylene sulfide: 40% glass reinforced	3—9
Phenylene Oxides: Glass fiber reinforced	4—6
Epoxy, Standard: Cast rigid	4.4
Polyester, Thermoplastic Moldings: Glass reinforced grade	<5
Polyester, Thermoplastic Moldings: Asbestos—filled grade	<5
Polyester, Thermoplastic: Glass reinforced self extinguishing	5
ABS Resins: Medium impact	5—20
ABS Resins: High impact	5—50
Polyacetal Homopolymer: 20% glass reinforced	7

Selecting Total Elongation of Polymers (Continued)

Polymer	Elongation (in 2 in.), (ASTM D638) (%)
Ceramic reinforced (PTFE)	10—200
Polyacetal Homopolymer: 22% TFE reinforced	12
Polyvinyl Chloride & Copolymers: Vinylidene chloride	15—30
Polyarylsulfone	15—40
6/6 Nylon: General purpose molding	15—60, 300
ABS Resins: Heat resistant	20
Nylon, Molded, Extruded Type 6: Cast	20
Olefin Copolymers, Molded: Ethylene butene	20
ABS Resins: Very high impact	20—50
Polyacetal Homopolymer: Standard	25
Acrylic Moldings: High impact grade	>25
Polyester, Thermoset Cast: Flexible	25—300
Nylon, Molded, Extruded Type 6: General purpose	30—100
ABS Resins: Low temperature impact	30—200
Polypropylene: High impact	30—>200
Polyacetal Copolymer: High flow	40
Phenylene Oxides: SE—100	50
Phenylene oxides (Noryl): Standard	50—100
Polyethylene, Type III: Melt Melt index 0.1—12.0	50—1,000
Phenylene Oxides: SE—1	60
Polyacetal Copolymer: Standard	60—75
Polyethylene, Type I: Melt index 200	80—100 (ASTM D412)

Selecting Total Elongation of Polymers (Continued)

Polymer	Elongation (in 2 in.), (ASTM D638) (%)
6/10 Nylon: General purpose	85—220
6/6 Nylon: General purpose extrusion	90—240
PVC–Acrylic Alloy: sheet	>100
Nylon, Type 11	100—120
Polypropylene: General purpose	100—600
Polyethylene, Type III: Melt index 1.5—15	100—700
Polycarbonate	110
ABS–Polycarbonate Alloy	110
Nylon, Type 12	120—350
Polytrifluoro chloroethylene (PTFCE)	125—175
Polyethylene, Type I: Melt index 6—26	125—675 (ASTM D412)
Chlorinated polyether	130
PVC–Acrylic Alloy: injection molded	150
Polyethylene, Type II: Melt index 20	200
Polyvinylidene— fluoride (PVDF)	200—300
Nylon, Molded, Extruded Type 6: Flexible copolymers	200—320
Polyethylene, Type II: Melt index 1.0—1.9	200—425
Polyvinyl Chloride & Copolymers: Nonrigid—general	200—450
Polyvinyl Chloride & Copolymers: Nonrigid—electrical	220—360
Polyester, Thermoplastic Moldings: General purpose grade	250
Fluorinated ethylene propylene (FEP)	250—330
Polytetrafluoroethylene (PTFE)	250—350

Selecting Total Elongation of Polymers (Continued)

Polymer	Elongation (in 2 in.), (ASTM D638) (%)
Polyester, Thermoplastic Moldings: General purpose grade	300
Olefin Copolymers, Molded: Polyallomer	300—400
Nylon, Type 8	400
Polyethylene, Type III: High molecular weight	400
Olefin Copolymers, Molded: Ionomer	450
Polyethylene, Type I: Melt index 0.3—3.6	500—725 (ASTM D412)
Olefin Copolymers, Molded: EEA (ethylene ethyl acrylate)	650
Olefin Copolymers, Molded: EVA (ethylene vinyl acetate)	650
Polyethylene, Type III: Melt index 0.2—0.9	700—1,000

Source: data compiled by J.S. Park *from* Charles T. Lynch, *CRC Handbook of Materials Science, Vol. 3*, CRC Press, Boca Raton, Florida, 1975 and *Engineered Materials Handbook, Vol.2*, Engineering Plastics, ASM International, Metals Park, Ohio, 1988.

SELECTING ELONGATION AT YIELD OF POLYMERS

Polymer	Elongation at Yield, (ASTM D638) (%)
Polystyrene: General purpose	1.0—2.3
Polystyrene: Glass fiber -30% reinforced	1.1
Polystyrene: Medium impact	1.2—3.0
Polyphenylene sulfide: 40% glass reinforced	1.25
Polystyrene: Glass fiber (30%) reinforced SAN	1.4—1.6
Polystyrene: High impact	1.5—2.0
Polyphenylene sulfide: Standard	1.6
Phenylene oxides (Noryl): Glass fiber reinforced	2—1.6
Polyacetal Copolymer: 25% glass reinforced	3
Polycarbonate	5
Nylon, Type 6: Cast	5
Polypropylene: Asbestos filled	5
6/6 Nylon: General purpose molding	5—25
6/6 Nylon: General purpose extrusion	5—30
6/10 Nylon: General purpose	5—30
Phenylene oxides (Noryl): Standard	5.6
Nylon, Type 12	5.8
Polyarylsulfone	6.5—13
Polypropylene: High impact	7—13
Polypropylene: General purpose	9—15
Polyacetal Homopolymer: Standard	12
Polyacetal Copolymer: Standard	12
Polyacetal Copolymer: High flow	12
Chlorinated polyether	15

Source: data compiled by J.S. Park *from* Charles T. Lynch, *CRC Handbook of Materials Science, Vol. 3*, CRC Press, Boca Raton, Florida, 1975 and *Engineered Materials Handbook, Vol.2*, Engineering Plastics, ASM International, Metals Park, Ohio, 1988.

Selecting Polymers:
Electrical Properties

SELECTING VOLUME RESISTIVITY OF POLYMERS

Polymer	Volume Resistivity (ASTM D257) ($\Omega \cdot$ cm)
Diallyl Phthalates; Molded: Dacron Filled	10^2—2.5×10^4
Diallyl Phthalates; Molded: Asbestos Filled	10^2—5×10^3
Diallyl Phthalates; Molded: Glass Fiber Filled	10^4—5×10^4
Diallyl Phthalates; Molded: Orlon Filled	6×10^4—6×10^6
Standard Epoxies: Cast Flexible	9.1×10^5—6.7×10^9
Standard Epoxies; Reinforced: High Strength Laminate	6.6×10^7—10^9
Molded Rubber Phenolic—Woodflour or Flock Filled	10^8—10^{11}
Phenolics; Molded: General: Woodflour and Flock Filled	10^9—10^{13}
Cellulose Acetate; Molded, Extruded; ASTM Grade: H6—1	10^{10}—10^{13}
Cellulose Acetate; Molded, Extruded; ASTM Grade: H4—1	10^{10}—10^{13}
Cellulose Acetate; Molded, Extruded; ASTM Grade: H2—1	10^{10}—10^{13}
Cellulose Acetate; ASTM Grade: MH—1, MH—2	10^{10}—10^{13}
Cellulose Acetate; ASTM Grade: MS—1, MS—2	10^{10}—10^{13}
Cellulose Acetate; Molded, Extruded; ASTM Grade: S2—1	10^{10}—10^{13}
Phenolics; Molded: High Shock: Chopped Fabric or Cord Filled	$>10^{10}$
Phenolics; Molded: Very High Shock: Glass Fiber Filled	10^{10}—10^{11}
Phenolics: Molded: Arc Resistant—Mineral Filled	10^{10}—10^{12}
Ureas; Molded: Cellulose Filled (ASTM Type 2)	5—8×10^{10}
Cellulose Acetate Butyrate; ASTM Grade: H4	10^{11}—10^{14}
Cellulose Acetate Butyrate; ASTM Grade: MH	10^{11}—10^{14}
Cellulose Acetate Butyrate; ASTM Grade: S2	10^{11}—10^{14}
Cellusose Acetate Propionate; ASTM Grade: 1	10^{11}—10^{14}
Cellusose Acetate Propionate; ASTM Grade: 3	10^{11}—10^{14}
Cellusose Acetate Propionate; ASTM Grade: 6	10^{11}—10^{14}

Selecting Volume Resistivity of Polymers (Continued)

Polymer	Volume Resistivity (ASTM D257) ($\Omega \cdot cm$)
Phenolics: Molded: Rubber Phenolic—Chopped Fabric Filled	10^{11}
Phenolics: Molded: Rubber Phenolic—Asbestos Filled	10^{11}
Ureas; Molded: Alpha—Cellulose filled (ASTM Type l)	$0.5—5 \times 10^{11}$
Melamines; Molded: Glass Fiber Filled	$1—7 \times 10^{11}$
Phenolics; Molded: Shock: Paper, Flock, or Pulp Filled	$1—50 \times 10^{11}$
Nylons: Type 8	1.5×10^{11}
Polyvinyl Chloride & Copolymers: Nonrigid—Electrical	$4—300 \times 10^{11}$
Melamines; Molded: Alpha Cellulose And Mineral Filled	10^{12}
Polyesters, Thermosets; Cast polyyester: Flexible	10^{12}
Melamines; Molded: Cellulose Electrical Filled	$10^{12}—10^{13}$
Reinforced Polyester: High Strength (Glass Fibers)	$1 \times 10^{12}—1 \times 10^{13}$
Reinforced Polyester: Heat & Chemical Resistant (Asbestos)	$1 \times 10^{12}—1 \times 10^{13}$
Polyvinyl Chloride & Copolymers: Nonrigid—General	$1—700 \times 10^{12}$
Polyesters, Thermosets; Cast polyyester: Rigid	10^{13}
PVC–Acrylic Alloy: PVC–Acrylic Sheet	$1—5 \times 10^{13}$
Nylons: Type 11	2×10^{13}
Nylons; Molded, Extruded; Type 6: General purpose	4.5×10^{13}
Alkyds; Molded: Putty (Encapsulating)	10^{14}
Alkyds; Molded: Rope (General Purpose)	10^{14}
Alkyds; Molded: Glass reinforced (heavy duty parts)	10^{14}
Acrylics; Moldings: Grades 5, 6, 8	$>10^{14}$
Alkyds; Molded: Granular (high speed molding)	$10^{14}—10^{15}$
Nylons: Type 12	$10^{14}—10^{15}$
6/6 Nylon: General purpose molding	$10^{14}—10^{15}$

Selecting Volume Resistivity of Polymers (Continued)

Polymer	Volume Resistivity (ASTM D257) ($\Omega \cdot cm$)
Polyacetal Copolymer: Standard	1×10^{14}
Polyacetal Copolymer: High Flow	1.0×10^{14}
Polyacetal Copolymer: 25% Glass Reinforced	1.2×10^{14}
High Performance Epoxies: Molded	$1.4—5.5 \times 10^{14}$
Woven Glass Fabric/ Silicone Laminate	$2—5 \times 10^{14}$
High Performance Epoxies: Cast, rigid	2.10×10^{14}
Nylons; Type 6: Cast	2.6×10^{14}
Nylons; Type 6: Glass fiber (30%) Reinforced	$2.8 \times 10^{14}—1.5 \times 10^{15}$
Polyester; Thermoplastic Moldings: Asbestos—Filled Grade	3×10^{14}
Thermoset Carbonate: Allyl Diglycol Carbonate	4×10^{14}
Polyphenylene sulfide: 40% Glass Reinforced	4.5×10^{14}
Polyvinylidene— fluoride (PVDF)	5×10^{14}
Polyacetal Homopolymer: 20% Glass Reinforced	5×10^{14}
Granular (Silica) Reinforced Silicones	5×10^{14}
Fibrous (Glass) Reinforced Silicones	9×10^{14}
Polyvinyl Chloride & Copolymers: Rigid—Normal Impact	$10^{14}—10^{16}$
Vinylidene chloride	$10^{14}—10^{16}$
Ceramic Reinforced (PTFE)	10^{15}
6/6 Nylon: General Purpose Extrusion	10^{15}
6/10 Nylon: General purpose	10^{15}
Acrylics; Cast Resin Sheets, Rods: General purpose, type II	$>10^{15}$
Acrylics; Cast Resin Sheets, Rods: General purpose, type I	$>10^{15}$
Polyethylenes; Molded, Extruded; Type II: Melt Index 20	$>10^{15}$
Polyethylenes; Molded, Extruded; Type II: Melt Index 1.0—1.9	$>10^{15}$

Selecting Volume Resistivity of Polymers (Continued)

Polymer	Volume Resistivity (ASTM D257) ($\Omega \cdot cm$)
Polyethylenes; Molded, Extruded; Type III: Melt Index 0.2—0.9	$>10^{15}$
Polyethylenes; Type III: Melt Melt Index 0.l—12.0	$>10^{15}$
Polyethylenes; Molded, Extruded; Type III: Melt Index 1.5—15	$>10^{15}$
Polyethylenes; Molded, Extruded; Type III: High Molecular Weight	$>10^{15}$
Olefin Copolymers; Molded: EVA (ethylene vinyl acetate)	0.15×10^{15}
Chlorinated Polyvinyl Chloride	1×10^{15}—2×10^{16}
Standard Epoxies: Molded	1—5×10^{15}
Polyacetal Homopolymer: Standard	1×10^{15}
ABS Resins; Molded, Extruded: High impact	1—4×10^{15}
ABS Resins; Molded, Extruded: Very high impact	1—4×10^{15}
ABS Resins; Molded, Extruded: Low temperature impact	1—4×10^{15}
ABS Resins; Molded, Extruded: Heat resistant	1—5×10^{15}
Polycarbonate (40% Glass Fiber Reinforced)	1.4×10^{15}
Polypropylene: Asbestos Filled	1.5×10^{15}
Polyester; Thermoplastic Moldings: General Purpose Grade	2×10^{15}
ABS Resins; Molded, Extruded: Medium impact	2—4×10^{15}
Olefin Copolymers; Molded: EEA (ethylene ethyl acrylate)	2.4×10^{15}
6/6 Nylon; Molded, Extruded: Glass Fiber Reinforced	2.6—5.5×10^{15}
Polymides: Unreinforced	4×10^{15}
PVC–Acrylic Alloy: PVC–Acrylic Injection Molded	5×10^{15}
Standard Epoxies: Cast Rigid	6.1×10^{15}
Reinforced Polyester Sheet Molding, General Purpose	6.4×10^{15}—2.2×10^{16}
Polymides: Glass Reinforced	9.2×10^{15}
Olefin Copolymers; Molded: Ionomer	10×10^{15}

Selecting Volume Resistivity of Polymers (Continued)

Polymer	Volume Resistivity (ASTM D257) ($\Omega \cdot$ cm)
Styrene Acrylonitrile (SAN)	$>10^{16}$
Epoxy Novolacs: Cast, rigid	$>10^{16}$
Olefin Copolymers; Molded: Polyallomer	$>10^{16}$
Polystyrenes; Molded: General Purpose	$>10^{16}$
Polystyrenes; Molded: Medium Impact	$>10^{16}$
Polystyrenes; Molded: High Impact	$>10^{16}$
Polyester; Thermoplastic Moldings: General Purpose Grade	$1\text{---}4 \times 10^{16}$
Chlorinated Polyether	1.5×10^{16}
Polypropylene: Glass Reinforced	1.7×10^{16}
Acrylics; Moldings: High Impact Grade	2.0×10^{16}
Polycarbonate	2.1×10^{16}
ABS–Polycarbonate Alloy	2.2×10^{16}
Polyester; Thermoplastic Moldings: Glass Reinforced Grades	$3.2\text{---}3.3 \times 10^{16}$
Polyarylsulfone	$3.2\text{---}7.71 \times 10^{16}$
Polyester Moldings: Glass Reinforced Self Extinguishing	3.4×10^{16}
Polystyrenes; Molded: Glass Fiber -30% Reinforced	3.6×10^{16}
Polypropylene: Flame Retardant	$4 \times 10^{16}\text{---}10^{17}$
Glass Fiber (30%) Reinforced SAN	4.4×10^{16}
Phenylene Oxides (Noryl): Standard	5×10^{16}
Phenylene Oxides: SE—100	10^{17}
Phenylene Oxides: SE—1	10^{17}
Phenylene Oxides: Glass Fiber Reinforced	10^{17}
Phenylene Oxides (Noryl): Glass Fiber Reinforced	10^{17}
Polypropylene: High Impact	10^{17}

Selecting Volume Resistivity of Polymers (Continued)

Polymer	Volume Resistivity (ASTM D257) ($\Omega \cdot cm$)
Polypropylene: General Purpose	$>10^{17}$
Polyethylenes; Molded, Extruded; Type I: Melt Index 0.3—3.6	10^{17}—10^{19}
Polyethylenes; Molded, Extruded; Type I: Melt Index 6—26	10^{17}—10^{19}
Polyethylenes; Molded, Extruded; Type I: Melt Index 200	10^{17}—10^{19}
Polytrifluoro Chloroethylene (PTFCE), Molded,Extruded	10^{18}
Polytetrafluoroethylene (PTFE), Molded,Extruded	$>10^{18}$
Fluorinated Ethylene Propylene (FEP)	$>2 \times 10^{18}$

Source: data compiled by J.S. Park *from* Charles T. Lynch, *CRC Handbook of Materials Science, Vol. 3*, CRC Press, Boca Raton, Florida, 1975 and *Engineered Materials Handbook, Vol.2*, Engineering Plastics, ASM International, Metals Park, Ohio, 1988.

SELECTING DISSIPATION FACTOR FOR POLYMERS @ 60 Hz

Polymer	Dissipation Factor (ASTM D150) @ 60 Hz
Polystyrenes; Molded: General purpose	0.0001–0.0003
Fluorocarbons; Molded,Extruded: Polytetrafluoroethylene (PTFE)	0.0002
Fluorocarbons; Molded,Extruded: Fluorinated ethylene propylene (FEP)	0.0003
Polystyrenes; Molded: Medium impact	0.0004–0.002
Polystyrenes; Molded: High impact	0.0004–0.002
Polyethylenes; Molded, Extruded: Type I: Melt index 0.3—3.6	<0.0005
Polyethylenes; Molded, Extruded: Type I: Melt index 6—26	<0.0005
Polyethylenes; Molded, Extruded: Type I: Melt index 200	<0.0005
Polyethylenes; Molded, Extruded: Type II: Melt index 20	<0.0005
Polyethylenes; Molded, Extruded: Type II: Melt index 1.0—1.9	<0.0005
Polyethylenes; Molded, Extruded: Type III: Melt index 0.2—0.9	<0.0005
Polyethylenes; Molded, Extruded: Type III: Melt Melt index 0.1—12.0	<0.0005
Polyethylenes; Molded, Extruded: Type III: Melt index 1.5—15	<0.0005
Polyethylenes; Molded, Extruded: Type III: High molecular weight	<0.0005
Olefin Copolymers; Molded: Polyallomer	>0.0005
Polypropylene: General purpose	0.0005–0.0007
Fluorocarbons; Molded,Extruded: Ceramic reinforced (PTFE)	0.0005–0.0015
Phenylene Oxides: SE—100	0.0007
Phenylene Oxides: SE—1	0.0007
Polypropylene: Flame retardant	0.0007–0.017
Phenylene oxides (Noryl): Standard	0.0008
Polycarbonate	0.0009
Phenylene Oxides: Glass fiber reinforced	0.0009
Olefin Copolymers; Molded: EEA (ethylene ethyl acrylate)	0.001

Selecting Dissipation Factor for Polymers @ 60 Hz (Continued)

Polymer	Dissipation Factor (ASTM D150) @ 60 Hz
Epoxy novolacs: Cast, rigid	0.001—0.007
Polypropylene: High impact	<0.0016
Polyarylsulfone	0.0017—0.003
Phenylene oxides (Noryl): Glass fiber reinforced	0.0019
Polypropylene: Glass reinforced	0.002
Silicones; Molded, Laminated: Granular (silica) reinforced silicones	0.002—0.004
ABS–Polycarbonate Alloy	0.0026
Polymides: Unreinforced	0.003
Olefin Copolymers; Molded: EVA (ethylene vinyl acetate)	0.003
Olefin Copolymers; Molded: Ionomer	0.003
ABS Resins; Molded, Extruded: Medium impact	0.003—0.006
Polyester;: Thermosets: Cast Rigid	0.003—0.04
Polymides: Glass Reinforced	0.0034
Standard Epoxies: General Purpose Glass Cloth Laminate	0.004-0.006
Diallyl Phthalates; Molded: Glass Fiber Filled	0.004—0.015 (Dry)
Diallyl Phthalates; Molded: Dacron Filled	0.004—0.016 (Dry)
Polyacetal Homopolymer: 20% glass reinforced	0.0047
Polyacetal Homopolymer: Standard	0.0048
Standard Epoxies: Cast Flexible	0.0048-0.0380
Polystyrenes; Molded: Glass fiber -30% reinforced	0.005
Polystyrenes; Molded: Glass fiber (30%) reinforced SAN	0.005
ABS Resins; Molded, Extruded: High impact	0.005—0.007
ABS Resins; Molded, Extruded: Low temperature impact	0.005—0.01
ABS Resins; Molded, Extruded: Very high impact	0.005—0.010

Selecting Dissipation Factor for Polymers @ 60 Hz (Continued)

Polymer	Dissipation Factor (ASTM D150) @ 60 Hz
Epoxies; High Performance Resins: Cast, Rigid	0.0055—0.0074
Polycarbonate (40% glass fiber reinforced)	0.006
Polystyrenes; Molded: Styrene acrylonitrile (SAN)	>0.006
Polypropylene: Asbestos filled	0.007
Nylons; Molded, Extruded; Type 6: Flexible Copolymers	0.007—0.010
Epoxies; High Performance Resins: Molded	0.0071—0.025
Standard Epoxies: Cast Rigid	0.0074
Reinforced Polyester Sheet molding compounds, general purpose	0.0087—0.04
Silicones; Molded, Laminated: Fibrous (glass) reinforced silicones	0.01
Cellulose Acetate Butyrate; Molded, Extruded; ASTM Grade: H4	0.01—0.04
Cellulose Acetate Butyrate; Molded, Extruded; ASTM Grade: MH	0.01—0.04
Cellulose Acetate Butyrate; Molded, Extruded; ASTM Grade: S2	0.01—0.04
Cellusose Acetate Propionate; Molded, Extruded; ASTM Grade: 1	0.01—0.04
Cellusose Acetate Propionate; Molded, Extruded; ASTM Grade: 3	0.01—0.04
Cellusose Acetate Propionate; Molded, Extruded; ASTM Grade: 6	0.01—0.04
Cellulose Acetate; Molded, Extruded; ASTM Grade: H4—1	0.01—0.06
Cellulose Acetate; Molded, Extruded; ASTM Grade: H2—1	0.01—0.06
Cellulose Acetate; Molded, Extruded; ASTM Grade: MH—1, MH—2	0.01—0.06
Cellulose Acetate; Molded, Extruded; ASTM Grade: MS—1, MS—2	0.01—0.06
Cellulose Acetate; Molded, Extruded; ASTM Grade: S2—1	0.01—0.06
Polyester;: Thermosets: Flexible	0.01—0.18
Chlorinated polyether	0.011
Standard Epoxies: Molded	0.011-0.018
Nylons; Molded, Extruded; 6/6 Nylon: General purpose molding	0.014—0.04

Selecting Dissipation Factor for Polymers @ 60 Hz (Continued)

Polymer	Dissipation Factor (ASTM D150) @ 60 Hz
Nylons; Type 6: Cast	0.015
Nylons; Molded, Extruded; 6/6 Nylon: Glass fiber reinforced	0.018—0.009
Chlorinated polyvinyl chloride	0.0189—0.0208
Alkyds; Molded: Rope (general purpose)	0.019
Fluorocarbons; Molded,Extruded: Polytrifluoro chloroethylene (PTFCE)	0.02
Silicones; Molded, Laminated: Woven glass fabric/ silicone laminate	0.02
Polyvinyl Chloride & Copolymers: Rigid—normal impact	0.020—0.03
Alkyds; Molded: Glass reinforced (heavy duty parts)	0.02—0.03
Phenolics; Molded; Very High Shock: Glass Fiber Filled	0.02—0.03
Nylons; Molded, Extruded; Type 6: Glass fiber (30%) reinforced	0.022—0.008
Diallyl Phthalates; Molded: Orlon Filled	0.023—0.015 (Dry)
Melamines; Molded: Cellulose Electrical Filled	0.026—0.192
Nylons; Molded, Extruded: Type 11	0.03
ABS Resins; Molded, Extruded: Heat resistant	0.030—0.040
Alkyds; Molded: Granular (high speed molding)	0.030—0.040
Alkyds; Molded: Putty (encapsulating)	0.030—0.045
Acrylics; Moldings: High impact grade	0.03—0.04
Thermoset Carbonate: Allyl diglycol carbonate	0.03—0.04
Polyvinyl Chloride & Copolymers: Vinylidene chloride	0.03—0.15
Ureas; Molded: Woodflour filled	0.035—0.040
Ureas; Molded: Alpha—cellulose filled (ASTM Type l)	0.035—0.043
PVC–Acrylic Injection Molded	0.037
Nylons; Molded, Extruded; 6/10 Nylon: General purpose	0.04
Acrylics; Moldings: Grades 5, 6, 8	0.04—0.06

Selecting Dissipation Factor for Polymers @ 60 Hz (Continued)

Polymer	Dissipation Factor (ASTM D150) @ 60 Hz
Ureas; Molded: Cellulose filled (ASTM Type 2)	0.042—0.044
Melamines; Molded: Unfilled	0.048—0.162
Fluorocarbons; Molded,Extruded: Polyvinylidene—fluoride (PVDF)	0.05
Diallyl Phthalates; Molded: Asbestos Filled	0.05—0.03 (Dry)
Acrylics; Cast Resin Sheets, Rods: General Purpose, Type I	0.05—0.06
Acrylics; Cast Resin Sheets, Rods: General Purpose, Type II	0.05—0.06
Polyvinyl Chloride & Copolymers; Molded, Extruded: Nonrigid—general	0.05—0.15
Phenolics; Molded; General: Woodflour & Flock Filled	0.05—0.30
Nylons; Molded, Extruded; Type 6: General Purpose	0.06—0.014
PVC–Acrylic Sheet	0.076
Polyvinyl Chloride & Copolymers: Nonrigid—electrical	0.08—0.11
Phenolics; Molded; Shock: Paper, Flock, or Pulp Filled	0.08—0.35
Phenolics; Molded; High Shock: Chopped Fabric or Cord Filled	0.08—0.45
Phenolics: Molded: Arc resistant—Mineral Filled	0.13—0.16
Melamines; Molded: Glass Fiber Filled	0.14—0.23
Rubber Phenolic—Asbestos Filled	0.15
Phenolics: Molded: Rubber Phenolic—Woodflour or Flock Filled	0.15—0.60
Nylons; Molded, Extruded: Type 8	0.19
Rubber Phenolic—Chopped Fabric Filled	0.5

Source: data compiled by J.S. Park *from* Charles T. Lynch, *CRC Handbook of Materials Science, Vol. 3*, CRC Press, Boca Raton, Florida, 1975 and *Engineered Materials Handbook, Vol.2*, Engineering Plastics, ASM International, Metals Park, Ohio, 1988.

SELECTING DISSIPATION FACTOR FOR POLYMERS @ 10^6 Hz

Polymer	Dissipation Factor (ASTM D150) @ 10^6 Hz
Polystyrenes; Molded: General purpose	0.0001–0.0005
Fluorocarbons; Molded,Extruded: Polytetrafluoroethylene (PTFE)	0.0002
Polypropylene: General purpose	0.0002–0.0003
Polypropylene: High impact	0.0002—0.0003
Molded,Extruded Fluorinated ethylene propylene (FEP)	0.0003
Polystyrenes; Molded: Medium impact	0.0004–0.002
Polystyrenes; Molded: High impact	0.0004–0.002
Fluorocarbons; Molded,Extruded: Ceramic reinforced (PTFE)	0.0005–0.0015
Polypropylene: Flame retardant	0.0006–0.003
Polyphenylene sulfide: Standard	0.0007
Silicones; Molded, Laminated: Granular (silica) reinforced silicones	0.001—0.004
Polyphenylene sulfide: 40% glass reinforced	0.0014—0.0041
Phenylene Oxides: Glass fiber reinforced	0.0015
Polypropylene: Asbestos filled	0.002
Polystyrenes; Molded: Glass fiber -30% reinforced	0.002
Silicones; Molded, Laminated: Woven glass fabric/ silicone laminate	0.002
Phenylene Oxides: SE—100	0.0024
Phenylene Oxides: SE—1	0.0024
Polypropylene: Glass reinforced	0.003
Phenylene oxides (Noryl): Standard	0.0034
Polyacetal Homopolymer: 20% glass reinforced	0.0036
Silicones; Molded, Laminated: Fibrous (glass) reinforced silicones	0.004
Polyacetal Homopolymer: Standard	0.0048
Phenylene oxides (Noryl): Glass fiber reinforced	0.0049

Selecting Dissipation Factor for Polymers @ 10^6 Hz (Continued)

Polymer	Dissipation Factor (ASTM D150) @ 10^6 Hz
ABS Resins; Molded, Extruded: Heat resistant	0.005—0.015
Polymides: Glass Reinforced	0.0055
Polyarylsulfone	0.0056—0.012
ABS–Polycarbonate Alloy	0.0059
Polyester;: Thermosets: Cast Rigid	0.006—0.04
Polycarbonate (40% glass fiber reinforced)	0.007
Polystyrenes; Molded: Styrene acrylonitrile (SAN)	0.007–0.010
Fluorocarbons; Molded,Extruded: Polytrifluoro chloroethylene (PTFCE)	0.007—0.010
ABS Resins; Molded, Extruded: High impact	0.007—0.015
ABS Resins; Molded, Extruded: Medium impact	0.008—0.009
ABS Resins; Molded, Extruded: Very high impact	0.008—0.016
ABS Resins; Molded, Extruded: Low temperature impact	0.008—0.016
Reinforced Polyester Sheet molding compounds, general purpose	0.0086—0.022
Polystyrenes; Molded: Glass fiber (30%) reinforced SAN	0.009
Diallyl Phthalates; Molded: Dacron Filled	0.009—0.017 (Wet)
Polycarbonate	0.01
Standard Epoxies: High Strength Laminate	0.010-0.017
Nylons; Molded, Extruded; Type 6: Flexible Copolymers	0.010—0.015
Acrylics; Moldings: High impact grade	0.01—0.02
Cellulose Acetate; Molded, Extruded; ASTM Grade: H4—1	0.01—0.10
Cellulose Acetate; Molded, Extruded; ASTM Grade: H2—1	0.01—0.10
Cellulose Acetate; Molded, Extruded; ASTM Grade: MH—1, MH—2	0.01—0.10
Cellulose Acetate; Molded, Extruded; ASTM Grade: MS—1, MS—2	0.01—0.10
Cellulose Acetate; Molded, Extruded; ASTM Grade: S2—1	0.01—0.10

Selecting Dissipation Factor for Polymers @ 10^6 Hz (Continued)

Polymer	Dissipation Factor (ASTM D150) @ 10^6 Hz
Chlorinated polyether	0.011
Polymides: Unreinforced	0.011
Diallyl Phthalates; Molded: Glass Fiber Filled	0.012—0.020 (Wet)
Standard Epoxies: Molded	0.013—0.020
Alkyds; Molded: Glass reinforced (heavy duty parts)	0.015—0.022
Epoxies; High Performance Resins: Glass Cloth Laminate	0.0158
Alkyds; Molded: Putty (encapsulating)	0.016—0.020
Nylons; Molded, Extruded; 6/6 Nylon: Glass fiber reinforced	0.017—0.018
Alkyds; Molded: Granular (high speed molding)	0.017—0.020
Nylons; Molded, Extruded; Type 6: Glass fiber (30%) reinforced	0.019—0.015
Chlorinated polyvinyl chloride	0.02
Nylons; Molded, Extruded: Type 11	0.02
Phenolics; Molded; Very High Shock: Glass Fiber Filled	0.02
Melamines; Molded: Glass Fiber Filled	0.020—0.03
Acrylics; Cast Resin Sheets, Rods: General Purpose, Type I	0.02—0.03
Acrylics; Cast Resin Sheets, Rods: General Purpose, Type II	0.02—0.03
Acrylics; Moldings: Grades 5, 6, 8	0.02—0.03
Cellulose Acetate Butyrate; Molded, Extruded; ASTM Grade: H4	0.02—0.05
Cellulose Acetate Butyrate; Molded, Extruded; ASTM Grade: MH	0.02—0.05
Cellulose Acetate Butyrate; Molded, Extruded; ASTM Grade: S2	0.02—0.05
Cellusose Acetate Propionate; Molded, Extruded; ASTM Grade: 1	0.02—0.05
Cellusose Acetate Propionate; Molded, Extruded; ASTM Grade: 3	0.02—0.05
Cellusose Acetate Propionate; Molded, Extruded; ASTM Grade: 6	0.02—0.05
Polyester;: Thermosets: Flexible	0.02—0.06

Selecting Dissipation Factor for Polymers @ 10^6 Hz (Continued)

Polymer	Dissipation Factor (ASTM D150) @ 10^6 Hz
Alkyds; Molded: Rope (general purpose)	0.023
Standard Epoxies: General Purpose Glass Cloth Laminate	0.024—0.026
Ureas; Molded: Cellulose filled (ASTM Type 2)	0.027—0.029
Melamines; Molded: Alpha Cellulose Filled	0.028
Ureas; Molded: Alpha—cellulose filled (ASTM Type 1)	0.028—0.032
Ureas; Molded: Woodflour filled	0.028—0.032
Epoxies; High Performance Resins: Cast, Rigid	0.029—0.028
Melamines; Molded: Alpha Cellulose Mineral Filled	0.030
Nylons; Molded, Extruded; Type 6: General Purpose	0.03—0.04
Phenolics; Molded; General: Woodflour & Flock Filled	0.03—0.07
Phenolics; Molded; Shock: Paper, Flock, or Pulp Filled	0.03—0.07
Phenolics; Molded; High Shock: Chopped Fabric or Cord Filled	0.03—0.09
PVC–Acrylic Injection Molded	0.031
Melamines; Molded: Unfilled	0.031—0.040
Standard Epoxies: Cast Rigid	0.032
Melamines; Molded: Cellulose Electrical Filled	0.032—0.12
Standard Epoxies: Cast Flexible	0.0369-0.0622
Nylons; Molded, Extruded; 6/6 Nylon: General purpose molding	0.04
Diallyl Phthalates; Molded: Orlon Filled	0.045—0.040 (Wet)
Nylons; Type 6: Cast	0.05

Selecting Dissipation Factor for Polymers @ 10^6 Hz (Continued)

Polymer	Dissipation Factor (ASTM D150) @ 10^6 Hz
Nylons; Molded, Extruded: Type 8	0.08
Rubber Phenolic—Chopped Fabric Filled	0.09
PVC–Acrylic Sheet	0.094
Phenolics: Molded: Arc resistant—Mineral Filled	0.1
Thermoset Carbonate: Allyl diglycol carbonate	0.1—0.2
Phenolics: Molded: Rubber Phenolic—Woodflour or Flock Filled	0.1—0.2
Rubber Phenolic—Asbestos Filled	0.13
Diallyl Phthalates; Molded: Asbestos Filled	0.154—0.050 (Wet)
Fluorocarbons; Molded,Extruded: Polyvinylidene—fluoride (PVDF)	0.184

Source: data compiled by J.S. Park *from* Charles T. Lynch, *CRC Handbook of Materials Science, Vol. 3*, CRC Press, Boca Raton, Florida, 1975 and *Engineered Materials Handbook, Vol.2*, Engineering Plastics, ASM International, Metals Park, Ohio, 1988.

SELECTING DIELECTRIC STRENGTH OF POLYMERS

Polymer	Dielectric Strength (Short Time, ASTM D149) (V / mil)
Polyvinyl Chloride & Copolymers: Nonrigid–electrical	24—500
Phenolics; Molded: High shock: chopped fabric or cord filled	200—350
Reinforced polyester moldings: High strength (glass fibers)	200—400
Phenolics; Molded: General: woodflour and flock filled	200—425
Phenolics; Molded: Rubber phenolic—chopped fabric filled	250
Melamines; Molded: Glass fiber filled	250 —300
Phenolics; Molded: Shock: paper, flock, or pulp filled	250—350
Phenolics; Molded: Rubber phenolic—woodflour or flock filled	250—375
Cellulose Acetate Butyrate; Molded, Extruded; ASTM Grade: H4	250—400
Cellulose Acetate Butyrate; Molded, Extruded; ASTM Grade: MH	250—400
Cellulose Acetate Butyrate; Molded, Extruded; ASTM Grade: S2	250—400
Cellulose Acetate; Molded, Extruded; ASTM Grade: H6—1	250—600
Cellulose Acetate; Molded, Extruded; ASTM Grade: H4—1	250—600
Cellulose Acetate; Molded, Extruded; ASTM Grade: H2—1	250—600
Cellulose Acetate; ASTM Grade: MH—1, MH—2	250—600
Cellulose Acetate; ASTM Grade: MS—1, MS—2	250—600
Cellulose Acetate; Molded, Extruded; ASTM Grade: S2—1	250—600
Polyvinylidene— fluoride (PVDF): Molded,Extruded	260
Silicones: Fibrous (glass) reinforced silicones	280 (in oil)
Epoxies; High performance resins: Molded	280—400 (step)
Polymides: Glass reinforced	300—310
Resins; Molded, Extruded: Very high impact	300—375
Ceramic reinforced (PTFE): Molded,Extruded	300—400
6/6 Nylon: Glass fiber Molybdenum disulfide filled	300—400

Selecting Dielectric Strength of Polymers (Continued)

Polymer	Dielectric Strength (Short Time, ASTM D149) (V / mil)
Polyesters: Cast Thermosets: Rigid	300—400
Polyesters: Cast Thermosets: Flexible	300—400
Ureas; Molded: Alpha–cellulose filled (ASTM Type l)	300—400
Ureas; Molded: Woodflour filled	300—400
Diallyl Phthalates; Molded: Asbestos filled	300—400 (wet)
Resins; Molded, Extruded: Low temperature impact	300—415
Diallyl Phthalates; Molded: Glass fiber filled	300—420 (wet)
Cellusose Acetate Propionate; Molded, Extruded' ASTM Grade: 1	300—450
Cellusose Acetate Propionate; Molded, Extruded' ASTM Grade: 3	300—450
Cellusose Acetate Propionate; Molded, Extruded' ASTM Grade: 6	300—450
Polystyrenes; Molded: High impact	300—650
Polypropylene: Glass reinforced	317—475
Nylons; Molded, Extruded: Type 8	340
Ureas; Molded: Cellulose filled (ASTM Type 2)	340—370
Phenolics; Molded: Rubber phenolic—asbestos filled	350
Reinforced polyester moldings: Heat & chemical resistant (asbestos)	350
Polyarylsulfone	350—383
Melamines; Molded: Cellulose electrical	350—400
Phenolics; Molded: Arc resistant—mineral filled	350—425
Diallyl Phthalates; Molded: Glass fiber filled	350—430 (dry)
Resins; Molded, Extruded: High impact	350—440
Diallyl Phthalates; Molded: Asbestos filled	350—450 (dry)
Diallyl Phthalates; Molded: Dacron filled	360—391 (wet)
Resins; Molded, Extruded: Heat resistant	360—400

Selecting Dielectric Strength of Polymers (Continued)

Polymer	Dielectric Strength (Short Time, ASTM D149) (V / mil)
Melamines; Molded: Alpha cellulose and mineral filled	375
Diallyl Phthalates; Molded: Orlon filled	375 (wet)
Phenolics; Molded: Very high shock: glass fiber filled	375—425
Diallyl Phthalates; Molded: Dacron filled	376—400 (dry)
Nylons; Molded, Extruded; Type 6: Cast	380
Silicones: Granular (silica) reinforced silicones	380 (in oil)
ABS Resins; Molded, Extruded: Medium impact	385
6/6 Nylon: General purpose molding	385
Nylons; Molded, Extruded; Type 6: General purpose	385—400
Polystyrenes; Molded: Glass fiber -30% reinforced	396
Acrylics; Moldings: Grades 5, 6, 8	400
Chlorinated polyether	400
Polycarbonate	400
PVC–Acrylic Alloy: PVC–acrylic injection molded	400
Phenylene Oxides: SE—100	400 (1/8 in.)
Diallyl Phthalates; Molded: Orlon filled	400 (dry)
Reinforced polyester: Sheet molding compounds, general purpose	400—440
Nylons; Molded, Extruded; Type 6: Glass fiber (30%) reinforced	400—450
6/6 Nylon; Molded, Extruded: Glass fiber reinforced	400—480
Acrylics; Moldings: High impact grade	400—500
Styrene acrylonitrile (SAN)	400—500
Polyester; Thermoplastic Moldings: General purpose grade	420—540
Nylons; Molded, Extruded: Type 11	425
Phenylene oxides (Noryl): Standard	425

Selecting Dielectric Strength of Polymers (Continued)

Polymer	Dielectric Strength (Short Time, ASTM D149) (V / mil)
Polystyrenes; Molded: Medium impact	>425
PVC–Acrylic Alloy: PVC–acrylic sheet	>429
Nylons; Molded, Extruded; Type 6: Flexible copolymers	440
Epoxy novolacs: Cast, rigid	444
Polypropylene: Asbestos filled	450
Acrylics; Cast Resin Sheets, Rods: General purpose, type II	450—500
Acrylics; Cast Resin Sheets, Rods: General purpose, type I	450—530
Polyphenylene sulfide: Standard	450—595
Polypropylene: High impact	450—650
6/10 Nylon: General purpose extrusion	470
Polycarbonate (40% glass fiber reinforced)	475
Phenylene oxides (Noryl): Glass fiber reinforced	480
Polyethylenes; Molded, Extruded; Type I: Melt index 0.3—3.6	480
Polyethylenes; Molded, Extruded; Type I: Melt index 6—26	480
Polyethylenes; Molded, Extruded; Type I: Melt index 200	480
Polyethylenes; Molded, Extruded; Type II: Melt index 20	480
Polyethylenes; Molded, Extruded; Type II: Melt index l.0—1.9	480
Polyethylenes; Molded, Extruded; Type III: Melt index 0.2—0.9	480
Polyethylenes; Molded, Extruded; Type III: Melt Melt index 0.l—12.0	480
Polyethylenes; Molded, Extruded; Type III: Melt index 1.5—15	480
Polyethylenes; Molded, Extruded; Type III: High molecular weight	480
Polypropylene: Flame retardant	485—700
Polyphenylene sulfide: 40% glass reinforced	490
ABS–Polycarbonate Alloy	500

Selecting Dielectric Strength of Polymers (Continued)

Polymer	Dielectric Strength (Short Time, ASTM D149) (V / mil)
Polyacetal Homopolymer: Standard	500
Polyacetal Homopolymer: 20% glass reinforced	500
Polyacetal Copolymer: Standard	500
Polyacetal Copolymer: High flow	500
Phenylene Oxides: SE—1	500 (1/8 in.)
Polystyrenes; Molded: General purpose	>500
Olefin Copolymers; Molded: Polyallomer	500—650
Glass fiber (30%) reinforced SAN	515
Olefin Copolymers; Molded: EVA (ethylene vinyl acetate)	525
Polytrifluoro chloroethylene (PTFCE): Molded,Extruded	530—600
Olefin Copolymers; Molded: EEA (ethylene ethyl acrylate)	550
Polyester; Thermoplastic Moldings: Glass reinforced grades	560—750
Polyacetal Copolymer: 25% glass reinforced	580
Polyester; Thermoplastic Moldings: Asbestos—filled grade	580
Polyester; Thermoplastic Moldings: General purpose grade	590
Polypropylene: General purpose	650 (125 mil)
Silicones: Woven glass fabric/ silicone laminate	725
Polyvinyl Chloride & Copolymers: Rigid–normal impact	725—1,400
Polyester; Thermoplastic Moldings: Glass reinforced self extinguishing	750
Nylons; Molded, Extruded: Type 12	840

Selecting Dielectric Strength of Polymers (Continued)

Polymer	Dielectric Strength (Short Time, ASTM D149) (V / mil)
Olefin Copolymers; Molded: Ionomer	1,000
Polytetrafluoroethylene (PTFE): Molded,Extruded	1,000—2,000
Phenylene Oxides: Glass fiber reinforced	1,020 (1/32 in.)
Chlorinated polyvinyl chloride	1,250—1,550
Fluorinated ethylene propylene(FEP): Molded,Extruded	2,100

Source: data compiled by J.S. Park *from* Charles T. Lynch, *CRC Handbook of Materials Science, Vol. 3*, CRC Press, Boca Raton, Florida, 1975 and *Engineered Materials Handbook, Vol.2*, Engineering Plastics, ASM International, Metals Park, Ohio, 1988.

SELECTING DIELECTRIC CONSTANTS OF POLYMERS
AT 60 Hz

Polymer	Dielectric Constant (ASTM D150) 60 Hz
Polytetrafluoroethylene (PTFE) (0.01 in thickness)	2.1
Fluorinated ethylene propylene(FEP) (0.01 in thickness)	2.1
Polypropylene: General purpose	2.20—2.28
Polypropylene: High impact	2.20—2.28
Polyethylenes; Molded, Extruded; Type I: Melt index 0.3—3.6	2.3
Polyethylenes; Molded, Extruded; Type I: Melt index 6—26	2.3
Polyethylenes; Molded, Extruded; Type I: Melt index 200	2.3
Polyethylenes; Molded, Extruded; Type II: Melt index 20	2.3
Polyethylenes; Molded, Extruded; Type II: Melt index 1.0—1.9	2.3
Polyethylenes; Molded, Extruded; Type III: Melt index 0.2—0.9	2.3
Polyethylenes; Molded, Extruded; Type III: Melt Melt index 0.1—12.0	2.3
Polyethylenes; Molded, Extruded; Type III: Melt index 1.5—15	2.3
Polyethylenes; Molded, Extruded; Type III: High molecular weight	2.3
Polyallomer	2.3
Polypropylene: Glass reinforced	2.3—2.5
Polyvinyl Chloride & Copolymers: Rigid—normal impact	2.3—3.7
Olefin Copolymers; Molded: Ionomer	2.4
Polystyrenes; Molded: General purpose	2.45—2.65
Polystyrenes; Molded: Medium impact	2.45—4.75
Polystyrenes; Molded: High impact	2.45—4.75
Polypropylene: Flame retardant	2.46—2.79
ABS Resins; Molded, Extruded: Low temperature impact	2.5—3.5
Polytrifluoro chloroethylene (PTFCE)	2.6—2.7
Styrene acrylonitrile (SAN)	2.6—3.4

Selecting Dielectric Constants of Polymers at 60 Hz (Continued)

Polymer	Dielectric Constant (ASTM D150) 60 Hz
Phenylene Oxides: SE—100	2.65
Phenylene Oxides: SE—1	2.69
ABS Resins; Molded, Extruded: Heat resistant	2.7—3.5
ABS–Polycarbonate Alloy	2.74
Polypropylene: Asbestos filled	2.75
Olefin Copolymers; Molded: EEA (ethylene ethyl acrylate)	2.8
ABS Resins; Molded, Extruded: Medium impact	2.8—3.2
ABS Resins; Molded, Extruded: High impact	2.8—3.2
ABS Resins; Molded, Extruded: Very high impact	2.8—3.5
Polyesters Cast Thermosets: Rigid	2.8—4.4
Ceramic reinforced (PTFE)	2.9—3.6
Phenylene Oxides: Glass fiber reinforced	2.93
Polyvinyl Chloride & Copolymers: Vinylidene chloride	3—5
Phenylene oxides (Noryl): Standard	3.06—3.15
Chlorinated polyvinyl chloride	3.08
Chlorinated polyether	3.1
Polystyrenes; Molded: Glass fiber -30% reinforced	3.1
Polyester; Thermoplastic Moldings: General purpose grade	3.1—3.3
Polyester; Thermoplastic Moldings: General purpose grade	3.16
Olefin Copolymers; Molded: EVA (ethylene vinyl acetate)	3.16
Polycarbonate	3.17
Polyesters Cast Thermosets: Flexible	3.18—7.0
Nylons; Molded, Extruded Type 6: Flexible copolymers	3.2—4.0
Nylons: Type 11	$3.3 \ (10^3 \ \text{Hz})$

Selecting Dielectric Constants of Polymers at 60 Hz (Continued)

Polymer	Dielectric Constant (ASTM D150) 60 Hz
Diallyl Phthalates; Molded: Orlon filled	3.3—3.9 (Dry)
Epoxy novolacs: Cast, rigid	3.34—3.39
Glass fiber (30%) reinforced Styrene acrylonitrile (SAN)	3.5
Diallyl Phthalates; Molded: Dacron filled	3.5—3.8 (Dry)
Acrylics; Moldings: Grades 5, 6, 8	3.5—3.9
Acrylics; Moldings: High impact grade	3.5—3.9
Polyester; Thermoplastic Moldings: Asbestos—filled grade	3.5—4.2
Acrylics; Cast Resin Sheets, Rods: General purpose, type I	3.5—4.5
Acrylics; Cast Resin Sheets, Rods: General purpose, type II	3.5—4.5
Diallyl Phthalates; Molded: Glass fiber filled	3.5—4.5 (Dry)
Cellulose Acetate Butyrate; Molded, Extruded; ASTM Grade: H4	3.5—6.4
Cellulose Acetate Butyrate; Molded, Extruded; ASTM Grade: MH	3.5—6.4
Cellulose Acetate Butyrate; Molded, Extruded; ASTM Grade: S2	3.5—64
Cellulose Acetate; Molded, Extruded; ASTM Grade: H6—1	3.5—7.5
Cellulose Acetate; Molded, Extruded; ASTM Grade: H4—1	3.5—7.5
Cellulose Acetate; Molded, Extruded; ASTM Grade: H2—1	3.5—7.5
Cellulose Acetate; Molded, Extruded; ASTM Grade: MH—1, MH—2	3.5—7.5
Cellulose Acetate; Molded, Extruded; ASTM Grade: MS—1, MS—2	3.5—7.5
Cellulose Acetate; Molded, Extruded; ASTM Grade: S2—1	3.5—7.5
Polyarylsulfone	3.51—3.94
Phenylene oxides (Noryl): Glass fiber reinforced	3.55
Nylons: Type 12	3.6 (10^3 Hz)
Polyacetal Homopolymer: Standard	3.7
Polyacetal Copolymer: Standard	3.7 (100 Hz)

Selecting Dielectric Constants of Polymers at 60 Hz (Continued)

Polymer	Dielectric Constant (ASTM D150) 60 Hz
Polyacetal Copolymer: High flow	3.7 (100 Hz)
Polyester Moldings: Glass reinforced self extinguishing	3.7—3.8
Cellusose Acetate Propionate; Molded, Extruded; ASTM Grade: 1	3.7—4.0
Cellusose Acetate Propionate; Molded, Extruded; ASTM Grade: 3	3.7—4.0
Cellusose Acetate Propionate; Molded, Extruded; ASTM Grade: 6	3.7—4.0
Polyester; Thermoplastic Moldings: Glass reinforced grades	3.7—4.2
Polycarbonate (40% glass fiber reinforced)	3.8
PVC–Acrylic Alloy: PVC–acrylic sheet	3.86
6/10 Nylon: General purpose	3.9
Polyacetal Copolymer: 25% glass reinforced	3.9 (100 Hz)
Silicones; Molded, Laminated: Woven glass fabric/ silicone laminate	3.9—4.2
High performance Epoxies: Cast, rigid	3.96—4.02
Nylons; Type 6: Cast	4
6/6 Nylon: General purpose molding	4
PVC–Acrylic Alloy: PVC–acrylic injection molded	4
Polyacetal Homopolymer: 20% glass reinforced	4
Nylons; Molded, Extruded Type 6: General purpose	4.0—5.3
Standard Epoxies: Cast rigid	4.02
Silicones; Molded, Laminated: Granular (silica) reinforced silicones	4.1—4.5
Polymides: Unreinforced	4.12
Silicones; Molded, Laminated: Fibrous (glass) reinforced silicones	4.34
Thermoset Carbonate: Allyl diglycol carbonate	4.4
Standard Epoxies: Molded	4.4-5.4
Epoxy novolacs: Glass cloth laminate	4.41—4.43

Selecting Dielectric Constants of Polymers at 60 Hz (Continued)

Polymer	Dielectric Constant (ASTM D150) 60 Hz
Standard Epoxies: Cast flexible	4.43-4.79
Diallyl Phthalates; Molded: Asbestos filled	4.5—5.2 (Dry)
Nylons; Molded, Extruded Type 6: Glass fiber (30%) reinforced	4.6—5.6
Polyester Thermosets: Sheet molding compounds, general purpose	4.62—5.0
High performance Epoxies: Molded	4.7—5.7
Polymides: Glass reinforced	4.84
Phenolics; Molded; General: woodflour and flock filled	5.0—9.0
Alkyds; Molded: Glass reinforced (heavy duty parts)	5.2—6.0
Standard Epoxies: General purpose glass cloth laminate	5.3-5.4
Alkyds; Molded: Putty (encapsulating)	5.4—5.9
Polyvinyl Chloride & Copolymers: Nonrigid—general	5.5—9.1
Phenolics; Molded; Shock: paper, flock, or pulp filled	5.6—11.0
Alkyds; Molded: Granular (high speed molding)	5.7—6.3
Polyvinyl Chloride & Copolymers: Nonrigid—electrical	6.0—8.0
Melamines; Molded: Cellulose electrical	6.2—7.7
Phenolics; Molded; High shock: chopped fabric or cord filled	6.5—15.0
Melamines; Molded: Glass fiber filled	7.0—11.1
Ureas; Molded: Alpha—cellulose filled (ASTM Type 1)	7.0—9.5
Ureas; Molded: Woodflour filled	7.0—9.5
Phenolics; Molded; Very high shock: glass fiber filled	7.1—7.2

Selecting Dielectric Constants of Polymers at 60 Hz (Continued)

Polymer	Dielectric Constant (ASTM D150) 60 Hz
Ureas; Molded: Cellulose filled (ASTM Type 2)	7.2—7.3
Alkyds; Molded: Rope (general purpose)	7.4
Phenolics; Molded: Arc resistant—mineral	7.4
Melamines; Molded: Unfilled	7.9—11.0
Phenolics; Molded: Rubber phenolic—woodflour or flock	9—16
Nylons: Type 8	9.3
Polyvinylidene— fluoride (PVDF) (0.125 in thickness)	10
Rubber phenolic—chopped fabric	15
Rubber phenolic—asbestos	15
6/6 Nylon; Molded, Extruded:Glass fiber reinforced	40—44

Source: data compiled by J.S. Park *from* Charles T. Lynch, *CRC Handbook of Materials Science, Vol. 3*, CRC Press, Boca Raton, Florida, 1975 and *Engineered Materials Handbook, Vol.2*, Engineering Plastics, ASM International, Metals Park, Ohio, 1988.

SELECTING DIELECTRIC CONSTANTS OF POLYMERS AT 10^6 Hz

Polymer	Dielectric Constant (ASTM D150) 10^6 Hz
Polypropylene: Glass reinforced	2—2.25
Polypropylene: General purpose	2.23—2.24
Polypropylene: High impact	2.23—2.27
ABS Resins; Molded, Extruded: Very high impact	2.4—3.0
ABS Resins; Molded, Extruded: Low temperature impact	2.4—3.0
Polystyrenes; Molded: Medium impact	2.4—3.8
Polystyrenes; Molded: General purpose	2.45—2.65
Polypropylene: Flame retardant	2.45—2.70
Acrylics; Moldings: High impact grade	2.5—3.0
Polystyrenes; Molded: High impact	2.5—4.0
Styrene acrylonitrile (SAN)	2.6—3.02
Polypropylene: Asbestos filled	2.6—3.17
Phenylene Oxides: SE—100	2.64
Phenylene Oxides: SE—1	2.68
ABS–Polycarbonate Alloy	2.69
Acrylics; Moldings: Grades 5, 6, 8	2.7—2.9
ABS Resins; Molded, Extruded: High impact	2.7—3.0
Acrylics; Cast Resin Sheets, Rods: General purpose, type I	2.7—3.2
Acrylics; Cast Resin Sheets, Rods: General purpose, type II	2.7—3.2
ABS Resins; Molded, Extruded: Medium impact	2.75—3.0
Standard Epoxies: Cast flexible	2.78-3.52
ABS Resins; Molded, Extruded: Heat resistant	2.8—3.2
Polyesters Cast Thermosets: Rigid	2.8—4.4
Chlorinated polyether	2.92

Selecting Dielectric Constants of Polymers at 10⁶ Hz (Continued)

Polymer	Dielectric Constant (ASTM D150) 10^6 Hz
Phenylene Oxides: Glass fiber reinforced	2.92
Polycarbonate	2.96
Polystyrenes; Molded: Glass fiber -30% reinforced	3
Nylons; Molded, Extruded Type 6: Flexible copolymers	3.0—3.6
Polyacetal Copolymer: Standard	3—7
Polyacetal Copolymer: High flow	3—7
Polyacetal Copolymer: 25% glass reinforced	3—9
Phenylene oxides (Noryl): Standard	3.03—3.10
Chlorinated polyvinyl chloride	3.2—3.6
Cellulose Acetate Butyrate; Molded, Extruded; ASTM Grade: H4	3.2—6.2
Cellulose Acetate Butyrate; Molded, Extruded; ASTM Grade: MH	3.2—6.2
Cellulose Acetate Butyrate; Molded, Extruded; ASTM Grade: S2	3.2—6.2
Cellulose Acetate; Molded, Extruded; ASTM Grade: H6—1	3.2—7.0
Cellulose Acetate; Molded, Extruded; ASTM Grade: H4—1	3.2—7.0
Cellulose Acetate; Molded, Extruded; ASTM Grade: H2—1	3.2—7.0
Cellulose Acetate; Molded, Extruded; ASTM Grade: MH—1, MH—2	3.2—7.0
Cellulose Acetate; Molded, Extruded; ASTM Grade: MS—1, MS—2	3.2—7.0
Cellulose Acetate; Molded, Extruded; ASTM Grade: S2—1	3.2—7.0
Polyphenylene sulfide: Standard	3.22—3.8
Nylons; Type 6: Cast	3.3
PVC–Acrylic Alloy: PVC–acrylic injection molded	3.4
Silicones; Molded, Laminated: Granular (silica) reinforced silicones	3.4 —4.3
Glass fiber (30%) reinforced Styrene acrylonitrile (SAN)	3.4—3.6
Cellusose Acetate Propionate; Molded, Extruded; ASTM Grade: 1	3.4—3.7

Selecting Dielectric Constants of Polymers at 10^6 Hz (Continued)

Polymer	Dielectric Constant (ASTM D150) 10^6 Hz
Cellusose Acetate Propionate; Molded, Extruded; ASTM Grade: 3	3.4—3.7
Phenylene oxides (Noryl): Glass fiber reinforced	3.41
Standard Epoxies: Cast rigid	3.42
PVC–Acrylic Alloy: PVC–acrylic sheet	3.44
6/10 Nylon: General purpose	3.5
Thermoset Carbonate: Allyl diglycol carbonate	3.5—3.8
6/6 Nylon; Molded, Extruded:Glass fiber reinforced	3.5—4.1
High performance Epoxies: Cast, rigid	3.53—3.58
Polyarylsulfone	3.54—3.7
Polycarbonate (40% glass fiber reinforced)	3.58
6/6 Nylon: General purpose molding	3.6
Nylons; Molded, Extruded Type 6: General purpose	3.6—3.8
Polyacetal Homopolymer: Standard	3.7
Cellusose Acetate Propionate; Molded, Extruded; ASTM Grade: 6	3.7—3.4
Diallyl Phthalates; Molded: Dacron filled	3.7—3.9 (Wet)
Polyesters Cast Thermosets: Flexible	3.7—6.1
Silicones; Molded, Laminated: Woven glass fabric/ silicone laminate	3.8—397
Polyphenylene sulfide: 40% glass reinforced	3.88
Nylons; Molded, Extruded Type 6: Glass fiber (30%) reinforced	3.9—5.4
Polymides: Unreinforced	3.96
Nylons: Type 8	4
Phenolics; Molded; General: woodflour and flock filled	4.0—7.0
Polyacetal Homopolymer: 20% glass reinforced	4—0
Standard Epoxies: Molded	4.1-4.6

Selecting Dielectric Constants of Polymers at 10^6 Hz (Continued)

Polymer	Dielectric Constant (ASTM D150) 10^6 Hz
Diallyl Phthalates; Molded: Orlon filled	4.1—3.4 (Wet)
Silicones; Molded, Laminated: Fibrous (glass) reinforced silicones	4.28
High performance Epoxies: Molded	4.3—4.8
Diallyl Phthalates; Molded: Glass fiber filled	4.4—4.6 (Wet)
Alkyds; Molded: Putty (encapsulating)	4.5—4.7
Alkyds; Molded: Glass reinforced (heavy duty parts)	4.5—5.0
Phenolics; Molded; Shock: paper, flock, or pulp filled	4.5—7.0
Phenolics; Molded; High shock: chopped fabric or cord filled	4.5—7.0
Polyester Thermosets: Sheet molding compounds, general purpose	4.55—4.75
Phenolics; Molded; Very high shock: glass fiber filled	4.6—6.6
Standard Epoxies: General purpose glass cloth laminate	4.7-4.8
Polymides: Glass reinforced	4.74
Standard Epoxies: High strength laminate	4.8-5.2
Alkyds; Molded: Granular (high speed molding)	4.8—5.1
Diallyl Phthalates; Molded: Asbestos filled	4.8—6.5 (Wet)
Phenolics; Molded: Arc resistant—mineral	5
Phenolics; Molded: Rubber phenolic—woodflour or flock	5
Rubber phenolic—chopped fabric	5
Rubber phenolic—asbestos	5
High performance Epoxies: Glass cloth laminate	5.1
Melamines; Molded: Cellulose electrical	5.2—6.0
Melamines; Molded: Alpha cellulose mineral filled	5.6
Melamines; Molded: Glass fiber filled	6.0—7.9
Melamines; Molded: Unfilled	6.3—7.3

Selecting Dielectric Constants of Polymers at 10^6 Hz (Continued)

Polymer	Dielectric Constant (ASTM D150) 10^6 Hz
Ureas; Molded: Cellulose filled (ASTM Type 2)	6.4—6.5
Ureas; Molded: Alpha—cellulose filled (ASTM Type 1)	6.4—6.9
Ureas; Molded: Woodflour filled	6.4—6.9
Melamines; Molded: Alpha cellulose filled	6.4—8.1
Alkyds; Molded: Rope (general purpose)	6.8

Source: data compiled by J.S. Park *from* Charles T. Lynch, *CRC Handbook of Materials Science, Vol. 3*, CRC Press, Boca Raton, Florida, 1975 and *Engineered Materials Handbook, Vol.2*, Engineering Plastics, ASM International, Metals Park, Ohio, 1988.

SELECTING ARC RESISTANCE OF POLYMERS

Polymer	Arc Resistance (ASTM D495) (seconds)
Rubber phenolic—asbestos filled	5—20
Phenolics; Molded; General: woodflour and flock filled	5—60
Phenolics; Molded; Shock: paper, flock, or pulp filled	5—60
Phenolics; Molded; High shock: chopped fabric or cord filled	5—60
Rubber phenolic—woodflour or flock filled	7—20
Rubber phenolic—chopped fabric filled	10—20
Polypropylene: Flame retardant	15—40
Polystyrenes; Molded: High impact	20—100
Polystyrenes; Molded: Medium impact	20—135
PVC–Acrylic Alloy: PVC–acrylic injection molded	25
Polystyrenes; Molded: Glass fiber -30% reinforced	28
Polyphenylene sulfide: 40% glass reinforced	34
Polymides: Glass reinforced	50—180
Phenolics; Molded; Very high shock: glass fiber filled	60
Polystyrenes; Molded: General purpose	60—135
Glass fiber (30%) reinforced SAN	65
Polyarylsulfone	67—81
Melamines; Molded: Cellulose electrical filled	70—135
Polypropylene: Glass reinforced	73—77
Phenylene Oxides: SE—100	75
Phenylene Oxides: SE—1	75
Standard Epoxies: Cast flexible	75—98
PVC–Acrylic Alloy: PVC–acrylic sheet	80
Polyester; Thermoplastic Moldings: Glass reinforced self extinguishing	80

Selecting Arc Resistance of Polymers (Continued)

Polymer	Arc Resistance (ASTM D495) (seconds)
Ureas; Molded: Woodflour filled	80—110
Ureas; Molded: Cellulose filled (ASTM Type 2)	85—110
Diallyl Phthalates; Molded: Orlon filled	85—115
Nylons; Molded, Extruded Type 6: Glass fiber (30%) reinforced	92—81
ABS–Polycarbonate Alloy	96
Standard Epoxies: Cast rigid	100
Ureas; Molded: Alpha—cellulose filled (ASTM Type 1)	100—135
Melamines; Molded: Unfilled	100—145
Styrene acrylonitrile (SAN)	100—150
Diallyl Phthalates; Molded: Dacron filled	105—125
Polyester; Thermoplastic Moldings: Asbestos—filled grade	108
Phenylene oxides (Noryl): Glass fiber reinforced	114
Polyesters Cast Thermosets: Rigid	115—135
Epoxy novolacs: Cast, rigid	120
6/6 Nylon; Molded, Extruded: General purpose molding	120
6/6 Nylon; Molded, Extruded: General purpose extrusion	120
6/10 Nylon: General purpose	120
Phenylene Oxides: Glass fiber reinforced	120
Polycarbonate	120 (tungsten electrode)
Polycarbonate (40% glass fiber reinforced)	120 (tungsten electrode)
Polypropylene: Asbestos filled	121—125
Phenylene oxides (Noryl): Standard	122
Polypropylene: High impact	123—140
Melamines; Molded: Alpha cellulose and mineral filled	125

Selecting Arc Resistance of Polymers (Continued)

Polymer	Arc Resistance (ASTM D495) (seconds)
Polyester; Thermoplastic Moldings: General purpose grade	125
Polypropylene: General purpose	125—136
Diallyl Phthalates; Molded: Asbestos filled	125—140
Diallyl Phthalates; Molded: Glass fiber filled	125—140
Polyesters Cast Thermosets: Flexible	125—145
Polyacetal Homopolymer: Standard	129
Polyester; Thermoplastic Moldings: Glass reinforced grades	130
Reinforced polyester moldings: High strength (glass fibers)	130—170
Standard Epoxies: General purpose glass cloth laminate	130—180
Reinforced polyester: Sheet molding compounds, general purpose	130—180
6/6 Nylon; Molded, Extruded: Glass fiber Molybdenum disulfide filled	135
Standard Epoxies: Molded	135—190
Polyacetal Copolymer: 25% glass reinforced	136
6/6 Nylon; Molded, Extruded: Glass fiber reinforced	148—100
Polymides: Unreinforced	152
Alkyds; Molded: Putty (encapsulating)	180
Alkyds; Molded: Rope (general purpose)	180
Alkyds; Molded: Granular (high speed molding)	180
Alkyds; Molded: Glass reinforced (heavy duty parts)	180
Phenolics; Molded: Arc resistant—mineral	180
High performance Epoxies: Molded	180—185
Melamines; Molded: Glass fiber filled	180—186
Thermoset Carbonate: Allyl diglycol carbonate	185
Polyacetal Homopolymer: 20% glass reinforced	188

Selecting Arc Resistance of Polymers (Continued)

Polymer	Arc Resistance (ASTM D495) (seconds)
Polyester; Thermoplastic Moldings: General purpose grade	190
Molded,Extruded Polytetrafluoroethylene (PTFE)	>200
Silicones; Molded, Laminated: Woven glass fabric/ silicone laminate	225—250
Polyacetal Copolymer: Standard	240
Polyacetal Copolymer: High flow	240
Silicones; Molded, Laminated: Fibrous (glass) reinforced silicones	240
Silicones; Molded, Laminated: Granular (silica) reinforced silicones	250—310
Molded,Extruded Polytrifluoro chloroethylene (PTFCE)	>360
Acrylics; Cast Resin Sheets, Rods: General purpose, type I	No track
Acrylics; Cast Resin Sheets, Rods: General purpose, type II	No track
Acrylic Moldings: Grades 5, 6, 8	No track
Acrylic Moldings: High impact grade	No track

Source: data compiled by J.S. Park *from* Charles T. Lynch, *CRC Handbook of Materials Science, Vol. 3*, CRC Press, Boca Raton, Florida, 1975 and *Engineered Materials Handbook, Vol.2*, Engineering Plastics, ASM International, Metals Park, Ohio, 1988.

Selecting Polymers: Optical Properties

SELECTING TRANSPARENCY OF POLYMERS

Polymer	Transparency (visible light) (ASTM D791) (%)
Alkyds; Molded: Putty (encapsulating)	Opaque
Alkyds; Molded: Rope (general purpose)	Opaque
Alkyds; Molded: Granular (high speed molding)	Opaque
Alkyds; Molded: Glass reinforced (heavy duty parts)	Opaque
Chlorinated polyether	Opaque
Chlorinated polyvinyl chloride	Opaque
Standard Epoxies: General purpose glass cloth laminate	Opaque
Standard Epoxies: High strength laminate	Opaque
Standard Epoxies: Filament wound composite	Opaque
High performance Epoxies: Molded	Opaque
High performance Epoxies: Glass cloth laminate	Opaque
Epoxy novolacs: Glass cloth laminate	Opaque
Melamines; Molded: Cellulose electrical	Opaque
6/6 Nylon; Molded, Extruded: Glass fiber reinforced	Opaque
6/6 Nylon; Molded, Extruded: Glass fiber Molybdenum disulfide filled	Opaque
6/6 Nylon; Molded, Extruded: General purpose extrusion	Opaque
6/10 Nylon: General purpose	Opaque
6/10 Nylon: Glass fiber (30%) reinforced	Opaque
ABS–Polycarbonate Alloy	Opaque
PVC–Acrylic Alloy: PVC–acrylic injection molded	Opaque
Polymides: Unreinforced	Opaque
Polymides: Glass reinforced	Opaque

Selecting Transparency of Polymers (Continued)

Polymer	Transparency (visible light) (ASTM D791) (%)
Reinforced polyester moldings: High strength (glass fibers)	Opaque
Reinforced polyester moldings: Heat & chemical resistsnt (asbestos)	Opaque
Reinforced polyester: Sheet molding compounds, general purpose	Opaque
Phenylene Oxides: SE—100	Opaque
Phenylene Oxides: SE—1	Opaque
Phenylene Oxides: Glass fiber reinforced	Opaque
Phenylene oxides (Noryl): Glass fiber reinforced	Opaque
Polypropylene: Asbestos filled	Opaque
Polypropylene: Glass reinforced	Opaque
Polypropylene: Flame retardant	Opaque
Polyphenylene sulfide: Standard	Opaque
Polyphenylene sulfide: 40% glass reinforced	Opaque
Polystyrenes; Molded: Medium impact	Opaque
Polystyrenes; Molded: High impact	Opaque
Polystyrenes; Molded: Glass fiber -30% reinforced	Opaque
Glass fiber (30%) reinforced Styrene acrylonitrile (SAN)	Opaque
Silicones; Molded, Laminated: Fibrous (glass) reinforced silicones	Opaque
Silicones; Molded, Laminated: Granular (silica) reinforced silicones	Opaque
Silicones; Molded, Laminated: Woven glass fabric/ silicone laminate	Opaque
Ureas; Molded: Cellulose filled (ASTM Type 2)	Opaque
Ureas; Molded: Woodflour filled	Opaque
PVC–Acrylic Alloy: PVC–acrylic sheet	Opaque

Selecting Transparency of Polymers (Continued)

Polymer	Transparency (visible light) (ASTM D791) (%)
Polypropylene: General purpose	Translucent—opaque
Polypropylene: High impact	Translucent—opaque
Polycarbonate (40% glass fiber reinforced)	Translucent
6/6 Nylon; Molded, Extruded: General purpose molding	Translucent
Polystyrenes; Molded: General purpose	Transparent
Styrene acrylonitrile (SAN)	Transparent
Ureas; Molded: Alpha—cellulose filled (ASTM Type 1)	21.8
Polycarbonate	75—85
Cellulose Acetate; Molded, Extruded; ASTM Grade: H6—1	75—90
Cellulose Acetate; Molded, Extruded; ASTM Grade: H4—1	75—90
Cellulose Acetate Butyrate; Molded, Extruded; ASTM Grade: H4	75—92
Cellulose Acetate; Molded, Extruded; ASTM Grade: H2—1	80—90
Cellulose Acetate; Molded, Extruded; ASTM Grade: MH—1, MH—2	80—90
Cellulose Acetate; Molded, Extruded; ASTM Grade: MS—1, MS—2	80—90
Cellulose Acetate Butyrate; Molded, Extruded; ASTM Grade: MH	80—92
Cellusose Acetate Propionate; Molded, Extruded; ASTM Grade: 1	80—92
Cellusose Acetate Propionate; Molded, Extruded; ASTM Grade: 3	80—92
Cellusose Acetate Propionate; Molded, Extruded; ASTM Grade: 6	80—92
Polytrifluoro chloroethylene (PTFCE) Molded, Extruded	80—92
Cellulose Acetate; Molded, Extruded; ASTM Grade: S2—1	80—95
Standard Epoxies: Molded	85
Cellulose Acetate Butyrate; Molded, Extruded; ASTM Grade: S2	85—95

Selecting Transparency of Polymers (Continued)

Polymer	Transparency (visible light) (ASTM D791) (%)
Thermoset Carbonate: Allyl diglycol carbonate	89—92
Acrylic Moldings: High impact grade	90
Standard Epoxies: Cast flexible	90
Acrylics; Cast Resin Sheets, Rods: General purpose, type I	91—92 (0.125 in.)
Acrylics; Cast Resin Sheets, Rods: General purpose, type II	91—92 (0.125 in.)
Acrylic Moldings: Grades 5, 6, 8	>92

Source: data compiled by J.S. Park *from* Charles T. Lynch, *CRC Handbook of Materials Science, Vol. 3*, CRC Press, Boca Raton, Florida, 1975 and *Engineered Materials Handbook, Vol.2*, Engineering Plastics, ASM International, Metals Park, Ohio, 1988.

SELECTING REFRACTIVE INDICES OF POLYMERS

Polymer	Refractive Index (ASTM D542) (n_D)
Fluorinated ethylene propylene(FEP) Molded, Extruded	1.34
Polytetrafluoroethylene (PTFE) Molded, Extruded	1.35
Polyvinylidene— fluoride (PVDF) Molded, Extruded	1.42
Polytrifluoro chloroethylene (PTFCE) Molded, Extruded	1.43
Cellusose Acetate Propionate; Molded, Extruded; ASTM Grade: 1	1.46—1.49
Cellusose Acetate Propionate; Molded, Extruded; ASTM Grade: 3	1.46—1.49
Cellusose Acetate Propionate; Molded, Extruded; ASTM Grade: 6	1.46—1.49
Cellulose Acetate Butyrate; Molded, Extruded; ASTM Grade: H4	1.46—1.49 (D543)
Cellulose Acetate Butyrate; Molded, Extruded; ASTM Grade: MH	1.46—1.49 (D543)
Cellulose Acetate Butyrate; Molded, Extruded; ASTM Grade: S2	1.46—1.49 (D543)
Cellulose Acetate; Molded, Extruded; ASTM Grade: H6—1	1.46—1.50
Cellulose Acetate; Molded, Extruded; ASTM Grade: H4—1	1.46—1.50
Cellulose Acetate; Molded, Extruded; ASTM Grade: H2—1	1.46—1.50
Cellulose Acetate; Molded, Extruded; ASTM Grade: MH—1, MH—2	1.46—1.50
Cellulose Acetate; Molded, Extruded; ASTM Grade: MS—1, MS—2	1.46—1.50
Cellulose Acetate; Molded, Extruded; ASTM Grade: S2—1	1.46—1.50
Acrylics; Cast Resin Sheets, Rods: General purpose, type II	1.485—1.495
Acrylics; Cast Resin Sheets, Rods: General purpose, type I	1.485—1.500
Acrylic Moldings: Grades 5, 6, 8	1.489—1.493
Acrylic Moldings: High impact grade	1.49

Selecting Refractive Indices of Polymers (Continued)

Polymer	Refractive Index (ASTM D542) (n_D)
Thermoset Carbonate: Allyl diglycol carbonate	1.5
Polyesters Cast Thermosets: Flexible	1.50—1.57
Polyethylenes; Molded, Extruded; Type I: Melt index 0.3—3.6	1.51
Polyethylenes; Molded, Extruded; Type I: Melt index 6—26	1.51
Polyethylenes; Molded, Extruded; Type I: Melt index 200	1.51
Polyethylenes; Molded, Extruded; Type II: Melt index 20	1.51
Polyethylenes; Molded, Extruded; Type II: Melt index 1.0—1.9	1.51
Polyesters Cast Thermosets: Rigid	1.53—1.58
Polyethylenes; Molded, Extruded; Type III: Melt index 0.2—0.9	1.54
Polyethylenes; Molded, Extruded; Type III: Melt Melt index 0.1—12.0	1.54
Polyethylenes; Molded, Extruded; Type III: Melt index 1.5—15	1.54
Styrene acrylonitrile (SAN)	1.565—1.569
Polycarbonate	1.586
Polystyrenes; Molded: General purpose	1.6
Polyvinyl Chloride & Copolymers: Vinylidene chloride	1.60—1.63
Standard Epoxies: Cast flexible	1.61
Standard Epoxies: Molded	1.61
Phenylene oxides (Noryl): Standard	1.63
Polyarylsulfone	1.651
Polyacetal Homopolymer: Standard	Opaque

Selecting Refractive Indices of Polymers (Continued)

Polymer	Refractive Index (ASTM D542) (n_D)
Polyacetal Homopolymer: 20% glass reinforced	Opaque
Polyacetal Homopolymer: 22% TFE reinforced	Opaque
Polyacetal Copolymer: Standard	Opaque
Polyacetal Copolymer: 25% glass reinforced	Opaque
Polyacetal Copolymer: High flow	Opaque
Polystyrenes; Molded: Medium impact	Opaque
Polystyrenes; Molded: High impact	Opaque
Polystyrenes; Molded: Glass fiber -30% reinforced	Opaque
Glass fiber (30%) reinforced Styrene acrylonitrile (SAN)	Opaque

Source: data compiled by J.S. Park *from* Charles T. Lynch, *CRC Handbook of Materials Science, Vol. 3*, CRC Press, Boca Raton, Florida, 1975 and *Engineered Materials Handbook, Vol.2*, Engineering Plastics, ASM International, Metals Park, Ohio, 1988.

Selecting Polymers: Chemical Properties

SELECTING WATER ABSORPTION OF POLYMERS

Polymer	Water Absorption in 24 hr (ASTM D570) (%)
Polytrifluoro chloroethylene (PTFCE); Molded, Extruded	0
Alkyds; Molded: Glass reinforced (heavy duty parts)	0.007—0.10
Fluorinated ethylene propylene(FEP)	<0.01
Polyethylenes; Molded, Extruded; Type I: Melt index 0.3—3.6	<0.01
Polyethylenes; Molded, Extruded; Type I: Melt index 6—26	<0.01
Polyethylenes; Molded, Extruded; Type I: Melt index 200	<0.01
Polyethylenes; Molded, Extruded; Type II: Melt index 20	<0.01
Polyethylenes; Molded, Extruded; Type II: Melt index 1.0—1.9	<0.01
Polyethylenes; Molded, Extruded; Type III: Melt index 0.2—0.9	<0.01
Polyethylenes; Molded, Extruded; Type III: Melt Melt index 0.1—12.0	<0.01
Polyethylenes; Molded, Extruded; Type III: Melt index 1.5—15	<0.01
Polyethylenes; Molded, Extruded; Type III: High molecular weight	<0.01
Polypropylene: High impact	<0.01—0.02
Polypropylene: General purpose	<0.01—0.03
Chlorinated polyether	0.01
Polytetrafluoroethylene (PTFE); Molded, Extruded	0.01
Polyvinyl Chloride & Copolymers: Vinylidene chloride	>0.1 (ASTM D635)
Polypropylene: Flame retardant	0.02—0.03
Polypropylene: Asbestos filled	0.02—0.04
Polypropylene: Glass reinforced	0.02—0.05
Silicones: Woven glass fabric/ silicone laminate	0.03—0.05
Polyvinylidene— fluoride (PVDF)	0.03—0.06
Polystyrenes; Molded: Medium impact	0.03—0.09
Polyvinyl Chloride & Copolymers: Rigid—normal impact	0.03—0.40 (ASTM D635)

Selecting Water Absorption of Polymers (Continued)

Polymer	Water Absorption in 24 hr (ASTM D570) (%)
High performance Epoxies; Glass cloth laminate	0.04—0.06
Standard Epoxies; High strength laminate	0.05
Standard Epoxies; General purpose glass cloth laminate	0.05—0.07
Standard Epoxies; Filament wound composite	0.05—0.07
Alkyds; Molded: Rope (general purpose)	0.05—0.08
Polystyrenes; Molded: High impact	0.05—0.22
PVC–Acrylic Alloy: PVC–acrylic sheet	0.06
Phenylene Oxides: Glass fiber reinforced	0.06
Polyester; Thermoplastic Moldings: Glass reinforced grades	0.06—0.07
Polyester; Moldings: Glass reinforced self extinguishing	0.07
Polyester; Thermoplastic Moldings: Glass reinforced grade	0.07
Phenylene Oxides: SE—100	0.07
Phenylene Oxides: SE—1	0.07
Polystyrenes; Molded: Glass fiber –30% reinforced	0.07
Polycarbonate (40% glass fiber reinforced)	0.08
Polyester; Thermoplastic Moldings: General purpose grade	0.08
Silicones; Molded, Laminated: Granular (silica) reinforced	0.08—0.1
Alkyds; Molded: Granular (high speed molding)	0.08—0.12
Polyester; Thermoplastic Moldings: General purpose grade	0.09
Melamines; Molded: Glass fiber filled	0.09—0.60
Polyester; Thermoplastic Moldings: Asbestos—filled grade	0.1
Alkyds; Molded: Putty (encapsulating)	0.10—0.15
Rubber phenolic—asbestos filled	0.10—0.50
Silicones; Molded, Laminated: Fibrous (glass) reinforced	0.1—0.15

Selecting Water Absorption of Polymers (Continued)

Polymer	Water Absorption in 24 hr (ASTM D570) (%)
Standard Epoxies; Cast rigid	0.1—0.2
Epoxy novolacs: Cast, rigid	0.1—0.7
Phenolics; Molded; Very high shock: glass fiber filled	0.1—1.0
Chlorinated polyvinyl chloride	0.11
High performance Epoxies; Molded	0.11—0.2
Polyesters: Cast Thermosets: Flexible	0.12—2.5
PVC–Acrylic Alloy: PVC–acrylic injection molded	0.13
Polycarbonate	0.15
Styrene acrylonitrile (SAN): Glass fiber (30%) reinforced	0.15
Polyester: Sheet molding compounds, general purpose	0.15—0.25
Phenylene oxides (Noryl): Glass fiber reinforced	0.18—0.22
Thermoset Carbonate: Allyl diglycol carbonate	0.2
6/10 Nylon: Glass fiber (30%) reinforced	0.2
Polymides: Glass reinforced	0.2
Polyacetal Homopolymer: 22% TFE reinforced	0.2
Ceramic reinforced (PTFE)	>0.2
Styrene acrylonitrile (SAN)	0.20—0.35
Polyesters: Cast Thermosets: Rigid	0.20—0.60
ABS Resins; Molded, Extruded: Medium impact	0.2—0.4
ABS Resins; Molded, Extruded: Heat resistant	0.2—0.4
Acrylics; Cast Resin Sheets, Rods: General purpose, type II	0.2—0.4
Acrylics; Moldings: High impact grade	0.2—0.4
ABS Resins; Molded, Extruded: High impact	0.2—0.45
ABS Resins; Molded, Extruded: Very high impact	0.2—0.45

Selecting Water Absorption of Polymers (Continued)

Polymer	Water Absorption in 24 hr (ASTM D570) (%)
ABS Resins; Molded, Extruded: Low temperature impact	0.2—0.45
Melamines; Molded: Unfilled	0.2—0.5
Polyvinyl Chloride & Copolymers: Nonrigid—general	0.2—1.0 (ASTM D635)
ABS–Polycarbonate Alloy	0.21
Polyacetal Copolymer: Standard	0.22
Polyacetal Copolymer: High flow	0.22
Phenylene oxides (Noryl): Standard	0.22
Polymides: Unreinforced	0.24—0.47
Nylons; Type 12	0.25
Polyacetal Homopolymer: Standard	0.25
Polyacetal Homopolymer: 20% glass reinforced	0.25
Ppolyester moldings: Heat & chemical resistant (asbestos)	0.25—0.50
Melamines; Molded: Cellulose electrical filled	0.27—0.80
Polyacetal Copolymer: 25% glass reinforced	0.29
Polystyrenes; Molded: General purpose	0.30—0.2
Acrylics; Cast Resin Sheets, Rods: General purpose, type I	0.3—0.4
Acrylics; Moldings: Grades 5, 6, 8	0.3—0.4
Melamines; Molded: Alpha cellulose and mineral filled	0.3—0.5
Standard Epoxies; Molded	0.3—0.8
Phenolics; Molded; General: woodflour and flock filled	0.3—0.8
Nylons; Type 11	0.4
6/10 Nylon: General purpose	0.4
Polyarylsulfone	0.4
Polyvinyl Chloride & Copolymers: Nonrigid—electrical	0.40—0.75 (ASTM D635)

Selecting Water Absorption of Polymers (Continued)

Polymer	Water Absorption in 24 hr (ASTM D570) (%)
Standard Epoxies; Cast flexible	0.4—0.1
Ureas; Molded: Alpha—cellulose filled (ASTM Type l)	0.4—0.8
Phenolics; Molded; Shock: paper, flock, or pulp filled	0.4—1.5
Phenolics; Molded; High shock: chopped fabric or cord filled	0.4—1.75
Nylons; 6/6 Nylon: Glass fiber Molybdenum disulfide filled	0.5—0.7
Phenolics; Molded; Arc resistant—mineral filled	0.5—0.7
Reinforced polyester moldings: High strength (glass fibers)	0.5—0.75
Rubber phenolic—woodflour or flock filled	0.5—2.0
Rubber phenolic—chopped fabric filled	0.5—2.0
Nylons; Type 6: Cast	0.6
Nylons; Molded, Extruded; 6/6 Nylon: Glass fiber reinforced	0.8—0.9
Nylons; Molded, Extruded; Type 6: Flexible copolymers	0.8—1.4
Nylons; Molded, Extruded; Type 6: Glass fiber (30%) reinforced	0.9—1.2
Cellulose Acetate Butyrate; ASTM Grade: S2	0.9—1.3
Cellulose Acetate Butyrate; ASTM Grade: MH	1.3—1.6
Cellusose Acetate Propionate; ASTM Grade: 3	1.3—1.8
Nylons; Molded, Extruded; Type 6: General purpose	1.3—1.9
Nylons; Molded, Extruded; 6/6 Nylon: General purpose molding	1.5
Nylons; Molded, Extruded; 6/6 Nylon: General purpose extrusion	1.5
Cellusose Acetate Propionate; ASTM Grade: 6	1.6

Selecting Water Absorption of Polymers (Continued)

Polymer	Water Absorption in 24 hr (ASTM D570) (%)
Cellusose Acetate Propionate; ASTM Grade: 1	1.6—2.0
Cellulose Acetate; Molded, Extruded; ASTM Grade: H4—1	1.7—2.7
Cellulose Acetate; Molded, Extruded; ASTM Grade: H2—1	1.7—2.7
Cellulose Acetate; ASTM Grade: MH—1, MH—2	1.8—4.0
Cellulose Acetate Butyrate; ASTM Grade: H4	2
Cellulose Acetate; ASTM Grade: MS—1, MS—2	2.1—4.0
Cellulose Acetate; Molded, Extruded; ASTM Grade: S2—1	2.3—4.0
Nylons; Type 8	9.5

Source: data compiled by J.S. Park *from* Charles T. Lynch, *CRC Handbook of Materials Science, Vol. 3*, CRC Press, Boca Raton, Florida, 1975 and *Engineered Materials Handbook, Vol.2*, Engineering Plastics, ASM International, Metals Park, Ohio, 1988.

SELECTING FLAMMABILITY OF POLYMERS

Polymer	Flammability (ASTM D635) (ipm)
Alkyds; Molded: Glass reinforced (heavy duty parts)	Nonburning
Alkyds; Molded: Putty (encapsulating)	Nonburning
Ceramic reinforced (PTFE)	Noninflammable
Chlorinated polyvinyl chloride	Nonburning
Fibrous (glass) reinforced silicones	Nonburning
Fluorinated ethylene propylene (FEP)	Noninflammable
Granular (silica) reinforced silicones	Nonburning
Polyphenylene sulfide: 40% glass reinforced	Non—burning
Polyphenylene sulfide: Standard	Non—burning
Polytetrafluoroethylene (PTFE); Molded,Extruded	Nonintlammable
Polytrifluoro chloroethylene (PTFCE); Molded,Extruded	Noninflammable
PVC–Acrylic Alloy: PVC–acrylic injection molded	Nonburning
PVC–Acrylic Alloy: PVC–acrylic sheet	Nonburning
Alkyds; Molded: Granular (high speed molding)	Self extinguishing
Alkyds; Molded: Rope (general purpose)	Self extinguishing
Chlorinated polyether	Self extinguishing
Epoxies; High performance resins: Cast, rigid	Self extinguishing
Epoxies; High performance resins: Glass cloth laminate	Self extinguishing
Epoxies; High performance resins: Molded	Self extinguishing
Melamines; Molded: Alpha cellulose and mineral filled	Self extinguishing
Melamines; Molded: Cellulose electrical filled	Self extinguishing

Selecting Flammability of Polymers (Continued)

Polymer	Flammability (ASTM D635) (ipm)
Melamines; Molded: Glass fiber filled	Self extinguishing
Melamines; Molded: Unfilled	Self extinguishing
Nylons; Molded, Extruded; Type 6: Cast	Self extinguishing
Nylons; Molded, Extruded; Type 6: General purpose	Self extinguishing
Nylons; Molded, Extruded; Type 8	Self extinguishing
Nylons; Molded, Extruded; Type 11	Self extinguishing
6/6 Nylon: General purpose extrusion	Self extinguishing
6/6 Nylon: General purpose molding	Self extinguishing
6/10 Nylon: General purpose	Self extinguishing
Phenolics: Molded: Arc resistant—mineral filled	Self extinguishing
Phenolics; Molded; General: woodflour and flock filled	Self extinguishing
Phenolics; Molded; High shock: chopped fabric or cord filled	Self extinguishing
Phenolics; Molded; Shock: paper, flock, or pulp filled	Self extinguishing
Phenolics; Molded; Very high shock: glass fiber filled	Self extinguishing
Phenylene oxides (Noryl): Glass fiber reinforced	Self extinguishing
Phenylene oxides (Noryl): Standard	Self extinguishing
Phenylene Oxides: Glass fiber reinforced	Self extinguishing
Phenylene Oxides: SE—1	Self extinguishing
Phenylene Oxides: SE—100	Self extinguishing
Polyarylsulfone	Self extinguishing

Selecting Flammability of Polymers (Continued)

Polymer	Flammability (ASTM D635) (ipm)
Polycarbonate	Self extinguishing
Polycarbonate (40% glass fiber reinforced)	Self extinguishing
Polyester; Moldings: Glass reinforced self extinguishing	Self extinguishing
Polypropylene: Flame retardant	Self extinguishing
Polyvinyl Chloride & Copolymers: Nonrigid—electrical	Self extinguishing
Polyvinyl Chloride & Copolymers: Nonrigid—general	Self extinguishing
Polyvinyl Chloride & Copolymers: Rigid—normal impact	Self extinguishing
Polyvinyl Chloride & Copolymers: Vinylidene chloride	Self extinguishing
Polyvinylidene— fluoride (PVDF)	Self extinguishing
Reinforced polyester moldings: High strength (glass fibers)	Self extinguishing
Reinforced polyester: Heat and chemical resistant (asbestos)	Self extinguishing
Reinforced polyester: Sheet molding compounds, general purpose	Self extinguishing
Rubber phenolic—asbestos filled	Self extinguishing
Rubber phenolic—chopped fabric filled	Self extinguishing
Rubber phenolic—woodflour or flock filled	Self extinguishing
Standard Epoxies: Filament wound composite	Self extinguishing
Standard Epoxies: High strength laminate	Self extinguishing
Standard Epoxies: Molded	Self extunguishing
Ureas; Molded: Alpha—cellulose filled (ASTM Type 1)	Self extinguishing
Ureas; Molded: Cellulose filled (ASTM Type 2)	Self extinguishing
Ureas; Molded: Woodflour filled	Self extinguishing

Selecting Flammability of Polymers (Continued)

Polymer	Flammability (ASTM D635) (ipm)
Standard Epoxies: General purpose glass cloth laminate	Slow burn to Self extinguishing
Polyester; Thermoset: Cast polyyester: Flexible	Slow burn to self extinguishing
Nylons; Molded, Extruded; Type 6: Glass fiber (30%) reinforced	Slow burn
6/6 Nylon: Glass fiber reinforced	Slow burn
6/6 Nylon: Glass fiber Molybdenum disulfide filled	Slow burn
6/10 Nylon: Glass fiber (30%) reinforced	Slow burn
Polyester; Thermoplastic Injection Moldings: General purpose grade	Slow burn
Polyester; Thermoplastic Injection Moldings: Glass reinforced grades	Slow burn
Polyester; Thermoplastic Moldings: General purpose grade	Slow burn
Polyester; Thermoplastic Moldings: Glass reinforced grade	Slow burn
Silicones; Woven glass fabric/ silicone laminate	0.12
Standard Epoxies: Cast rigid	0.3-0.34
Thermoset Carbonate: Allyl diglycol carbonate	0.35
Cellulose Acetate Butyrate; Molded, Extruded; ASTM Grade: H4	0.5—1.5
Cellulose Acetate Butyrate; Molded, Extruded; ASTM Grade: MH	0.5—1.5
Cellulose Acetate Butyrate; Molded, Extruded; ASTM Grade: S2	0.5—1.5
Cellusose Acetate Propionate; Molded, Extruded; ASTM Grade: 1	0.5—1.5
Cellusose Acetate Propionate; Molded, Extruded; ASTM Grade: 3	0.5—1.5
Cellusose Acetate Propionate; Molded, Extruded; ASTM Grade: 6	0.5—1.5
Polystyrenes; Molded: High impact	0.5—1.5
Cellulose Acetate; Molded, Extruded; ASTM Grade: H6—1	0.5—2.0
Cellulose Acetate; Molded, Extruded; ASTM Grade: H4—1	0.5—2.0

Selecting Flammability of Polymers (Continued)

Polymer	Flammability (ASTM D635) (ipm)
Cellulose Acetate; Molded, Extruded; ASTM Grade: H2—1	0.5—2.0
Cellulose Acetate; Molded, Extruded; ASTM Grade: MH—1, MH—2	0.5—2.0
Cellulose Acetate; Molded, Extruded; ASTM Grade: MS—1, MS—2	0.5—2.0
Cellulose Acetate; Molded, Extruded; ASTM Grade: S2—1	0.5—2.0
Polystyrenes; Molded: Medium impact	0.5—2.0
Nylons; Molded, Extruded; Type 6: Flexible copolymers	Slow burn, 0.6
Polypropylene: General purpose	0.7—1
Polyacetal Homopolymer: 20% glass reinforced	0.8
Polyacetal Homopolymer: 22% TFE reinforced	0.8
Polystyrenes; Molded: Styrene acrylonitrile (SAN)	0.8
Polyester; Thermoset: Cast polyyester: Rigid	0.87 to self extinguishing
ABS–Polycarbonate Alloy	0.9
Polyacetal Copolymer: 25% glass reinforced	1
Polypropylene: High impact	1
Polypropylene: Asbestos filled	1
Polypropylene: Glass reinforced	1
Polyethylenes; Molded, Extruded; Type I: Melt index 0.3—3.6	1
Polyethylenes; Molded, Extruded; Type I: Melt index 6—26	1
Polyethylenes; Molded, Extruded; Type I: Melt index 200	1
Polyethylenes; Molded, Extruded; Type II: Melt index 20	1
Polyethylenes; Molded, Extruded; Type II: Melt index 1.0—1.9	1
Polyethylenes; Molded, Extruded; Type III: Melt index 0.2—0.9	1
Polyethylenes; Molded, Extruded; Type III: Melt Melt index 0.1—12.0	1
Polyethylenes; Molded, Extruded; Type III: Melt index 1.5—15	1

Selecting Flammability of Polymers (Continued)

Polymer	Flammability (ASTM D635) (ipm)
Polyethylenes; Molded, Extruded; Type III: High molecular weight	1
ABS Resins; Molded, Extruded: Low temperature impact	1.0—1.5
Polystyrenes; Molded: General purpose	1.0—1.5
ABS Resins; Molded, Extruded: Medium impact	1.0—1.6
Polyacetal Homopolymer: Standard	1.1
Polyacetal Copolymer: Standard	1.1
Polyacetal Copolymer: High flow	1.1
ABS Resins; Molded, Extruded: High impact	1.3—1.5
ABS Resins; Molded, Extruded: Very high impact	1.3—1.5
ABS Resins; Molded, Extruded: Heat resistant	1.3—2.0

Source: data compiled by J.S. Park *from* Charles T. Lynch, *CRC Handbook of Materials Science, Vol. 3*, CRC Press, Boca Raton, Florida, 1975 and *Engineered Materials Handbook, Vol.2*, Engineering Plastics, ASM International, Metals Park, Ohio, 1988.

Index

Leading Phrase **[Key Word]** *Trailing Phrase* ....*Page*

A

Leading Phrase **[Key Word]** *Trailing Phrase* ....*Page*

C

Leading Phrase **[Key Word]** *Trailing Phrase* ....*Page*

C (Continued)

Leading Phrase **[Key Word]** *Trailing Phrase* ....*Page*

D (Continued)

Leading Phrase **[Key Word]** *Trailing Phrase* ....*Page*

E

Leading Phrase **[Key Word]** *Trailing Phrase* ....*Page*

E (Continued)

G (Continued)

H

I

M (Continued)

P (Continued)

P (Continued)

Leading Phrase **[Key Word]** *Trailing Phrase* ....*Page*

P (Continued)

R

Leading Phrase **[Key Word]** *Trailing Phrase* ....*Page*

S (Continued)

Leading Phrase **[Key Word]** *Trailing Phrase* ....*Page*

T (Continued)

Leading Phrase **[Key Word]** *Trailing Phrase* ....*Page*

T (Continued)

T (Continued)

Leading Phrase **[Key Word]** *Trailing Phrase* ....*Page*

T (Continued)

U

V

Leading Phrase **[Key Word]** *Trailing Phrase* ....*Page*

Y (Continued)

1

6